T0155902

Lecture Notes in Computer Science 13389

More information about this series at https://link.springer.com/bookseries/558

Silvia Chiusano · Tania Cerquitelli ·
Robert Wrembel (Eds.)

Advances in Databases and Information Systems

26th European Conference, ADBIS 2022
Turin, Italy, September 5–8, 2022
Proceedings

 Springer

Editors
Silvia Chiusano ⓘ
Politecnico di Torino
Turin, Italy

Tania Cerquitelli ⓘ
Politecnico di Torino
Turin, Italy

Robert Wrembel ⓘ
Poznań University of Technology
Poznań, Poland

ISSN 0302-9743   ISSN 1611-3349 (electronic)
Lecture Notes in Computer Science
ISBN 978-3-031-15739-4   ISBN 978-3-031-15740-0 (eBook)
https://doi.org/10.1007/978-3-031-15740-0

This Springer imprint is published by the registered company Springer Nature Switzerland AG
The registered company address is: Gewerbestrasse 11, 6330 Cham, Switzerland

Preface

This year ADBIS – the European Conference on Advances in Databases and Information Systems – celebrated its 26th anniversary.

The first ADBIS conference was held in Saint Petersburg, Russia (1997). Since then, ADBIS has taken place annually, with previous editions held in Poznan, Poland (1998); Maribor, Slovenia (1999); Prague, Czech Republic (2000); Vilnius, Lithuania (2001); Bratislava, Slovakia (2002); Dresden, Germany (2003); Budapest, Hungary (2004); Tallinn, Estonia (2005); Thessaloniki, Greece (2006); Varna, Bulgaria (2007); Pori, Finland (2008); Riga, Latvia (2009); Novi Sad, Serbia (2010); Vienna, Austria (2011); Poznan, Poland (2012); Genoa, Italy (2013); Ohrid, North Macedonia (2014); Poitiers, France (2015); Prague, Czech Republic (2016); Nicosia, Cyprus (2017); Budapest, Hungary (2018); Bled, Slovenia (2019); Lyon, France (2020); and Tartu, Estonia (2021).

The official ADBIS portal – http://adbis.eu – provides up to date information on all ADBIS conferences, committees, publications, and issues related to the ADBIS community.

The 26th ADBIS conference was held in Turin, Italy, during September 5–8, 2022, as a hybrid event. It received significant attention from both the research and industrial communities, as 90 papers were submitted to the conference. In total, 280 authors from 32 different countries submitted their research contributions to ADBIS 2022. The submitted papers had, on average, 3.1 authors each, and most of them were the outcome of international cooperation. The papers were reviewed by an international Program Committee (PC) consisting of 85 members.

The Program Committee selected 23 regular research papers for inclusion in this volume (an acceptance rate of 25%). The selected papers span a wide spectrum of topics related to the ADBIS conference from different areas of research in database and information systems, including graph processing, time series and data streams, data quality, OLAP, advanced querying, performance, and machine learning. The Program Committee also selected 28 short papers (an acceptance rate of 42%), which were included in CCIS, volume 1652.

ADBIS 2022 featured the following four keynote speakers:

- Sihem Amer-Yahia (CNRS, University of Grenoble Alpes, France) - AI-Powered Data-driven Education
- Daniele Quercia (King's College London and Nokia Bell Labs, UK) - Insider Stories: Analyzing Stress, Depression, and Staff Welfare at Major US Companies from Online Reviews
- Carlo Curino (Microsoft, USA) - Tensor Query Processing: Neural Network $$ to speed up Databases and Classical ML!

– Bruno Lepri (Bruno Kessler Foundation, Italy) - Understanding and rewiring cities and societies: a computational social science perspective

ADBIS 2022 was also accompanied by the following tutorials:

– Mirjana Ivanović (University of Novi Sad, Serbia) - AI approaches in processing and using data in personalized medicine
– Rosa Meo (University of Turin, Italy) - Explainable, Interpretable, Trustworthy, Responsible, Ethical, Fair, Verifiable AI... What's Next?
– Johann Gamper (Free University of Bozen-Bolzano) - What's New in Temporal Databases?
– Stefano Rizzi (University of Bologna, Italy) - OLAP and NoSQL: Happily Ever After

Thanks to the reputation of ADBIS, selected best papers of ADBIS 2022 will be invited for a special issue of the following Q1 journals: Information Systems (Elsevier) and Information Systems Frontiers (Springer). Therefore, the PC chairs would like to express their sincere gratitude to the Information Systems Editors-in-Chief: Dennis Shasha, Gottfried Vossen, and Matthias Weidlich, as well as the Information Systems Frontiers Editors-in-Chief: Ram Ramesh and H. Raghav Rao, for their approval of these special issues.

Finally, we would like to thank everyone who contributed to ADBIS 2022:

– the authors for submitting their research papers to the conference;
– the keynote speakers and tutorial presenters who honored us with their insightful talks;
– members of the Program Committee and external reviewers for dedicating their time and expertise to build the conference program;
– members of the ADBIS Steering Committee for their trust and support, and especially its chair Yannis Manolopoulos;
– all members of the Organizing Committee; and
– our partners:

  • Politecnico di Torino for hosting and supporting the event;
  • the Department of Control and Computer Engineering and the SmartData center at Politecnico di Torino for supporting the event; and
  • Springer for publishing the proceedings and constant support for the conference over years.

The ADBIS 2022 Organizing Committee supported diversity and inclusion by offering some grants, supporting a few researchers to participate in the conference and become part of the ADBIS community. All grants were assigned based on the underrepresented community, gender, and role/position. The grants included:

– two free regular registrations, assigned to researchers from Argentina and Brazil,
– three regular registration fee discounts of 200 Euros, assigned to researchers from Estonia, Lebanon, and Italy, and

– four regular registration fee discounts of 150 Euros, assigned to researchers from Croatia, France, and Italy.

July 2022

Silvia Chiusano
Tania Cerquitelli
Robert Wrembel

Organization

General Chair

Silvia Chiusano   Polytechnic University of Turin, Italy

Program Committee Chairs

Tania Cerquitelli   Polytechnic University of Turin, Italy
Robert Wrembel   Poznan University of Technology, Poland

Workshop Chairs

Kjetil Nørvåg   Norwegian University of Science and Technology, Norway
Barbara Catania   University of Genoa, Italy

Doctoral Consortium Chairs

Genoveva Vargas Solar   CNRS, LIRIS, France
Ester Zumpano   University of Calabria, Italy

Local Organization Committee Chairs

Luca Cagliero   Polytechnic University of Turin, Italy
Paolo Garza   Polytechnic University of Turin, Italy
Bartolomeo Vacchetti   Polytechnic University of Turin, Italy
Giovanni Malnati   Polytechnic University of Turin, Italy

Publicity Chair

Oscar Romero   Polytechnic University of Catalonia - BarcelonaTech, Spain

Proceedings Chair

Khalid Belhajjame   Paris Dauphine University - PSL, France

Special Issue Chair

Ladjel Bellatreche   ISAE-ENSMA, France

Diversity and Inclusion Chair

Jérôme Darmont   University of Lyon 2, France

Steering Committee

Andreas Behrend   TH Köln, Germany
Ladjel Bellatreche   ISAE-ENSMA, France
Maria Bielikova   Kempelen Institute of Intelligent Technologies,
                                   Slovakia
Barbara Catania   University of Genoa, Italy
Jérôme Darmont   University of Lyon 2, France
Johann Eder   Universität Klagenfurt, Austria
Johann Gamper   Free University of Bozen-Bolzano, Italy
Tomáš Horváth   Eötvös Loránd University, Hungary
Mirjana Ivanović   University of Novi Sad, Serbia
Marite Kirikova   Riga Technical University, Latvia
Manuk Manukyan   Yerevan State University, Armenia
Raimundas Matulevicius   University of Tartu, Estonia
Tadeusz Morzy   Poznan University of Technology, Poland
Kjetil Nørvåg   Norwegian University of Science and Technology,
                                   Norway
Boris Novikov   National Research University, Higher School of
                                   Economics, Saint Petersburg, Russia
George Papadopoulos   University of Cyprus, Cyprus
Jaroslav Pokorný   Charles University in Prague, Czech Republic
Oscar Romero   Polytechnic University of Catalonia -
                                   BarcelonaTech, Spain
Sergey Stupnikov   Russian Academy of Sciences, Russia
Bernhard Thalheim   University of Kiel, Germany
Goce Trajcevski   Iowa State University, USA
Valentino Vranić   Slovak University of Technology in Bratislava,
                                   Slovakia
Tatjana Welzer   University of Maribor, Slovenia
Robert Wrembel   Poznan University of Technology, Poland
Ester Zumpano   University of Calabria, Italy

Program Committee

Alberto Abelló	Polytechnic University of Catalonia - BarcelonaTech, Catalonia
Cristina D. Aguiar	University of São Paulo, Brasil
Syed Muhammad Fawad Ali	IBM, Germany
Bernd Amann	Sorbonne University, France
Witold Andrzejewski	Poznan University of Technology, Poland
Daniele Apiletti	Polytechnic University of Turin, Italy
Costin Badica	University of Craiova, Romania
Sylvio Barbon Junior	University of Trieste, Italy
Andreas Behrend	University of Bonn, Germany
Khalid Belhajjame	Paris Dauphine University - PSL, France
Ladjel Bellatreche	ISAE-ENSMA, France
András Benczúr	Eötvos Loránd University, Hungary
Fadila Bentayeb	University of Lyon 2, France
Maria Bielikova	Kempelen Institute of Intelligent Technologies, Slovakia
Sandro Bimonte	INRAE, France
Pawel Boiński	Poznan University of Technology, Poland
Zoran Bosnić	University of Ljubljana, Slovenia
Omar Boussaid	University of Lyon 2, France
Drazen Brdjanin	University of Banja Luka, Bosnia and Herzegovina
Damiano Carra	University of Verona, Italy
Jérôme Darmont	University of Lyon 2, France
Claudia Diamantini	Marche Polytechnic University, Italy
Christos Doulkeridis	University of Piraeus, Greece
Johann Eder	University of Klagenfurt, Austria
Markus Endres	University of Passau, Germany
Javier A. Espinosa-Oviedo	University of Lyon, France
Georgios Evangelidis	University of Macedonia, Greece
Flavio Ferrarotti	Software Competence Centre Hagenberg, Austria
Alessandro Fiori	Polytechnic University of Turin, Italy
Flavius Frasincar	Erasmus University Rotterdam, The Netherlands
Johann Gamper	Free University of Bozen-Bolzano, Italy
Matteo Golfarelli	University of Bologna, Italy
Marcin Gorawski	Silesian University of Technology, Poland
Jānis Grabis	Riga Technical University, Latvia
Francesco Guerra	University of Modena and Reggio Emilia, Italy
Giancarlo Guizzardi	Federal University of Espirito Santo, Brazil
Tomas Horvath	Eötvös Loránd University, Hungary
Mirjana Ivanović	University of Novi Sad, Serbia

Stefan Jablonski	University of Bayreuth, Germany
Aida Kamišalić	University of Maribor, Slovenia
Zoubida Kedad	University of Versailles, France
Attila Kiss	Eötvös Loránd University, Hungary
Julius Köpke	University of Klagenfurt, Austria
Dejan Lavbič	University of Ljubljana, Slovenia
Yurii Litvinov	St. Petersburg State University, Russia
Audrone Lupeikiene	Vilnius University, Lithuania
Federica Mandreoli	University of Modena and Reggio Emilia, Italy
Yannis Manolopoulos	Open University of Cyprus, Cyprus
Patrick Marcel	University of Tours, France
Sara Migliorini	University of Verona, Italy
Angelo Montanari	University of Udine, Italy
Lia Morra	Polytechnic University of Turin, Italy
Tadeusz Morzy	Poznan University of Technology, Poland
Boris Novikov	National Research University, Higher School of Economics, Russia
Kjetil Nørvåg	Norwegian University of Science and Technology, Norway
Andreas Oberweis	Karlsruhe Institute of Technology, Germany
George Papadopoulos	University of Cyprus, Cyprus
András Pataricza	Budapest University of Technology and Economics, Hungary
Jan Platos	Technical University of Ostrava, Czech Republic
Jaroslav Pokorný	Charles University in Prague, Czech Republic
Giuseppe Polese	University of Salerno, Italy
Alvaro E. Prieto	University of Extremadura, Spain
Elisa Quintarelli	University of Verona, Italy
Miloš Radovanović	University of Novi Sad, Serbia
Franck Ravat	University of Toulouse, France
Stefano Rizzi	University of Bologna, Italy
Oscar Romero	Polytechnic University of Catalonia, Spain
Gunter Saake	University of Magdeburg, Germany
Kai-Uwe Sattler	TU Ilmenau, Germany
Milos Savic	University of Novi Sad, Serbia
Claudia Steinberger	University of Klagenfurt, Austria
Sergey Stupnikov	Russian Academy of Sciences, Russia
Bernhard Thalheim	University of Kiel, Germany
Goce Trajcevski	Iowa State University, USA
Raquel Trillo-Lado	University of Zaragoza, Spain
Genoveva Vargas Solar	CNRS, LIRIS, France

Goran Velinov	Ss. Cyril and Methodius University, North Macedonia
Peter Vojtas	Charles University in Prague, Czech Republic
Isabelle Wattiau	ESSEC and Cnam, France
Marek Wojciechowski	Poznan University of Technology, Poland
Vladimir Zadorozhny	University of Pittsburgh, USA
Ester Zumpano	University of Calabria, Italy

Additional Reviewers

Stylianos Argyrou
Paul Blockhaus
Miroslav Blšták
Andrea Brunello
Gabriel Campero Durand
Andrea Chiorrini
Francesco Del Buono
Chiara Forresi
Marco Franceschetti
Matteo Francia
Verena Geist

Joseph Giovanelli
Nicolas Labroche
Christos Mettouris
Simone Monaco
Federico Motta
Uchechukwu Njoku
Thomas Photiadis
Matúš Pikuliak
Nicola Saccomanno
Vladimir A. Shekhovtsov
Emanuele Storti

Abstracts of the Keynote Talks

Toward AI-Powered Data-Driven Education

Sihem Amer-Yahia

CNRS, Univ. Grenoble Alpes, France

Abstract. Educational platforms are increasingly becoming AI-driven. Besides providing a wide range of course filtering options, personalized recommendations of learning material and teachers are driving today's research. While accuracy plays a major role in evaluating those recommendations, many factors must be considered including learner retention, throughput, upskilling ability, equality of learning opportunities, and satisfaction. This creates a tension between learner-centered and platform-centered approaches. I will describe research at the intersection of data-driven recommendations and education theory. This includes multi-objective algorithms that leverage collaboration and affinity in peer learning, studying the impact of learning strategies on platforms and people, and automating the generation of sequences of courses. I will end the talk with a discussion of the central role data management systems could play in enabling holistic educational experiences.

Insider Stories: Analyzing Stress, Depression, and Staff Welfare at Major US Companies from Online Reviews

Daniele Quercia

King's College London, and Nokia Bell Labs in Cambridge, UK

Abstract. We mined 440K company reviews published during twelve successive years on GlassDoor, and developed state-of-the-art deep-learning frameworks to accurately extract mentions of:

1. Stress [1, 2]. There are two types of stress: distress refers to harmful stimuli, while eustress refers to healthy, euphoric stimuli that create a sense of fulfillment and achievement. Telling the two types of stress apart is challenging, let alone quantifying their impact across corporations. We scored each company to be either a low stress, passive, negative stress, or positive stress company. We found that (former) employees of positive stress companies tended to describe high-growth and collaborative workplaces in their reviews, and that such companies' stock evaluations grew, on average, 5.1 times in 10 years (2009–2019) as opposed to the companies of the other three stress types that grew, on average, 3.7 times in the same time period. We also found that the four stress scores aggregated every year – from 2008 to 2020 – closely followed the unemployment rate in the U.S.: a year of positive stress (2008) was rapidly followed by several years of negative stress (2009–2015), which peaked during the Great Recession (2009–2011).
2. Internal Sustainability Efforts (ISEs) [3], which reflect whether a company supports gender equality, diversity, and general staff welfare. Commitment to ISEs manifested itself not only at micro-level (companies scoring high in ISEs enjoyed high stock growth) but also at macro-level (states hosting these companies were economically wealthy and equal, and attracted the so-called creative class).

References

1. Pressure Test: Good Stress for Company Success. https://arxiv.org/abs/2107.12362
2. Sen, I., et al.: Depression at Work: Exploring Depression in Major US Companies from Online Reviews. In: Proceedings of the ACM on Human-Computer Interaction, 2022
3. Insider Stories: Analyzing Internal Sustainability Efforts of Major US Companies from Online Reviews. https://arxiv.org/abs/2205.01217

Tensor Query Processing: Neural Network $$ to speed up Databases and Classical ML!

Carlo Curino

Microsoft, USA

Abstract. Massive market interest in AI has driven unprecedented investments in Special HW and runtimes for Neural Networks. Tensor computations are emerging as the de-facto API for all these special HW and runtimes. In this talk, we show how we can automatically transform and optimize relational queries and Classical ML pipelines into tensor computations, and run on special hardware. Interestingly the performance we obtain significantly outperform classical systems and even custom-build GPU DBMSs. At the same time, this approach retains very low engineering costs, thanks to a minute code footprint (<10 k LoC) and free portability—as we piggyback on tensor runtimes getting ported to all the new HW coming out. We conclude touching on further research directions that emerge once both queries and ML models are uniformly represented as tensors computations.

Understanding and Rewiring Cities and Societies: A Computational Social Science Perspective

Bruno Lepri

Bruno Kessler Foundation, Italy

Abstract. The almost universal adoption of mobile phones, the exponential growth in the usage of Internet services and social media platforms, and the proliferation of digital payment systems, wearable devices, and connected objects has led to the existence of unprecedented amounts of data about human behavior. Thus, we live in an unprecedented historic moment where the availability of vast amounts of behavioral data, combined with advances in machine learning, are enabling us to build predictive computational models of human behavior. In my talk, I will show examples of how those computational models of human behavior can be used to better understand and to design more efficient companies, cities, and societies, For example, I will present some works where we have leveraged mobile phone data, credit card transactions, Google Street View images, and social media data in order (i) to infer how vital and livable a city is, (ii) to find the urban conditions that magnify and influence urban life, (iii) to study their relationship with societal outcomes such as urban crime and segregation, and (iv) to model the impact of migrations and pandemic shocks such as COVID-19, etc. Finally, I will also discuss key human-centric requirements for a positive disruption of these novel approaches including a fundamental renegotiation of user-centric data ownership and management, the development of tools and participatory infrastructures towards increased algorithmic transparency and accountability, and the creation of living labs for experimenting and co-creating data-driven policies.

Contents

On Line Analytical Processing

Advanced Querying

Performance

Keynote Talk and Tutorials

Understanding and Rewiring Cities

Bruno Lepri$^{(\boxtimes)}$, Simone Centellegher, and Marco De Nadai

Fondazione Bruno Kessler, Trento, Italy
{lepri,centellegher,denadai}@fbk.eu

Abstract. Nowadays, the ever increasing digitization of our societies is producing an unprecedented amount of data about human behavior. At the same time, advances in machine learning and complex systems enable us to build explanatory and/or predictive computational models of human behavior. Interestingly, these data and models can also be used to better understand the factors associated with specific neighborhoods' outcomes such as vitality, safety perception, crime levels, innovation, segregation, traffic congestion, etc., and to design more efficient policymakers' interventions. In particular, leveraging census data, mobile phone traces, information from OpenStreetMap, and street view images, we describe a set of studies where we (i) infer how vital and livable a city is; (ii) find urban appearance conditions that magnify and influence urban life; (iii) study the relationship of urban conditions with societal outcomes such as urban crime levels; and (iv) model the impact of pandemic shocks such as COVID-19 and related non-pharmaceutical interventions on human behavior.

1 Introduction

Cities have always played an essential role for innovation, economic prosperity and diversity [1]. Quantitative evidence from many empirical studies points to an acceleration of economic and social life with the population size of cities [2]. More specifically, these gains apply to a wide variety of quantities, including gross domestic product (GDP), wages, patents, violent crime, the spreading of contagious diseases, and the number of human interactions [3–6].

Supposedly, this acceleration is linked to the ability of human beings of learning from each other [7,8]. Indeed, cities foster this greatest talent to flourish by supplying diverse high quality amenities and places where to meet each other and socialize [9], and access to people with different background and skills [10].

However, these empirical studies on the acceleration of emergent phenomena such as economic output, innovation, crime, etc. [3–5] have considered cities as a whole, thus neglecting the evidence that extremely diverse neighborhoods coexist within the same city. Instead, several studies in urban demography and urban sociology have shown that cities have always been economically and ethnically spatially segregated [11–16], as well as unevenly affected by crime (e.g., crime hotspots), etc. [17,18]. Until recently, however, empirical studies and advancements on these neighborhood-level differences were limited due to the high cost

© Springer Nature Switzerland AG 2022
S. Chiusano et al. (Eds.): ADBIS 2022, LNCS 13389, pp. 3–10, 2022.
https://doi.org/10.1007/978-3-031-15740-0_1

of census and surveys' data. For this reason, the digitization of our societies and the resulting explosion of data sources (e.g., mobile phones, social media, credit card transactions, etc.) and analytical methods and tools have started to deeply revolutionize the study of cities [19,20]. In particular, nowadays we can passively observe and even predict many aspects of human mobility and social behavior and interactions in cities [20,21].

Here, we describe our approach of studying neighborhoods' characteristics and their relationship with people behaviors combining novel sources of automatically collected data (e.g., data on the presence of Point Of Interests from OpenStreetMap, street view images, mobile phone traces, etc.) and a mixture of methodologies ranging from more traditional statistical models to machine learning and complex systems' approaches. The proposed methodological approach and the variety of obtained results are relevant to researchers within a broad range of fields, from urban computing, computational social science, and complex systems to urban-planning, urban sociology and criminology, as well as to policymakers.

2 Neighborhoods' Characteristics and Urban Vitality

According to the urban activist Jane Jacobs, in her most influential book "The Death and Life of Great American Cities" [22], there exist four conditions that promote life in a city: (i) *mixed land uses*, for which districts should provide multiple primary functions to attract people for different purposes; (ii) *small blocks*, to promote contacts between people; (iii) *buildings diversity (age and form)*, to mix high-rent and low-rent tenants; and (iv) sufficient *dense concentration* of people and buildings.

Exploiting the large amount of data collected from mobile phones as well as census data and data from OpenStreetMap (OSM) [23], we were able to empirically test these conditions, overcoming the extensive difficulties in collecting data for entire cities (previously collected with surveys). In particular, we used mobile phone data to extract information for urban vitality, and census data and data from OpenStreetMap to compute proxies of urban diversity, as per Jacobs's four conditions, in six Italian cities (i.e., Bologna, Florence, Milan, Palermo, Rome, and Turin).

Our results suggest that Jacobs's four conditions for maintaining a vital urban life hold for Italian cities despite different structural and socio-economic conditions from the American cities originally described by Jacobs. We also find that vibrant Italian districts have a dense concentration of office workers, third places (e.g., restaurants, pubs, general stores) within walking distance, smaller streets, and historical buildings.

The findings, the developed methodology and the variety of structural features, closely linked to district activity, could be used by municipalities and regulators to assess the vitality of a neighborhood. Moreover, the methodology explained in [23] could be used to monitor and quantify regulatory interventions and potentially provide suggestions for missing structural features in a particular neighborhood.

3   How Safety Perception Influences Vitality

In her book, Jane Jacobs has also introduced the *natural surveillance hypothesis* [22], which suggests that citizens can contribute to maintaining the safety of their neighbourhoods through a continued natural and informal surveillance. However, Jacobs argued that neighbourhoods and buildings need certain physical qualities to support natural surveillance, such as well-lit streets and buildings with street-facing windows. Jacobs' idea that the physical quality of a neighbourhood can enhance its safety was later expanded by Oscar Newman's *defensible space theory* [24]. The defensible space theory expands on the idea of natural surveillance by suggesting that neighbours will be more likely to protect an area when clear physical demarcations are separating what is considered public and private property [22,24]. Examples of architectural markers of defensible space are archways in the entrance of building complexes or staircases in the entrance of townhouses. These archways and staircases serve an aesthetic purpose and signal the boundary between a city's public space and the private and semi-private spaces that neighbours are expected to watch and defend.

In our work [25], we investigate whether safer-looking neighbourhoods are more likely to experience more human activity and therefore experience more natural surveillance. Using mobile phone data as a proxy for human activity for the cities of Rome and Milan, and scores of perceived safety, estimated using a well-known Convolutional Neural Network (AlexNet [26]) trained on a ground-truth dataset of Google Street View images scored using a crowdsourced safety visual perception survey[1], we find that (i) safer-looking neighbourhoods are more active than what is expected from their population density, employee density, and distance to the city centre; and (ii) there exist a positive correlation between safety appearance and human activity for females and people over 50 years old, and a negative correlation for people under 30 years old. This suggests that safety perception depends on the demographic of the population.

The neural network allowed us also to identify and understand the urban features that contribute to the safe appearance of a neighbourhood. To do so, we occlude a portion of the images in input to the Convolutional Neural Network and compute how the safety prediction changes depending on the part occluded.

Through this approach, we find that greenery and street-facing windows contribute to a positive appearance of safety (in agreement with Oscar Newman's defensible space theory [24]). Our results suggest that urban appearance modulates levels of human activity and, consequently, a neighbourhood's rate of natural surveillance [25].

---

[1] http://pulse.media.mit.edu.

4   The Interplay of Neighborhoods' Socio-Economic Conditions, Urban Environment, People Behaviors, and Crime

As seen in the previous section, natural surveillance, namely the mechanisms by which residents can contribute to maintaining the safety of their neighbourhoods, finds its roots in urban planning, where specific aspects of urban physical characteristics [22,24] are related to urban security. In this work [27], we investigate which factors are at play with urban crime, particularly how crime levels are associated with some aspects of social disorganization, the characteristics of built environment, and the mobility routines of people. We are not interested in predicting crime, mainly looking at the few places with the highest number of crimes (i.e., crime hotspots), but in shedding light on the diverse factors at play with urban crime. Previous studies focused on just a subset of static factors at a time in a single city. This limits our understanding of the complex urban interplay between crime, people, places, culture and human mobility. We address the need for a comprehensive study that explores crime theories across multiple cities of the world, analyzing the cities of Bogotá, Boston, Los Angeles and Chicago. The four cities differ in cultural, economic, historical and geographical aspects.

Using data sources such as mobile phone records and OpenStreetMap (OSM), we have developed a Bayesian hierarchical model that considers proxy variables for the social disorganization theory (i.e., economic disadvantage, ethnic heterogeneity, and residential stability) [28,29], the built environment and the mobility routines of people. Taking this into account, we extract social disorganization variables from census data, we compute proxies for the built environment (e.g., land use mix, small blocks, building mix, building density, Walkscore, etc.) from both census and geographical data, while proxies of human mobility are extracted from mobile phone traces.

We found that the neighborhoods' built environment characteristics affect crime, which can be instrumental, especially in cities where census variables are challenging to collect. Moreover, we have compared two alternative and almost static definitions of neighbourhood effect: the social disorganization theory and the Jane Jacobs' theory on urban vitality conditions. Then, we modelled them jointly with the dynamics of people extracted from mobile phone data and we showed that the best description of crime requires modelling the socio-economic conditions, the built environment and the people mobility all together.

Given the cultural and historical differences between the four analyzed cities, our analysis shows a great variability of results, and there is not a model that "fits it all" that can learn from one city and can be used to easily study crime in another city.

The resulting framework is potentially reproducible at scale and could be used to analyze crime in different cities. In addition, one could use the insights to make recommendations for policies and initiatives that could be the most effective in improving urban citizens' security.

5   The Impact of COVID-19 Pandemic on Human Behavior

As we have seen in the previous sections, leveraging mobile phone data enable us to study several aspects of cities such as their vitality, safety and criminality. In [30] we leverage longitudinal GPS mobility data of hundreds of thousands of anonymous individuals to empirically show and quantify the dramatic disruption in people's mobility habits and social behavior due to the diffusion of the COVID-19 pandemic and to the enacted non-pharmaceutical interventions (NPIs) such as physical distancing mandates, closures of business venues, stay-at-home orders, etc.

In order to process raw GPS data and to give them a semantic meaning, we have computed stop locations defined as places where a person stays for at least 5 min. within a distance of 65 meters [30]. Moreover, when possible we associate each stop location to the nearest Point of Interest (POI), extracted from OpenStreetMap (OSM), where a POI represents a public location that people use for business or recreational activities.

With this data source, we have explored and characterized individuals' mobility trajectories and we have shown how individuals changed their patterns of visits, their routine behaviors and their person-to-person contact activity over time. During the COVID-19 pandemic, individuals dramatically reduced the number of visits to POIs, which only partially recovered the pre-pandemic levels, while the duration of visits to POIs remained, after NPI's relaxation, significantly shorter than in the pre-pandemic period. This finding suggests that people were less willing to spend time in POIs, reasonably to minimise social contacts in public venues. The reductions in the number of visits are also heterogeneous. POIs categories such as *Arts & Entertainment* and *Nightlife Spot* were severely impacted by the pandemic. Even inside the *Shop & Service* there were differences, where essential shops such as supermarkets faced a lower reduction in the number of visits than the non-essential shops.

By only looking at aggregated mobility, we have just a partial view of changes in human behavior during the pandemic. To better understand the complexity of the individuals' mobility changes, we focused our attention on mobility motifs and routines that characterize the chronological sequence of where an individual goes.

We first transform the individual's chronological sequence of visits to places into a sequence of symbols (e.g., *Food, Residential, Workplace*) and then apply the Sequitur algorithm [31] to generate a hierarchical compressed representation of the original sequence. Overall, human routines during the COVID-19 pandemic got shorter and more predictable with respect to the pre-pandemic period. Moreover, the dramatic change in people's behavior also emerges from the similarity between the characteristic routine of different individuals. Applying agglomerative hierarchical clustering [32] of routines before and during the pandemic, we observe that clusters became larger, which means that mobility routines simplify and people's behavior gets more homogeneous. For example, by inspecting the everyday routines in the two biggest clusters, we observe that individuals limit their mobility to *Residential* ↔ *Residential*, *Residential* ↔ *Shops & Services* and *Shops & Services* ↔ *Shops & Services* routines.

Finally, we have also observed a risk adaptation factor, which increases the people's mobility over time regardless of the stringency of non-pharmaceutical interventions.

In sum, our results show that the policy interventions and people strategies to minimize the risk of infection have profoundly reshaped individuals' routines and habits, changing how they experience places and social interactions during the pandemic. These findings should inform policymakers in designing interventions to support individuals and commercial activities that experienced the major disruption during the pandemic.

6   Looking Ahead

As we have seen in our studies described above, the life of a city and of its neighborhoods is deeply associated with the urban structure and the urban appearance. However, it is still unclear how to design neighborhoods to become more vital and safer, to reduce traffic congestion, to be resilient to disruptive shocks such as a pandemic and the related policy restrictions, etc. A possible direction to explore is resorting to the usage of Generative Adversarial Networks (GANs) [33]. GANs have been successfully applied in computer vision for transferring style from an image to another [34], for image super-resolution [35], or to learn the mapping between different visual domains [36]. Interestingly, novel GAN architectures that use land use constraints, satellite imagery and/or street view images can be adopted to generate realistic conditioned urban images where a given attribute has to satisfy a specific value. More precisely, this sketched approach could generate a new image that modifies some specific information (e.g., the presence of a given building or of a set of buildings, the presence of parks, etc.) conditioned to some constraints (e.g., lowering crime levels, enhancing attractiveness, etc.).

A similar tool, based on a GAN-framework, may help policymakers and citizens to anticipate the consequences of urban changes as well as it may suggest to urban planners the specific intervention which would produce the desired change to a specific attribute (e.g., lowering traffic congestion, increasing safety perception, etc.)

References

1. Glaeser, E.L.: Triumph of the City: How Our Greatest Invention Makes Us Richer, Smarter, Greener, Healthier, and Happier. Penguin Press, New York (2011)
2. Milgram, S.: The experience of living in cities. Science **167**, 1461–1468 (1970)
3. Bettencourt, L.M.A., Lobo, J., Helbing, D., Kühnert, C., West, G.B.: Growth, innovation, scaling, and the pace of life in cities. Proc. Natl. Acad. Sci. **104**(17), 7301–7306 (2007)
4. Bettencourt, L.M.A., West, G.B.: A unified theory of urban living. Nature **467**(7318), 912–913 (2010)

5. Bettencourt, L.M.A.: The origin of scaling in cities. Science **340**(6139), 1438–1441 (2013)
6. Schläpfer, M., Bettencourt, L.M.A., Grauwin, S., Raschke, M., Claxton, R., Smoreda, Z., West, G.B., Ratti, C.: The scaling of human interactions with city size. J. R. Soc. Interf. **11**, 20130789 (2014)
7. Boyd, R., Richerson, P.J., Henrich, J.: The cultural niche: why social learning is essential for human adaptation. Proc. Natl. Acad. Sci. **108** (supplement_2), 10918–10925 (2011)
8. Pan, W., Ghoshal, G., Krumme, C., Cebrian, M., Pentland, A.: Urban characteristics attributable to density-driven tie formation. Nat. Commun. **4**, 1961 (2013)
9. Glaeser, E.L., Kolko, J., Saiz, A.: Consumer city. J. Econ. Geogr. **1**, 27–50 (2001)
10. Florida, R.: The Rise of the Creative Class - Revisited: Revised and Expanded. Basic Books (2014)
11. Logan, J.R., Stults, B.J., Farley, R.: Segregation of minorities in the metropolis: two decades of change. Demography **41**(1), 1–22 (2004)
12. Goldsmith, W., Blakely, E.: Separate Societies: Poverty and Inequality in US Cities. Temple University Press, Philadelphia (2010)
13. Cassiers, T., Kesteloot, C.: Socio-spatial inequalities and social cohesion in European cities. Urban Stud. **49**(9), 1909–1924 (2012)
14. Iceland, J., Weinberg, D., Hughes, L.: The residential segregation of detailed Hispanic and Asian groups in the United States: 1980–2010. Demograph. Res. **31**, 593–624 (2014)
15. Florida, R.: The New Urban Crisis: How Our Cities Are Increasing Inequality, Deepening Segregation, and Failing the Middle Class, and What We Can Do About it. Basic Books (2017)
16. Muster, S., Marcińczak, S., Van Ham, M., Tammaru, T.: Socioeconomic segregation in European capital cities: increasing separation between poor and rich. Urban Geogr. **38**(7), 1062–1083 (2017)
17. Sampson, R.J.: Great American City: Chicago and the Enduring Neighborhood Effect, University of Chicago Press, Chicago (2012)
18. Lee, Y., Eck, J.E., SooHyun, O., Martinez, N.N.: How concentrated is crime at places? A systematic review from 1970 to 2015. Crime Sci. **6**(6), (2017)
19. Barthelemy, M.: The Structure and Dynamics of Cities: Urban Data Analysis and Theoretical Modeling. Cambridge University Press, Cambridge (2016)
20. Zheng, Y.: Urban Computing. MIT Press, London (2019)
21. Zheng, Y., Capra, L., Wolfson, O., Hai, Y.: Urban computing: concepts, methodologies, and applications. ACM Trans. Intell. Syst. Technol. **5**(3), 1–55- (2014)
22. Jacobs, J.: The Death and Life of American Cities. Random House, New York (1961)
23. De Nadai, M., Staiano, J., Larcher, R., Sebe, N., Quercia, D., Lepri, B.: The death and life of great Italian cities: a mobile phone data perspective. In: Proceedings of The 25th International Conference On World Wide Web, pp. 413–423 (2016)
24. Newman, O.: Defensible Space. Macmillan, New York (1972)
25. De Nadai, M., et al.: Are safer looking neighborhoods more lively? A multimodal investigation into urban life. In: Proceedings of the 24th ACM International Conference on Multimedia, pp. 1127–1135 (2016)
26. Krizhevsky, A., Sutskever, I., Hinton, G.E. ImageNet classification with deep convolutional neural networks. In: Advances in Neural Information Processing Systems (NIPS), p. 25 (2012)

27. De Nadai, M., Xu, Y., Letouzé, E., González, M., Lepri, B.: Socio-economic, built environment, and mobility conditions associated with crime: a study of multiple cities. Sci. Rep. **10**, 1–12 (2020)
28. Sampson, R.J., Byron Groves, W.: Community structure and crime: testing social-disorganization theory. Am. J. Sociol. **94**(4), 774–802 (1989)
29. Sampson, R.J., Raudenbush, S.W., Earl, F.: Neighborhoods and violent crime: a multilevel study of collective efficacy. Science **277**(5328), 918–924 (1997)
30. Lucchini, I., et al.: Living in a pandemic: changes in mobility routines, social activity and adherence to COVID-19 protective measures. Sci. Rep. **11**, 1–12 (2021)
31. Nevill-Manning, C., Witten, I.: Identifying hierarchical structure in sequences: a linear-time algorithm. J. Artif. Intell. Res. **7**, 67–82 (1997)
32. Hastie, T., Tibshirani, R., Friedman, J.: The Elements of Statistical Learning, 2nd edn. Springer, New York (2009). https://doi.org/10.1007/978-0-387-84858-7
33. Goodfellow, I., et al.: Generative adversarial nets. In: Advances in Neural Information Processing Systems (NIPS), pp. 2672–2680 (2014)
34. Isola, P., Zhu, J.-Y., Zhou, T., Efros, A.A.: Image-to-image translation with conditional adversarial networks. In: Proceedings of the IEEE Conference on Computer Vision and Pattern Recognition (CPVR), pp. 5967–5976 (2017)
35. Ledig, C., et al.: Photo-realistic single image super-resolution using a Generative Adversarial Network. In: Proceedings of the IEEE Conference on Computer Vision and Pattern Recognition (CPVR), pp. 4681–4690 (2017)
36. Zhu, J.-Y., Park, T., Isola, P., Efros, A.A.: Unpaired image-to-image translation using cycle-consistent adversarial networks. In: Proceedings of the IEEE International Conference on Computer Vision (ICCV) (2017)

AI Approaches in Processing and Using Data in Personalized Medicine

Mirjana Ivanovic[1]($\boxtimes$) (iD), Serge Autexier[2] (iD), and Miltiadis Kokkonidis[3]

[1] Faculty of Sciences, University of Novi Sad, Novi Sad, Serbia
`mira@dmi.uns.ac.rs`
[2] German Research Center for Artificial Intelligence (DFKI), Bremen, Germany
`serge.autexier@dfki.de`
[3] Netcompany-Intrasoft S.A., Luxembourg, Luxembourg
`Miltiadis.KOKKONIDIS@netcompany-intrasoft.com`

Abstract. Nowadays, more and more people suffer from serious diseases and doctors and patients need sophisticated medical and health support. Accordingly, prominent health stakeholders have recognized the importance of development of such services to make patients' life easier. Such support requires the collection of patients' complex data. Holistic patient's data must be properly aggregated, processed, analyzed, and presented to the doctors/caregivers to recommend adequate treatment and actions to improve patient's health related parameters. Advanced artificial intelligence techniques offer the opportunity to analyze such big data, consume them, and derive new knowledge to support (personalized) medical decisions. New approaches like those based on advanced machine/deep learning, federated learning, transfer learning, explainable artificial intelligence open new paths for more quality use of health and medical data in future. In this paper, we will present some crucial aspects and examples of application of artificial intelligence approaches in (personalized) medical decisions.

Keywords: Artificial intelligence · Machine learning · Personalised medicine · Cancer treatment · Quality of life parameters

1 Introduction

We are witnesses of more and more sick population and it is necessary to take care of development of sophisticated multi-disciplinary approaches for medical diagnoses and treatments [7, 21]. Consequently, development of helpful medical services is getting crucial traction in medical innovation. The importance of improvements of patients' health related quality of life (QoL) parameters are also widely recognized in therapies and follow-ups of serious diseases survivors. Cancer patients experience a serious disruption of QoL parameters (fatigue, pain, psychological difficulties, appetite loss, sexual problems and so on). Additionally, they experience also "usual" problems like the majority of the population (anxiety, stress, sleep disorders, and so on) during active oncological treatment period.

© Springer Nature Switzerland AG 2022
S. Chiusano et al. (Eds.): ADBIS 2022, LNCS 13389, pp. 11–24, 2022.
https://doi.org/10.1007/978-3-031-15740-0_2

To support the development of sophisticated software services that can help patients to successfully cope with everyday activities it is necessary to find ways to collect and properly integrate wide spectra of complex patient data (apart from traditional clinical data also data collected from smart wearable devices, nutritional data, and so on). Health-related data should be aggregated in such forms that obtain adequate, useful, and reliable conclusions after processing. Results achieved after data processing should be presented to the doctors/clinicians in understandable and friendly form [7].

Modern, emergent approaches in collecting, processing, and analyzing patient's data support more appropriate interventions, and usually more tailored and personalized treatments [4, 10]. In this paper we will present the current state-of the-art in developing medical and clinical platforms, discuss crucial aspects and functionalities, and present characteristic examples of applications of artificial intelligence approaches in (personalized) medical decisions [23, 24].

The rest of the paper is organized as follows. In Sect. 2, different sources of patients' medical and health-related big data are briefly discussed. Section 3 considers some emergent artificial intelligence approaches that support quality medical decisions. characteristic medical decision support systems are presented in Sect. 4. Concluding remarks and future trends in processing big medical data are pointed out in the last section.

2   Different Sources of Patients' Medical Big Data - Collection and Processing

Electronic Health Record (EHR) is usually seen as basic source of information for any patient. It keeps data of several important aspects of a patient (like clinical information, diagnoses, medication, ...). For more reliable follow-ups of patient's health, it is necessary to consider also other data sources and in modern medical data processing they also can include so called Patient Health Record (PHR). A PHR usually contains the same (or similar) kinds of information as an EHR but it is managed by patient. For better analysis and use of all information collected for a patient it is necessary to combine them but also if possible, to incorporate patient's data from other diverse and multiple sources. For example, the CrowdHEALTH project [5] is oriented to the combination of patient's data from various sources to benefit from community knowledge and form Holistic Health Record (HHR). As a result of this project an integrated holistic platform is developed and it incorporates big data management mechanisms to support the logical pipeline of data management: acquisition, cleaning, integration, modelling, analysis, information extraction and interpretation [13].

Further improvement steps are oriented towards advanced approaches based on the integration of HHR and Social HHR (SHR). HHR as an extension of EHR usually contains data like physical activity, nutrition, environmental conditions, information collected from variety of sensors, social care information, and so on. SHR covers patient's aspects of social life and usually contains information about different social aspects and activities like: relationships, particular events, experiences, etc..

After identifying and considering different patient's data sources the next step in medical systems/platforms/frameworks is to find adequate ways and techniques to better acquire, manage, model, and process this data in order to achieve as much as possible, high-quality outcomes and results. Such result, based on huge amounts of data, should be exploited and presented in a user friendly way to doctors/clinicians, caregivers, or even to patients. The main intention in such systems is to try to achieve satisfactory level of tailored decisions to achieving better patient's QoL parameters.

Another interesting approach oriented towards use of complex patient's data and processing it by application of modern AI/ML approaches is developing under BD4QoL project [2]. The focus in the project is on implementation of a personalized management of head and neck cancer survivorship by providing doctors and survivors with an unobtrusive, privacy compliant, real-time monitoring. Such complex supportive environment should offer personalized interventions based on integration of Big Data-driven AI algorithms and models. Patient's traditional health data will be integrated with data collected from mobile and wearable devices for real-time assessment of patient's QoL.

Measuring cancer patient's QoL is about understanding the impact of cancer and how well people are living after their diagnosis and treatment. This includes a wide range of concerns, such as people's emotional or social wellbeing, finances, and ongoing physical problems, such as tiredness, sleep disorders, and pain.

We can conclude that there is trend in medical decision support systems to integrate "traditional" medical and health data sources with novel ones that include data from smart and wearable devices, IoT and sensors generated data, open data, environmental data, etc. Integration of multiscale/multimodal big health data is a challenging task in intelligent big data processing. Heterogeneous data should be aggregated in such a way to enable to generate meaningful conclusions to be presented to doctors in user friendly way.

The rapid development of information communication technologies, applications of IoT and pervasive smart environments in our everyday life promotes the frequent use of different smart wearable devices for monitoring and measuring some health parameters [8]. Constant improvements and development of such devices impose that they should satisfy specific requirements. So, for standard healthcare intervention 5 main features of wearable devices are detected in [16] to ease data collection: "(1) wireless mobility; (2) interactivity and intelligence; (3) sustainability and durability; (4) simple operation and miniaturization; and (5) wearability and portability."

If we concentrate on cancer patients, then it is evident that in several last decades the number of cancer survivors are increasing. Thus, there is a need to develop medical systems with tailored, personalized services that will help in improving patients' QoL parameters. So, it is important to include in patient's medical records their personal experiences. So, in contemporary medical decision support systems a number of questionnaires/tools to measure cancer patient's individual views of his/her health status should be considered. PROMs (Patient Reported Outcome Measures) and PREMs (Patient Reported Experience Measures) are widely used to check patient's perceptions from two aspects: health and experiences after receiving treatment/care. The QoL parameters are getting very important for cancer survivors. Therefore, the research activities should be oriented towards obtaining reliable and early predictors and QoL parameters over time, and improve treatment decisions and follow-up strategies. Medical systems/platforms/frameworks should support the utilization of big HHR and datasets, powerful AI/ML approaches that facilitate the integration of QoL instruments (like

PROMs and PREMs), implementation of a user-centered communication interface, and personalized support.

Modern societies are getting more and more "smart". Smart environments equipped with sensors, mobile and wearable intelligent device have a potential to positively influence patient's QoL. Especially, wearable devices play an important primary role to establish and maintain a connection between patients and doctors which offers a great potential to support the quality of medical treatment and recommendations. Additionally acquired complex data generated in smart environments and with wearable devices offers numerous opportunities in medicine and healthcare for the development of more powerful mobile health applications [18] or the development of complex IoT sensing-based health monitoring systems [6, 14].

3 Emergent Artificial Intelligence Approaches for Supporting Quality Medical Decisions

With appropriate medical treatment and support more and more people suffering from different critical diseases are living and normally go about their everyday routine activities. Also, more than ever people are living with and beyond cancer. Receiving adequate treatment tailored to their needs patients can keep and even increase their positive experience and QoL.

Different personalized services that support humans in their numerous activities are a modern approach in software development. Personalization is the process of tailoring specific service to reach the needs of individuals or groups with similar attitudes. Such an approach is also crucial in medical and healthcare domains. Personalized medical services for patients with similar needs are adequate therapies, decisions, interventions, and recommendations adjusted to their specific health status [1, 7]. Therapeutic strategy for "the right person at the right time" can support improvement and efficiency of the treatment, reduce possible side effects and increase the QoL.

QoL parameters are getting essential for cancer survivors. They influence the development of adequate services for person-centered monitoring and follow-up planning. Complex patient's data collected from multiple sources (EHR, PHR, data from wearable, smart devices, patient's reported outcomes, etc.) should be processed to be used in improving personalized treatment. Powerful AI/ML approaches are essential instruments for quality data processing that lead to better predictions, interventions and good health status. However, before applying AI/ML techniques diverse data must be prepared in an adequate way (i.e. aggregated, processed, analyzed). Contemporary AI approaches: (deep) ML [7], explainable AI (XAI), image processing (IP), natural language processing (NLP), agent technologies [10], robots, and so on, immensely influence the development of medical systems/platforms/frameworks. Contemporary AI approaches as federated learning (closely related to cloud/edge concepts), high possibilities of neural network architectures combined with transfer learning (e.g., repurposing features of established models explored to address data heterogeneity) offer great capability for developing powerful medical applications. Existing but also newly developing medical systems should utilize patients' big datasets integrated with QoL instruments, make more power and reliable AI/ML engines, support more friendly user-centered communication, visualize results of AI predictive models and make them more understandable to doctors (employing different XAI techniques), improve personalized support, and so on. Such holistic systems should: improve post-treatment patients' health status and QoL; follow-up the patients to meet their needs and make their everyday life bearable; but also help in predicting the status of new patients.

Many large projects focus on cancer patients and better QoL parameters based on their available complex data. Considering some of them (e.g. https://oncorelie f.eu/, https://www.gatekeeper-project.eu/, http://www.bd2decide.eu/, https://ascape-pro ject.eu/, etc.), we outline a "typical Health AI system". Such a system is composed of several subsystems each containing various components devoted to specific task. Three groups of such subsystems can be distinguished.

1) Data Management subsystem that is responsible for secured patients' data collection from multiple sources usually taking care of anonymization [22]. They are also focused on the aggregation of heterogeneous data, their transformation in some of widely used clinical data standards which address different aspects [21] (like SNOMED Clinical Terms, openEHR archetype, FAIR) but also to prepare them in formats appropriate for AI/ML processing. The essential tasks of this subsystem are focused on multiple sources data collection and its preparation for AI/ML processing. During these activities privacy preservation of patients' data must be guaranteed, and its preprocessing, harmonization, and semantical alignment is needed based on variety of services: Data Collection service, Privacy preserving Service, Data Curation/Filtering services, Data Harmonization services, and some others.

2) AI/ML subsystem is tightly related to the Data Management subsystem. This subsystem, for which depending on the nature of the data suitable AI/ML methods are selected to be applied, is responsible for comprehensive data processing and analyses of computed results. Based on a wide variety of techniques after data processing important and influential features/parameters are discovered, characteristic patterns of behaviors are noticed, powerful predictive models are generated. Predictive models, based on available patients' datasets, produce quality predictions, interventions, treatment recommendations that should be presented to doctors/caregivers/patients.

This subsystem usually includes Big Data analytics and modelling, and it is the central in a medical system and represents the logical link between the data management part and interface part. The AI/ML subsystem uses a variety of ML algorithms (for feature selection, outlier detection, classification, regression, and so on) based on available modern ML frameworks (like TensorFlow, Mahout, etc.). Depending on the application domain of a platform and particular disease, ML approaches can also include Medical Imaging Analytics using Recurrent (RNN) and Convolutional Neural Network (CNN) architectures, communication supported by virtual companions [10].

3) Intelligent/Smart Interface subsystem usually is the connection between patient's data and results achieved by AI/ML subsystem. Depending on the main aim of a medical system this subsystem can generate different forms of interfaces for doctors, caregivers, but also for patients. To extract and represent insights of the patient's health status and conditions the interface is usually implemented using powerful AI techniques (like XAI, data visualization, agent technologies and so on). Interface uses generated AI/ML models and suggests adequate, personalised treatments, interventions, various actions, activities, nutrition and so on. In some advanced medical systems AI/ML models can also be downloaded to patients' devices, locally analyze patients' data and recommend appropriate actions for improving their QoL.

In typical medical systems various predictive models are generated to support personalized medical decisions. To make results of predictive models more understandable to doctors and other end-users recently XAI methods are using (like Shapley Additive exPlanations, LIME, Anchors, Textual Explanations of Visual Models, Integrated Gradients) [9], and different ways of data visualization adjusted to different dashboards, smartphones and similar devices are applied. Additionally, the devices that end-users use in communication with a platform should allow for full access to relevant health-related data, support regular updates and obtain information about the patient's QoL parameters and effects of suggested interventions.

Depending on general organization and use of specific medical systems/platform/frameworks, it can incorporate and consider patients' datasets from arbitrary number of hospitals, train and lately use AI/ML predictive models considering common knowledge gained from all available datasets. Such types of systems adopt federate style of data processing, models training and using achieved powerful AI/ML predictive models. Federated learning (FL) as rather new ML approach creates an ML pipeline which significantly reduces the risk of data privacy being compromised. Federated ML is based on existence of multiple clients (edge nodes), that work together to train a single model organised and stored on single server i.e. cloud. In case of medical systems edge nodes usually represent hospitals.

An FL system has two actors: multiple edge nodes and the server. The server coordinates the training process between all edge nodes that participate in the construction of a global model. Each edge node receives a copy of the global model to be trained and updates it based on available local data. When training phase is finished all edge nodes are participating in the training send their updated model's weights back to the server to synchronize them and produce unique global model.

In such an approach sensitive patient's data remains decentralized and FL keeps the data at its local edge nodes (hospitals) and transfers only models' updates to the main server. Predictive models created and trained on local nodes'/edges' data are participating in creating global/centralized federated models. This is succeeded by distributing the model architecture and initial weights to all edge nodes participating in producing a global/federated model. Furthermore, edge nodes train their copy of the global model on local data. When training is finished achieving satisfactory results of the training, updated weights are sent back to the FL server (cloud) to contribute to creating new or updating existing common global/central model.

The central AI/ML component in a medical system is the main source of predictive models trained on a number of datasets from local edge nodes/hospitals, and the models are constantly updating when new data appears and getting more and more reliable and with higher prediction power.

4   Medical Decision Support Systems

In this section we will briefly present two characteristic medical systems based on contemporary technological achievements that incorporate abovementioned concepts and approaches.

4.1   Smart Ambient Intelligent Living Environments

Development of 4G and upcoming 5G and 6G networks have significant influences on development medical services and decision support systems/platforms/frameworks. Accordingly, a range of more and more sophisticated smart, intelligent devices support monitoring patients' daily activities and follow changes of their health related parameters. In such a way more and more patients are living in technologically interconnected worlds. Technological advancements and innovations can significantly support patients to efficiently cope with everyday activities [10].

Ambient Assisted Living (AAL) and Ambient intelligent (AmI) environments facilitate patients in their living space. They incorporate intelligent and flexible services to patients acting in their living space like: Sensors, Networks, Pervasive Ubiquitous Computing and AI, Unobtrusive Human-Computer Interfaces [20].

AAL and AmI encompass monitoring services that supports patients in their everyday activities and living habits, also suggesting them possible actions that can improve their QoL and wellbeing. Main functionalities of a comprehensive ALL environment are described in [3, 17] for a patients' residence or house. Numerous sensors are located at different places, such as sensors for light control, home automation control, presence sensor, medication control, and others to collect patients' data and monitor their daily activities and behaviors (see Fig. 1).

To propose patients' personalized predictions, treatments, recommendations or even possibility of prevention some serious diseases a wide variety of data are collected from such smart environments (like nutrients, physical activities, the microbiome, toxin exposure) and processed using advanced AI/ML methods.

The availability of the smart devices and wearable sensor technology [25] are prominent in a fast accumulation of patient's sensitive and complex health data. Emergent AI/ML techniques are promising in processes of mapping such big data into adequate, personalized health predictions.

4.2   Intelligent System for Supporting Cancer Patients

There are multiple challenges to be addressed by an AI system that aims to enhance clinical practice. A good AI engine with excellent analytical performance characteristics is not sufficient. Matters such as user experience, integration, security, privacy, etc. must also be addressed. The EU-funded research and innovation project ASCAPE (https://ascape-project.eu/) presents an interesting proposition covering all aforementioned aspects of a system that aims to be in a position to enhance clinical practice.

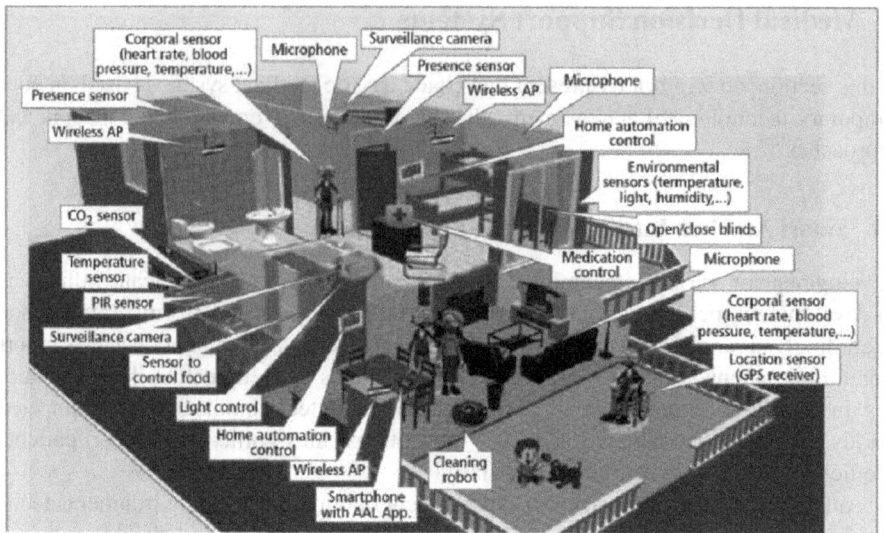

Fig. 1. AAL for elderly people's residence or house (from [17]).

Specifically, ASCAPE aims to provide doctors with an AI-powered tool that monitors and predicts the progression of QoL metrics corresponding to overall QoL and specific issues for a specific patient and offers suggestions for interventions that could improve outcomes. The ASCAPE personalized visualisations widget presents the patient's overall QoL timeline, various QoL issues timelines, a spider chart depicting the latest recorded and the predicted values for the various QoL issues and a list of interventions ASCAPE deems relevant, allowing the doctor to get an overview of the patient QoL and the history of interventions without a litany of interactions. The default view provides both recorded data and predictions for the case that any currently active interventions remain so. Doctors can see how different choices of interventions affect the predictions for the patient's overall QoL and all QoL issues simply by clicking on it. This is a simple interaction producing a predictable response from the system. The system also offers shortcuts, including one where the "ASCAPE-Proposed interventions" are selected.

ASCAPE, unlike the majority of similar clinically-targeted AI-focused research projects, paid particular attention to providing an easy pathway for integration with existing systems. Part of this effort relates to the user interface already discussed. The widget discussed and likewise the widget showing a summary of the current and predicted QoL issues status can easily be embedded into existing Health Information Systems (HIS) doctors are already using. This has the desirable consequence that doctors will not have to log in to yet another IT system and navigate to the patient again. ASCAPE makes integration a priority, point we will return to when discussing the ASCAPE architecture. Another priority for ASCAPE is that hospitals on the one hand maintain control of their patient data and on the other are able to collaborate on building AI models capturing knowledge from multiple hospitals' patients. For this it relies on two different technologies: Federated ML (FL) and ML on homomorphic encrypted (HE) data (Fig. 2, https://ascape-project.eu/marketing-material/ascapeframework-and-technical-innovations).

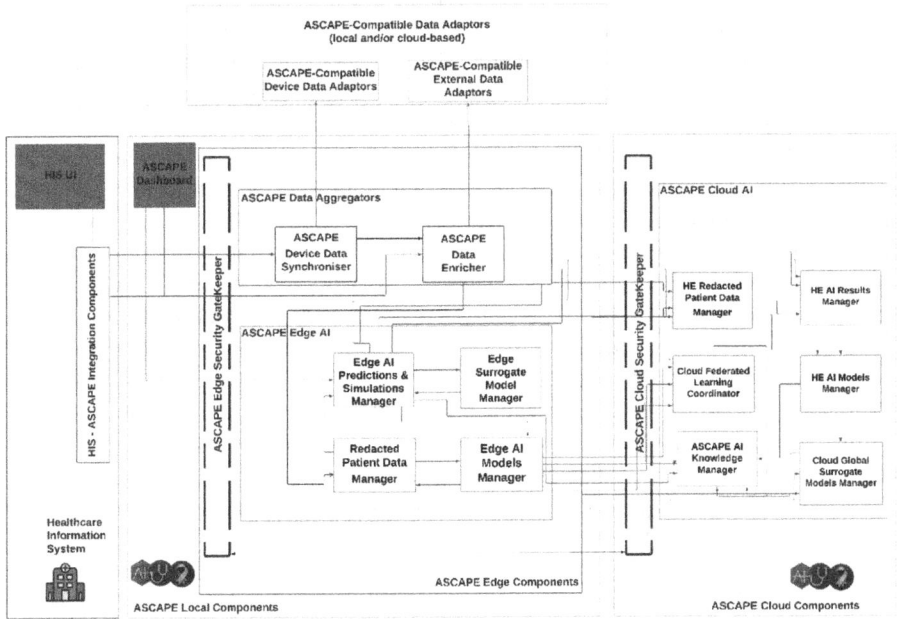

Fig. 2. The ASCAPE framework architecture.

HIS-ASCAPE Integration Components. These components allow an existing HIS to send its data (including both EHR and QoL questionnaire data) to ASCAPE and ideally also integrate the ASCAPE widgets and supporting backend code that provides the HIS with ASCAPE functionality identical to the stand-alone ASCAPE Dashboard's making the latter redundant and offering doctors the benefits of ASCAPE.

The ASCAPE Dashboard. A web application doctors may use, if ASCAPE is not sufficiently integrated into the HIS, in order to access ASCAPE functionality including AI-assisted monitoring of their patients' QoL status and recording information about proposed interventions, registering and de-registering a patient's wearable device.

The ASCAPE Edge Components. Installed locally at each hospital, these components collaborate with the HIS and the Dashboard on one end and, if so configured, with the ASCAPE Cloud which coordinates privacy-compliant collaborative model training with all participating hospitals and provides collaboratively training predictive models to all hospitals. Note that all edge node components that interact with the ASCAPE cloud, any interactions are initiated from the edge node towards the cloud in order to fit as best as possible to firewall settings in place at hospitals IT environments.

A. Edge Gatekeeper - A component that provides TLS/SSL termination, access control and an additional layer of pseudonymization.

B. Data Aggregators - Components that provide support to the task of sending to the ASCAPE Edge Node additional patient-related data not collected by the HIS but rather by ASCAPE-compatible Data Adaptors deployed locally or remotely; currently a FitBit data adapter and a weather service data adapter are included.

C. Redacted Patient Data Manager - The component responsible for storage of all patient data received from the HIS or the data aggregators within the Edge Node. It furthermore includes the extraction of patient specific inference requests and training datasets for the different target variables to train AI/ML models. The inference requests are forwarded to the Edge AI Predictions & Simulations Manager. The training datasets undergo privacy enhancing methods such as outlier detection and differential privacy and are then sent to the Edge AI Models Manager and Edge Surrogate Models Manager as well as after homomorphically encrypted to the ASCAPE cloud. Finally, the component is responsible for providing stored patient-related data to the HIS and/or the Dashboard.

D. Edge AI Models Manager - The component responsible for training local and global models with local data in collaboration with the ASCAPE Cloud, as well as for analytically evaluating models and choosing the ones that best fit local data. For each training dataset received from the Redacted Patient Data Manager several types of models are trained both on local data only as well in federated manner orchestrated by the Cloud Federated Learning Coordinator. For classification tasks, support vector machine classifiers, Naïve Bayes, K-nearest neighbors', Decision Tree, and Random Forest classifiers are trained, and for regression tasks Linear, Ridge, Lasso, Elastic Net, Kernel Ridge, Support Vector Machine, Random Forest, K-nearest neighbors', and AdaBoost regressions are used. All locally trained models are stored in the component as well as any global model obtained from the ASCAPE cloud. The quality of the models is evaluated over the locally available datasets, using appropriate metrics.

E. Edge Surrogate Models Manager - The component responsible for training local surrogate models (linear regression and decision trees) and for training global surrogate models for global predictive models with using the local data in collaboration with the ASCAPE Cloud. Surrogate models are trained to make the same predictions as the primary models (of the Edge AI Models Manager) but due to their nature (e.g. decision tree models) lend themselves to being used for explaining these predictions.

F. Edge AI Predictions & Simulations Manager - The component that uses the locally available models (local models, global models from via federated learning) as well as the Homomorphic Encrypted models at the ASCAPE cloud to produce QoL-related predictions and intervention suggestions to the HIS and/or the Dashboard. The used models are those with the best evaluation over the local data and the predictions from the HE models are obtained by sending encrypted patient-specific inference requests to the ASCAPE cloud and decrypting locally the received encrypted prediction. Furthermore, the component is responsible to compute feature attributions in form of Shapley Values to allow to visualize the impact of the different features on the predicted target values.

In addition to computing predictions and explanations, the component also pre-computes intervention suggestion: the goal is use the predictive capabilities of trained models and interventions of any kind for the patient and selected by the medical partners to provide for each patient with suggestions of interventions that have a positive effect on the predicted value. This is performed by simulations estimating the treatment effect of interventions and provide that information for retrieval by the ASCAPE Dashboard to show it to the doctors treating the patients, which can then take a decision.

The ASCAPE Cloud Components. The component allows privacy-preserving ML technologies on the ASCAPE Cloud: (i) the coordination and storage components for FL, (ii) the training, storage components for model training on HE data and encrypted predictions, and (iii) the components used for collaborative surrogate model training.

A. Cloud Gatekeeper - A component that provides TLS/SSL termination and controls which Edge Nodes may collaborate with the ASCAPE Cloud.

B. Cloud Federated Learning Coordinator - This component coordinates the federated training of global predictive models based on the patient data available at each participating edge node. The same type of models as locally are trained in federated manner for classification and regression tasks. The federated training is initiated by the edge nodes. If an edge needs a specific model and no global model is available in Cloud Knowledge Manager, it starts training locally and sends it as a first instance to the Cloud Federated Learning Coordinator. If a global model is available, the edge node updates it with its local training data and submits it again to the cloud (incremental FL mode). If more than one edge node wants to train a model, this component switches to semi-concurrent mode, where training happens in several rounds by collecting the trained or updated model from each edge node, creating an aggregated model.

C. Cloud Knowledge Manager - This component stores all available final global models on the cloud, from which they can be retrieved by the edge nodes. This way new edge nodes entering the federation can benefit from models previously trained on data from all other edge nodes.

D. HE Redacted Patient Manager - This component receives and stores the HE training datasets from all edge nodes. The training datasets can be identified regarding cancer type and target variables and are combined to a single HE dataset for each cancer type and target variable. These aggregated datasets are then forwarded to the HE AI Models Manager for training global HE predictive models.

E. HE AI Models Manager - This component stores all models trained on the aggregated HE datasets. They can be retrieved by the HE AI Results Manager to provide encrypted predictions on encrypted inference requests submitted from the edge.

F. HE AI Results Manager - The HE AI Results receives all encrypted inference requests for predictions from the different edge nodes. Based on the type, it retrieves the corresponding model from the HE AI Models Manager. If the model is not yet available, it waits until the model is available. The encrypted prediction is stored in the component in order to be retrieved by the edge node that submitted the request. The inference requests can be of different kinds: of course, any inference request in the edge node is also submitted to this component. However, during the computation of SHAPLEY values and the training of surrogate models further requests are created by the edge components and submitted to this component in order to determine these for the HE models.

G. Cloud Global Surrogate Models Manager - This component coordinates all activities to train global surrogate models. The training is initiated as soon as an edge node requests a surrogate model which is not yet trained. The Cloud Global Surrogate Models Manage then initiates the training both for linear regression and decision tree models. Meanwhile, the Edge Surrogate Model Manager creates the local training for the surrogate models by taking the local training dataset used for the global model, but labelling it using the predictions of the global model.

The training of linear regression surrogate models essentially works like the federated learning of normal models. Training of decision tree surrogate models is more involved, as separate training or update of models and aggregating via averaging is not possible. The decision tree with the best overall score across all datasets is used as the resulting surrogate model.

5 Conclusion

Growth of population and rapid technological development offer a variety of possibilities for implementing sophisticated and highly personalized medical services nowadays but in the future as well. Development of more and more powerful AI/ML algorithms, image processing, efficient big data processing, natural language processing, virtual and augmenter reality (VR/AR), IoT, agent technologies and other [11], offer a significant shift in medical and health domains [15].

All these possibilities direct medical research and practice in prominent directions [16]: more reliable and precise health analytics and predictive modeling [7], power data visualization techniques, tailored therapies, recommendations and interventions, personal user-friendly interfaces for communication [9] between different participants and stakeholders.

Avatars, metaverse [16], holographic construction [12] are newest concepts that have a high potential and can influence future development of holistic, sophisticated medical systems. In spite the fact that current achievements in these areas are sporadically used in medical systems it can be expected that they will have great influence and increase quality and functionality of medical systems in the future.

Ongoing and future research in the health domain needs extensive interdisciplinary and multidisciplinary collaborations. Important aspect of future medical systems should take care of patients' cognitive and emotional behavior and support adequate modelling in such systems. In this area agent technologies, holograms, AR/VR and metaverse definitely will play an essential role.

For the future development of complex integrated medical systems, it is also necessary to take care of development of other systems devoted to: **1. Planning and resource management, 2. Data management systems, 3. Decision support systems/knowledge base systems, 4. Remote care/self care systems.**

However, the near future is not so optimistic [19]. There are a lot of problems like diverse, limited, and distributed patients' data sources, satisfactory but not fully reliable AI/ML models, rather slow big data processing mechanisms, integration of wide variety of multiple AI services, personalized medicine limitations and so on.

Acknowledgments. This research was supported by the ASCAPE project. The ASCAPE project has received funding from the European Union's Horizon 2020 research and innovation programme under grant agreement No 875351.

References

1. Burmester, G.R.: Rheumatology 4.0: big data, wearables and diagnosis by computer. Ann. Rheum. Dis. **77**(7), 963–965 (2018)
2. H2020 project. https://www.bd4qol.eu/wps/portal/site/big-data-for-quality-of-life
3. Cicirelli, G., Marani, R., Petitti, A., Milella, A., D'Orazio, T.: Ambient assisted living: a review of technologies, methodologies and future perspectives for HealthyAging of population. Sensors **21**, 3549 (2021). https://doi.org/10.3390/s21103549
4. Claeys, A., Vialatte, J.S.: Advances in genetics: towards a Precision Medicine? Technological, social and ethical scientific issues of personalised medicine [Les progrès de la génétique: versune médecine de précision? Les enjeux scientifiques, technologiques, sociaux et éthiques de la médecine personnalisée] (2014)
5. Gallos, P., et al.: CrowdHEALTH: big data analytics and holistic health records. Stud. Health Technol. Inform. **258**, 255–256 (2019)
6. Hassanalieragh, M., et al.: Health monitoring and management using Internet-of-Things (IoT) sensing with cloud-based processing: opportunities and challenges. In: 2015 IEEE International Conference on Services Computing, pp. 285–292. IEEE (2015)
7. He, J., Baxter, S.L., Xu, J., Xu, J., Zhou, X., Zhang, K.: The practical implementation of artificial intelligence technologies in medicine. Nat. Med. **25**, 30–36 (2019)
8. Hiremath, S., Yang, G., Mankodiya, K.: Wearable internet of things: concept, architectural components and promises for person-centered healthcare. In: 2014 4th International Conference on Wireless Mobile Communication and Healthcare-Transforming Healthcare Through Innovations in Mobile and Wireless Technologies (MOBIHEALTH), pp. 304–307. IEEE (2014)
9. Holzinger, A., Saranti, A., Molnar, C., Biecek, P., Samek, W.: Explainable AI methods-a brief overview. In: Holzinger, A., Goebel, R., Fong, R., Moon, T., Müller, K.R., Samek, W. (eds.) Extending Explainable AI Beyond Deep Models and Classifiers, pp. 13–38. Springer, Cham (2022). https://doi.org/10.1007/978-3-031-04083-2_2
10. Ivanović, M., Ninković, S.: Personalized HealthCare and agent technologies. In: Jezic, G., Kusek, M., Chen-Burger, Y.-H., Howlett, R.J., Jain, L.C. (eds.) KES-AMSTA 2017. SIST, vol. 74, pp. 3–11. Springer, Cham (2018). https://doi.org/10.1007/978-3-319-59394-4_1
11. Ivanovic, M., Balaz, I.: Influence of artificial intelligence on personalized medical predictions, interventions and quality of life issues. In: ICSTCC 2020 - 24th International Conference on System Theory, Control and Computing, ICSTCC 2020, Sinaia, Romania, pp. 445–450. IEEE (2020). ISBN 978-1-7281-9809-5
12. Kairouz, P., et al.: Advances and open problems in federated learning. arXiv preprint arXiv: 1912.04977 (2019)
13. Kyriazis, D., et al.: Crowdhealth: holistic health records and big data analytics for health policy making and personalized health. Inform. Empowers Healthcare Transform. **238**, 19 (2017)
14. Autexier, S., Lüth, C., Drechsler, R.: Das Bremen Ambient Assisted Living Lab und darüber hinaus – Intelligente Umgebungen, smarte Services und Künstliche Intelligenz in der Medizin für den Menschen. In: Pfannstiel, M.A. (ed.) Künstliche Intelligenz im Gesundheitswesen. Springer, Wiesbaden (2022). https://doi.org/10.1007/978-3-658-33597-7_40
15. Lahiri, C.; Pawar, S.; Mishra, R.: Precision medicine and future of cancer treatment. Precis. Cancer Med. **2**, 33 (2019)
16. Lee, L.H., et al.: All one needs to know about metaverse: a complete survey on technological singularity, virtual ecosystem, and research agenda. arXiv preprint arXiv:2110.05352 (2021)
17. Lloret, J., Canovas, A., Sendra, S., Parra, L.: A smart communication architecture for ambient assisted living. IEEE Commun. Mag. **53**, 26–33 (2015)

18. Lv, Z., Chirivella, J., Gagliardo, P.: Bigdata oriented multimedia mobile health applications. J. Med. Syst. **40**(5), 1–10 (2016)
19. NHS England website. https://www.england.nhs.uk/cancer/living/. Accessed 20 May 2022
20. Salih, A., Abraham A.: Ambient Intelligence Assisted Healthcare Monitoring. LAP LAMBERT Academic Publishing, p. 192 (2016)
21. Schulz, S., Stegwee, R., Chronaki, C.: Standards in healthcare data. In: Kubben, P., Dumontier, M., Dekker, A. (eds.) Fundamentals of Clinical Data Science, pp. 19–36. Springer, Cham (2019). https://doi.org/10.1007/978-3-319-99713-1_3
22. Siddique, M., Mirza, M.A., Ahmad, M., Chaudhry, J., Islam, R.: A survey of big data security solutions in healthcare. In: Beyah, R., Chang, B., Li, Y., Zhu, S. (eds.) SecureComm 2018. LNICSSITE, vol. 255, pp. 391–406. Springer, Cham (2018). https://doi.org/10.1007/978-3-030-01704-0_21
23. Tyler, N.S., Mosquera-Lopez, C.M., Wilson, L.M., et al.: An artificial intelligence decision support system for the management of type 1 diabetes. Nat. Metab. **2**, 612–619 (2020)
24. Venne, J., et al.: International consortium for personalized medicine: an international survey about the future of personalized medicine. Pers. Med. **17**(2), 89–100 (2020)
25. Wu, M., Luo, J.: Wearable technology applications in healthcare: a literature review. Online J. Nurs. Inform **23**(3) (2019)

Explainable, Interpretable, Trustworthy, Responsible, Ethical, Fair, Verifiable AI... What's Next?

Rosa Meo$^{(\boxtimes)}$[iD], Roberto Nai[iD], and Emilio Sulis[iD]

University of Torino, Turin, Italy
{rosa.meo,roberto.nai,emilio.sulis}@unito.it
https://www.cs.unito.it/do/home.pl

Abstract. Artificial Intelligence plays an increasingly important role in many knowledge fields: computer science, technology, and other sciences such as health care, one of its most compelling applications. Artificial Intelligence has impacted arts, linguistics, law, sociology, society, and everyday lives. We are demanding many properties from the products of Artificial Intelligence: users of their application fields need trust and ask for fairness, accountability, and privacy. We overview the desired properties and recall the technology that enables Artificial Intelligence to satisfy them.

Keywords: Explainable AI · Trustworthy AI · Fairness · Ethics · Accountability

1 Introduction

Artificial Intelligence (AI) refers to computational systems whose actions and decisions resemble human intelligence, including functions typically associated with intelligence, such as learning, problem-solving, planning, and acting rationally, as defined by Russell and Norvig [18]. We interpret the term AI broadly to include closely related areas such as machine learning (ML). Systems that heavily use AI, have had a significant impact in domains that include healthcare, transportation, finance, social networking, e-commerce, and education. These "intelligent" systems have almost pervaded all the areas of our modern society. This growing societal impact has brought a set of risks and concerns, including the mistakes that AI systems can make. As a response, researchers are trying to design and deploy a new generation of systems that are trustworthy, i.e., meritable of trust from human beings and more robust to errors in software, resilient to cyber-attacks, and secure, in presence of incomplete scenarios.

The ingredients for a trustworthy Artificial Intelligence (AI) are manifolds. This is related to the deployment of an AI product: sometimes the output of an AI system is used to support the decision-making, and in this case, end-users will need to trust the outcomes of the artificial model. Other times the system is

© Springer Nature Switzerland AG 2022
S. Chiusano et al. (Eds.): ADBIS 2022, LNCS 13389, pp. 25–34, 2022.
https://doi.org/10.1007/978-3-031-15740-0_3

used to inform the user about the inner structure of the instances coming from the application domain. In these cases, the end-user needs to be convinced that the system has grasped a meaningful organization of the application domain examples.

Systems whose outcomes cannot be well-interpreted are difficult to trust, especially in sectors, such as healthcare or self-driving cars, in which the impact of an erroneous decision has moral and fairness implications [15]. This need for models that are trustworthy, fair, robust with respect to missing data, high-performing in the real-world applications led to the revival of eXplainable Artificial Intelligence (XAI) [13]. This field focuses on the understanding and interpretation of AI systems' behavior. The popularity of the search term "Explainable Artificial Intelligence" in the last five years, as measured by Google Trends, is illustrated in Fig. 1. The noticeable spike in recent years reflects also the increased research output of the same period.

XAI is not a monolithic concept: it reflects several related notions. The *explainability* and *interpretability* terms are often usually used interchangeably [5,14]. However, while they are very closely related, some works identify differences among related concepts [16]. We will distinguish them in the following. XAI has numerous applications: model validation, model debugging, and knowledge discovery [11]. The obtained explanations should show whether a machine learning model is grounded upon the possible biases in the training data or show when the learned models ignore important parts of the input data and instead rely on irrelevant ones. They could show that the flaws of the models could be caused by flaws in the training data.

As the demand for more explainable machine learning models with interpretable predictions rises, so does the need for methods that can help to achieve these goals. XAI is centered on the challenge of demystifying the black boxes but also implies *Responsible AI* as it can help to produce transparent models. Responsible AI takes into account societal values and moral and ethical considerations. Responsible AI has three main concepts: *Accountability, Responsability, Transparency*; these are called the A.R.T. of AI [9]. Finally, XAI is a part of a new generation of AI technologies called the *third wave AI* [21]. One of the objectives of this ambitious "wave" is to precisely generate models than can explain themselves.

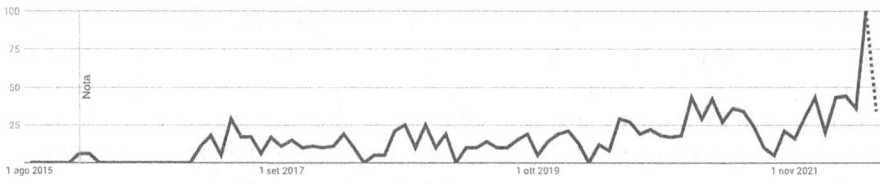

Fig. 1. Google Trends popularity index of the term "Explainable Artificial Intelligence" over the last five years (2017–2022).

2 Explainability

Explainability is more related to the techniques thought to convince the end-user about the validity of the model outcomes. The most common methods are providing post hoc explanations or recalling from the domain similar instances to the given one in input [11]. These post hoc explanations are local, and specific to single instances and can be model-agnostic or specific to the single method. The model agnostic ones treat the model to be explained as a black-box and assume the predictions of the global model can be approximated as the application of many interpretable white-box models, valid locally, in a small neighborhood of each input. Then, they sample the feature space in the neighborhood of each instance to prepare a training set that is passed to train a white-box model, such as a sparse linear model (Lasso), or if the local behavior is non-linear using if-then rules. Another approach is to determine the importance of each feature on the model by measuring the impact of features' perturbations on the output score. The results may be interpreted as counterfactual explanations, that describe a causal relationship between the input X and the output Y. They have the form: "If input X had not occurred, output Y would not have occurred".

Explanation approaches, designed for a specific type of model, leverage on the characteristics of the model to explain them. For instance, for Deep Neural Networks (DNN) we need to treat their structure as a white box and describe their components. There are three methods: back-propagation methods (top-down) compute the gradient of specific outputs with respect to the input and back-propagate it to derive the contribution of each feature. This method can be efficiently implemented in software libraries (PyTorch or TensorFlow) as a modified gradient function but can give noisy explanatory results. Perturbation methods work bottom-up (with mask perturbations in an optimization frame-work) and learn a perturbation mask that preserves the contribution of each feature and can be trained by an additional DNN. The intermediate methods either transform the representations at the higher layers of the DNN into a synthetic image together with an encoding of the target object in a mask, or they adopt a prediction's decomposition through the additive contribution of the hidden vectors in the DNN corresponding to each input (e.g., a word in the textual input to a Recurrent Neural Network). Therefore, each component of the decomposition quantifies the contribution of each input to the DNN output.

3 Interpretability

One of the most popular definitions of interpretability is the one of Doshi-Velez and Kim, who, in their work [10], define it as "the ability to explain or to present in understandable terms to a human". Interpretability is more focused on the task of exploration of the model properties with the goal of providing transparency to humans. For instance, clarifying the meaning of the components of a black-box model, like a deep neural network or a Support Vector Machine with the goal of understanding the model. The most common technique is to put aside an obscure model a "white-box" model, trained on the same instances. The latter model incorporates interpretability directly into its structure. This is the case of logical models (decision tree or rule-based model), linear models (that accompany features with coefficients whose magnitude informs their impact on the model outcome), attention model (for natural language, referred to as the words in the context).

One of the more interesting goals of learning an interpretation of a black box model is to understand the representations of the input (images) captured by the Deep Neural Network (DNN) model like a Convolutional one (CNN). Here we refer to the CNN internal network nodes because we know they encode artifacts learned from the input images. One of the most effective methods is finding the inputs that best activate neurons at a specific layer [11]. The optimization should be regularized using natural image priors produced by a generative model (GAN). Instead of directly optimizing the image, these methods optimize the latent space codes of the GAN to find an image that activates a given neuron. The visualization results provide several interesting observations. The neurons from the first layer to the last layer learn representations at several levels of abstraction, from general to task-specific. The second interesting learned issue is that a neuron is multifaceted, i.e., could respond to different images, semantically related to the same concept (i.e. faces). CNN learns distributed code for objects and learns objects by the representation of their parts that can be shared across different categories [11].

Based on the above, interpretability is mostly connected with the intuition behind the outputs of a model [1] and the idea that the more interpretable a machine learning system is, the easier it is to identify cause-and-effect relationships within the system inputs and outputs. Doshi-Velez and Kim [10] proposed the following classification of evaluation methods for interpretability: application-grounded, human-grounded, and functionally-grounded; Fig. 2 shows the taxonomy proposed. Application-grounded evaluation concerns itself with how the results of the interpretation process affect the human, domain expert, and end-user in terms of a specific and well-defined task or application. Human-grounded evaluation is similar to application-grounded evaluation; however, there are two main differences: first, the tester, in this case, does not have to be a domain expert, but can be any human end-user, and secondly, the end goal is not to evaluate a produced interpretation with respect to its fitness for a specific application, but rather to test the quality of the produced interpretation in a more general setting and measure how well the general notions are captured. Functionally grounded evaluation does not require any experiments that involve humans but instead uses formal, well-defined mathematical definitions of interpretability to evaluate the quality of an interpretability method. This type of evaluation usually follows the other two types of evaluation: once a class of models has already passed some interpretability criteria via human-grounded or application-grounded experiments, then mathematical definitions can be used to further rank the quality of the interpretability models.

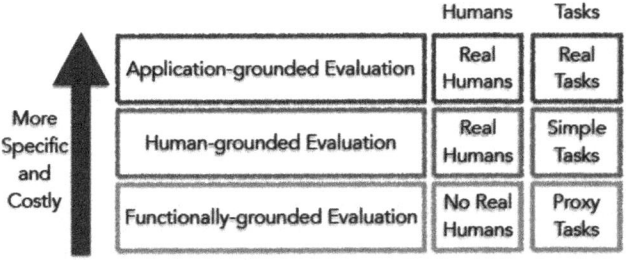

Fig. 2. Taxonomy of evaluation approaches for interpretability [10].

4   The Problem of Bias

AI explanations might reveal that decisions are influenced by factors that do not align with explicit organizational policies. Amazon canceled a plan to use AI to identify the best job candidates for technology positions upon discovering the models were biased against women because the training data consisted predominantly of males, reflecting historic hiring practices [7]. The example above explains that biases in AI mean biases in predictions. The ethical consequences of algorithmic decision-making by AI systems are a great concern. The emergence of biases in AI-led decision-making has seriously affected the adoption of AI. In order to build an unbiased system, a strong sense of justice needs to be in place to help decision makers act fairly without having any prejudice and favoritism [4].

The survey presented in [20] discusses the different sources of bias. Figure 3 shows a taxonomy of the sources of bias according to the authors.

As it is well-known, the knowledge discovery process in AI stems from a pipeline composed of different steps: data source cleaning, integration, feature selection or feature construction, model training, selection, validation, and finally, outcome presentation. All these steps might be the source of some bias. Some might be due to the users/analysts insufficient knowledge/preparation that comes out under the multiple forms of employing a sampling bias, showing a capture bias, a device bias, a measurement bias, or a negative set bias (insufficient examples for the negative class), or a confirmation bias (that leads to ignoring some relevant issues in the domain). All these examples of bias could lead to unsuitable choices in the data preparation.

Other biases could come from the presence of an ill-posed domain problem: the framing effect bias is sometimes due to the need to formulate the problem so that the experimental measured results could reflect some business objective. Another source of mistakes in the AI model is the confounding bias, that exists when an omitted feature is not included in the training data: this makes it impossible to measure the correlation between causes and effects. Another example is the inclusion of a proxy feature that is the source of indirect discrimination (e.g., zip code could be correlated to the ethnic condition, a sensitive feature that should be omitted to avoid discrimination based on ethnic conditions).

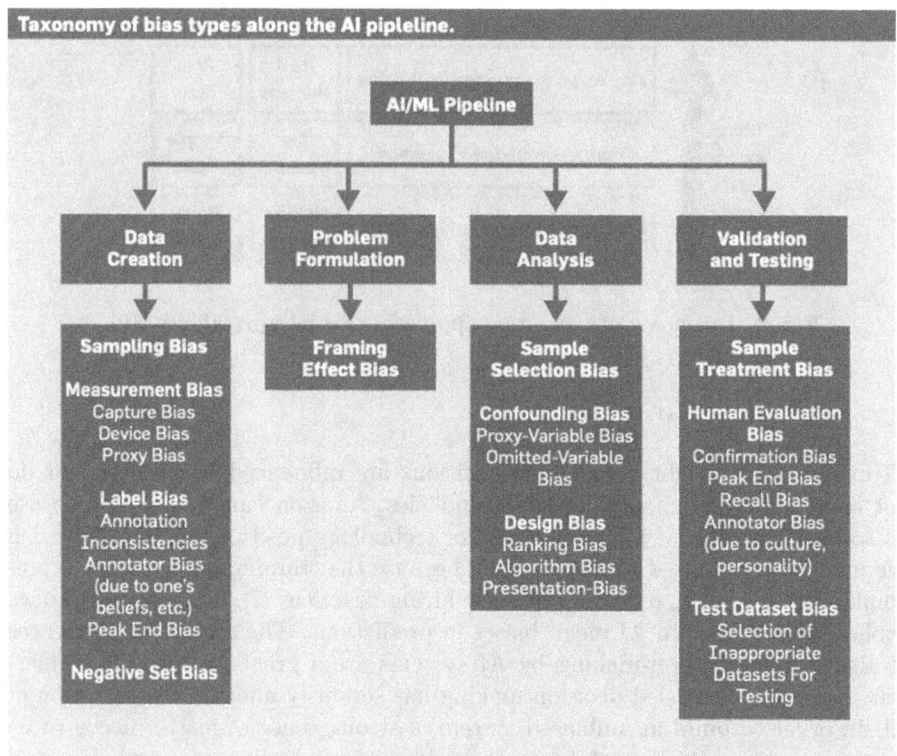

Fig. 3. Taxonomy of biases [20].

Other biases are algorithms biases: influence on the model outcome by how they explore the hypothesis and evaluate constraints, or how they present their results in a ranking to the users waiting for feedback. The users could not be impartial or could not be ready in their evaluation due to a recall bias. Finally, the deployment of the AI model might in turn influence the studied scenario and alter it.

Some effects of the biases could have serious effects on people's discrimination and on give serious doubts about the fair application of some AI models. These are discussed in Sect. 5.

5   Fairness

Because machine learning systems are increasingly adopted in real-life applications [1], any inequities or discrimination that are promoted by those systems have the potential to directly affect human lives [1]. Machine Learning Fairness is a sub-domain of machine learning interpretability that focuses solely on the social and ethical impact of machine learning algorithms by evaluating them in terms of impartiality and discrimination [6]. Traditionally, the fairness of a machine learning system has been evaluated by checking the model predictions and errors across certain demographic segments, for example, groups of a specific ethnicity or gender. In terms of dealing with a lack of fairness, a number of techniques have been developed both to remove bias from training data and from model predictions and to train models that learn to make fair predictions in the first place.

One of the proposed methods to control the bias in data is by maintaining diversity in examples' collection. It could allow the models to achieve statistical parity among the represented categories (or groups, among which some minorities exist and should be protected from discrimination). There are several measures for accounting fairness of treatment of an AI model to groups of people.

Statistical parity accounts for the parity of the rates of the favorable outcomes produced by the AI model when applied to exemplars coming from the unprivileged group and the privileged group.

Equal Opportunity accounts for the true positive rates between the unprivileged group and the privileged group.

Average Odds takes into consideration the odds between the false positive rates and true positive rates in the two groups.

Disparate impact compares the rates of the favorable outcome in the two groups by considering their ratio.

Unfortunately, not all of them could be applicable at the same time, because their satisfaction could depend on the distribution of the categories in the population and on the possible existence of some correlation between the sensitive attribute and the target.

Some software tools and standards of behavior in the data analysis and AI model development already exist [2] and are promoted by the big software vendors and by the European Community [17] and should be adopted by the software developers and the business development teams to verify the existence of some disparities in treatment by the AI model.

6   Verification

According to [19] the next generation of AI systems shall be verified with techniques similar to the formal methods used in computer science to test integrated circuits, debug software architectures, and Cyber-Physical Systems (CPS). It is composed of specifications, systems design that adheres to these specifications, verification by algorithmic search, specifications testing, simulation, and model checking. In Fig. 4 we show the tasks involved in the verification of a complex AI-based CPS, where a modular approach is essential for scalability but not yet easily reached. Finally, correct-by-construction design methods hold promise for achieving verified AI, but they are in their infancy and are still premature. Figure 4 summarizes the five challenge areas for verified AI. For each area, the current promising approaches are organized into three principles, depicted as nodes. Edges between nodes show the dependency among the principles, with the color denoting a common thread. The authors of [19] developed open-source tools, VerifAI [8] and Scenic [12] which implement the techniques based on the principles described.

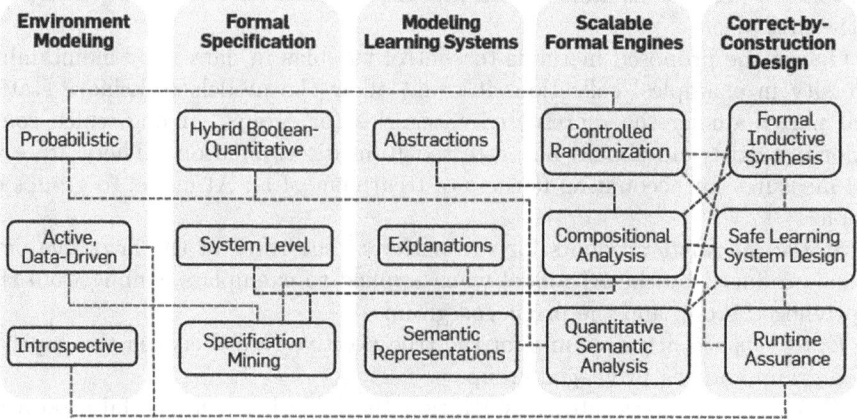

Fig. 4. Summary of the five challenge areas for verified AI, the corresponding principles proposed to address them, and their connections and dependencies [19].

For example, runtime assurance (fifth challenge) relies on introspective and data-driven environment modeling to extract monitorable assumptions and environment models (from the first challenge). Similarly, to perform system-level analysis (second challenge), we require compositional reasoning and abstraction. Some AI components may require specifications to be mined, while others are generated correct-by-construction via formal inductive synthesis (from the fifth challenge).

7   Accountability

The terms accountability, responsibility, and liability are closely related but carry different meanings. According to the OECD group of experts on AI [17], "accountability" implies ethical, moral, or in terms of management practices, codes of conduct. It guides the individuals' or organizations' actions and allows them to explain the reasons for which the actions were taken. From the viewpoint of moral principles, accountable systems are related to the concepts that guide "moral machines" [3] AI systems that are proposed and designed in a large-scale crowd-sourced experiment conducted by MIT researchers in 2018. The aim of the experiment is to collect and study the ethical and moral principles that should guide autonomous driving cars to take their decisions, in front of moral dilemmas. "Liability" generally refers to adverse legal implications arising from a person's or an organization's actions. "Responsibility" can also have ethical or moral expectations and refers to a causal link between an actor and an outcome.

Given these meanings, the term "accountability" best captures the essence of the moral principles behind the decisions of autonomous systems. In this context, "accountability" refers to the expectation that organizations or individuals will ensure the proper functioning of the AI systems that they design, develop, operate or deploy, throughout their lifecycle. For proving this, through their actions and the decision-making process they should provide documentation on the key decisions throughout the AI system lifecycle or they should conduct or allow auditing. From these viewpoints, accountability is related to systems that can be verified, as described in Sect. 6.

8   Conclusions

We provided a summary of the overview of the rapidly evolving field of explainable and interpretable AI. While many application areas of the AI systems need trust and fairness and demand responsible principles to guide the automated decisions, other applications like Cyber-Physical systems and autonomous driving need also the principles of the formal methods for obtaining verifiable systems, to guarantee software security also against cyber-attacks.

References

1. Adadi, A., Berrada, M.: Peeking inside the black-box: a survey on explainable artificial intelligence (XAI). IEEE Access **6**, 52138–52160 (2018)
2. AI Fairness 360. IBM (2022). https://aif360.mybluemix.net
3. Awad, E., et al.: The moral machine experiment. Nature **563**(7729), 59–64 (2018). https://doi.org/10.1038/s41586-018-0637-6
4. Bennetot, A., et al.: A practical tutorial on explainable AI techniques. arXiv preprint arXiv:2111.14260 (2021)
5. Bojarski, M., et al.: Explaining how a deep neural network trained with end-to-end learning steers a car. arXiv preprint arXiv:1704.07911 (2017)
6. Chouldechova, A., Roth, A.: The frontiers of fairness in machine learning. arXiv preprint arXiv:1810.08810 (2018)
7. Dastin, J.: Amazon scraps secret AI recruiting tool that showed bias against women. In: Ethics of Data and Analytics, pp. 296–299. Auerbach Publications (2018)
8. Dreossi, T., et al.: VERIFAI: a toolkit for the formal design and analysis of artificial intelligence-based systems. In: Dillig, I., Tasiran, S. (eds.) CAV 2019. LNCS, vol. 11561, pp. 432–442. Springer, Cham (2019). https://doi.org/10.1007/978-3-030-25540-4_25
9. Dignum, V.: Responsible artificial intelligence: designing AI for human values (2017)
10. Doshi-Velez, F., Kim, B.: Towards a rigorous science of interpretable machine learning. arXiv preprint arXiv:1702.08608 (2017)
11. Du, M., Liu, N., Hu, X.: Techniques for interpretable machine learning. Commun. ACM **63**(1), 68–77 (2019). https://doi.org/10.1145/3359786

12. Fremont, D.J., Dreossi, T., Ghosh, S., Yue, X., Sangiovanni-Vincentelli, A.L., Seshia, S.A.: Scenic: a language for scenario specification and scene generation. In: Proceedings of the 40th ACM SIGPLAN Conference on Programming Language Design and Implementation, PLDI 2019, pp. 63–78. Association for Computing Machinery, New York (2019). https://doi.org/10.1145/3314221.3314633

13. Gunning, D., Aha, D.: Darpa's explainable artificial intelligence (XAI) program. AI Mag. **40**(2), 44–58 (2019)

14. Koh, P.W., Liang, P.: Understanding black-box predictions via influence functions. In: International Conference on Machine Learning, pp. 1885–1894. PMLR (2017)

15. Linardatos, P., Papastefanopoulos, V., Kotsiantis, S.: Explainable AI: a review of machine learning interpretability methods. Entropy **23**(1), 18 (2020)

16. Lipton, Z.C.: The mythos of model interpretability: in machine learning, the concept of interpretability is both important and slippery. Queue **16**(3), 31–57 (2018)

17. OECD Policy Observatory: From principles to practice: tools for implementing trustworthy AI. OECD (2022). https://oecd.ai/en/tools

18. Russell, S.J., Norvig, P.: Artificial Intelligence: A Modern Approach, 3rd edn. Pearson, Hoboken (2009)

19. Seshia, S.A., Sadigh, D., Sastry, S.S.: Toward verified artificial intelligence. Commun. ACM **65**(7), 46–55 (2022). https://doi.org/10.1145/3503914

20. Srinivasan, R., Chander, A.: Biases in AI systems. Commun. ACM **64**(8), 44–49 (2021). https://doi.org/10.1145/3464903

21. Xu, W.: Toward human-centered AI: a perspective from human-computer interaction. Interactions **26**(4), 42–46 (2019)

OLAP and NoSQL: Happily Ever After

Stefano Rizzi$^{(\boxtimes)}$ ⓘ

DISI, University of Bologna, Viale Risorgimento 2, 40136 Bologna, Italy
stefano.rizzi@unibo.it

Abstract. NoSQL databases are preferred to relational ones for storing heterogeneous data with variable schema and structure. However, their schemaless nature adds complexity to analytical applications, in which a single OLAP analysis often involves large sets of data with different schemas. In this tutorial we describe the main approaches to enable OLAP on NoSQL data. We start from schema-on-read approaches, where data are left unchanged in their structure until they are accessed by the user, so they are put into multidimensional form at query time. Specifically, we show how this enables a form of approximated OLAP that embraces the inherent variety of schemaless data. Then we move to schema-on-write approaches, where a fixed multidimensional structure is forced onto data, which are loaded into a data warehouse to be then queried. In particular, we introduce multi-model data warehouses as a way to store data in multidimensional form and, at the same time, let each piece of data be natively represented through the most appropriate NoSQL model.

Keywords: NoSQL databases · OLAP · Multi-model databases

1 Introduction and Motivation

In recent years, NoSQL databases have been progressively eroding the predominance of relational databases [17]. A NoSQL database provides a mechanism for storage and retrieval of data that is modeled differently from the tabular relations used in relational databases; the particular suitability of a given type of NoSQL database (key-value, columnar, document-based, or graph-based) depends on the business problem it must address. Among the potential benefits of NoSQL databases, we mention better performance scaling, no ACID transactions, and no need for a unique schema. Indeed, NoSQL databases adopt a *schemaless* representation for data: schema is a "soft" concept and the instances referring to the same concept can be stored using different local schemas. Hence, these databases are preferred to relational ones for storing heterogeneous data with variable schemas and structural forms, such as those located in data lakes. Typical schema variants within a collection consist in missing or additional fields, in different names or types for a field, and in different structures for instances [15].

The growing use of NoSQL databases has resulted in vast amounts of semi-structured data holding precious information, which could be profitably

© Springer Nature Switzerland AG 2022
S. Chiusano et al. (Eds.): ADBIS 2022, LNCS 13389, pp. 35–44, 2022.
https://doi.org/10.1007/978-3-031-15740-0_4

integrated into existing business intelligence (BI) systems [1]. On-Line Analytical Processing (OLAP) is the querying paradigm normally used in the context of BI to analyze data stored in data warehouses, and it has been recognized to be an effective way for running analytics over big NoSQL data as well [12]. The OLAP paradigm entails dynamic analyses that read a huge quantity of data to compute a set of numbers that quantitatively describe a given business phenomenon. It assumes that data follow the *multidimensional model* [18], whose main concepts are *facts* (i.e., business phenomena such as sales), *dimensions* (coordinates used to analyze a fact, e.g., store, product, and date), *measures* (quantitative attributes that describe fact occurrences, e.g., sales revenue), and *hierarchies* (sequences of attributes that group dimension members at increasing levels of aggregation). OLAP comes in sessions, i.e., sequences of queries each obtained from the other by applying one OLAP operator (mainly, roll-up, drill-down, and slice-and-dice). Unfortunately, although the absence of a unique schema in NoSQL data grants flexibility to operational applications, it adds complexity to OLAP applications, in which a single analysis often involves large sets of data with different (and often conflicting) schemas.

In this tutorial we explore the most promising directions for enabling OLAP analyses on NoSQL data, distinguishing between the two approaches that can be followed (see Fig. 1 for an intuition): schema-on-write and schema-on-read [11]. *Schema-on-write* approaches force a (fixed) multidimensional structure in data, load them into a data warehouse using an ETL (Extract, Transform, and Load) process, then let these data be queried by users via OLAP tools. We discuss these approaches in Sect. 2. *Schema-on-read* approaches leave data unchanged in their structure until they are accessed by the user. The multidimensional schema is not devised at design time and forced in a data warehouse, but decided by every single user at querying time; clearly, this requires OLAP queries to be rewritten over NoSQL data sources. These approaches are the subject of Sect. 3. Finally, in Sect. 4 we draw the conclusions.

Fig. 1. In schema-on-read approaches (top), the user has a multidimensional view of data stored in their native (heterogeneous) form; in schema-on-write approaches (bottom), data are put into multidimensional form and stored

2 Schema-on-Read Approaches

In these approaches, source data are left unchanged in their own model and structure, to be directly queried in an OLAP fashion by the end-user without putting them in multidimensional form. Rather than being devised at design time, a multidimensional schema for accessing data is decided at querying time; while this requires OLAP queries to be rewritten over data sources on-the-fly and thus might give performance problems, it entails higher querying flexibility, simpler ETL, and lower effort for evolution. Schema-on-read approaches to enable OLAP on NoSQL data ground their roots into techniques for (i) schema discovery from XML/JSON documents, which deal with heterogeneity, quality, versioning, similarity, and comprehensiveness to produce unified schemas, schema matches, and skeleton schemas [3,22,24]; (ii) schema matching for XML/JSON documents using clustering or machine learning, in some cases considering a context [5,14]; (iii) multidimensional design from XML/JSON/columnar data, possibly by detecting and chasing functional dependencies [13,23].

In this tutorial we focus on two schema-on-read approaches, namely, Graph OLAP and Approximate OLAP; for other examples of schema-on-read approaches, see [2,11,19].

2.1 Graph OLAP

Given a graph-structured dataset, *Graph OLAP* [9] aims at returning a multidimensional view of it to enable efficient OLAP analyses. Source data are seen as a collection of *network snapshots*, each including some informational attributes (e.g., month and socialNetwork) and one graph (e.g., one where nodes are users and edges represent their interactions); both nodes and edges of this graph may be described by attributes (e.g., name is an attribute of user nodes, numberOfMessages is an attribute of collaboration edges). The multidimensional view of graph data provided by Graph OLAP relies on the two pillars of the multidimensional model, namely, dimensions and measures. Two types of dimensions are distinguished:

- *Informational dimensions* correspond to informational attributes and organize snapshots into groups based on different perspectives, where each group corresponds to a cube cell. Hierarchies can be defined on these dimensions, for instance, socialNetworks → all and month → year → all.
- *Topological dimensions* correspond to node/edge attributes and operate on individual network snapshots. Hierarchies can be defined on these dimensions too, e.g., user → nation → all.

As to measures, they are computed starting from numerical node/edge attributes by aggregating them in two different ways: (i) in *informational OLAP*, aggregation is done by grouping snapshots with identical values of informational dimensions; (ii) in *topological OLAP*, aggregation is done by grouping nodes with identical values of topological dimensions inside individual networks. An example is shown in Fig. 2.

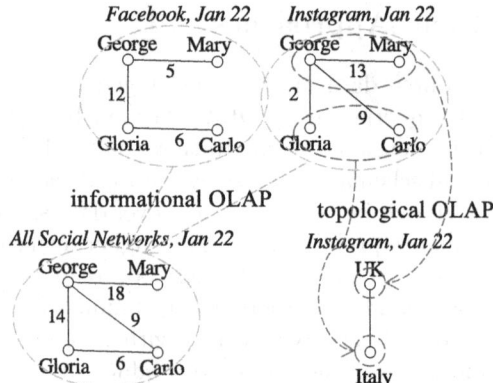

Fig. 2. Informational and topological OLAP in the Graph OLAP approach: in the first case, interactions are grouped on all social networks, in the second one, they are grouped by user's nation

2.2 Approximate OLAP

The basic idea of *Approximate OLAP* [16] is to enable multidimensional querying of document data with variable schemas, embracing data heterogeneity as an inherent source of information wealth in schemaless sources. Both inter-schema and intra-schema variety are considered; aimed at pursuing an inclusive approach to integration, OLAP querying is carried out on a "soft" schema where each source attribute is present to some extent. The approach encompasses four phases (see Fig. 3 for an example):

1. *Schema extraction*, whose goal is to identify the set of distinct, tree-like local schemas that occur inside a collection of documents.
2. *Schema integration*, which relies on inter-schema mappings and schema integration techniques to determine a tree-like global schema that gives the user a single and comprehensive description of the contents of the collection.
3. *FD enrichment*. An OLAP-compliant multidimensional view of the document data is obtained from the global schema by building a *dependency graph*, i.e., a graph that represents functional dependencies between the document fields; these dependencies are either inferred from the structure of the schema or determined (in approximate form) by analyzing the documents.
4. *Querying*. Here, the user can formulate OLAP queries on the dependency graph and execute them on the documents. To this end, each query is translated to the query language of the underlying document-oriented DBMS and reformulated into multiple queries, one for each local schema in the collection; the results presented to the user are obtained by merging the results of the single local queries. To make users aware of the impact of schema variety, a set of indicators describing the quality and reliability of the query result are computed.

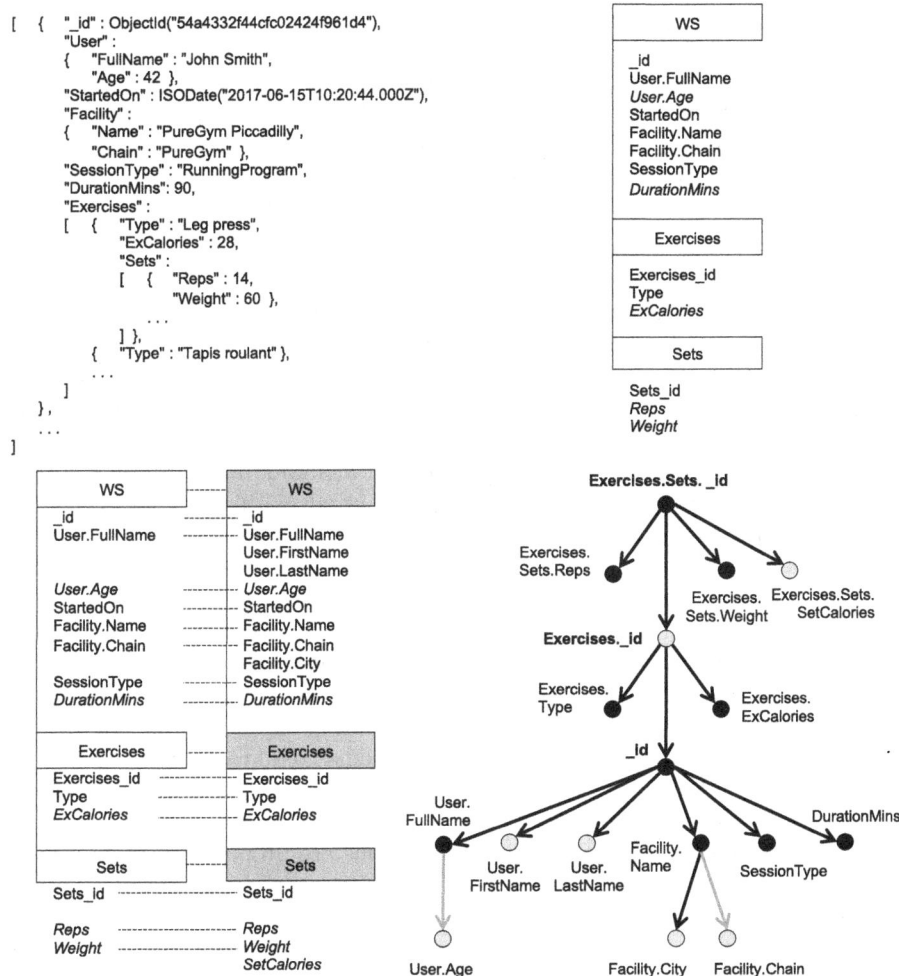

Fig. 3. Steps in approximate OLAP (adapted from [16]): a JSON document (top-left), its local schema (top-right, numerical fields in italics), its mappings with the global schema (bottom-left), and its dependency graph (bottom-right, grey arcs represent approximate functional dependencies discovered on documents)

3   Schema-on-Write Approaches

In these approaches, source data are moved into a data warehouse; this requires that they are put into multidimensional form to be then queried in an OLAP fashion by the user. The multidimensional schema is decided at design time and forced onto data at the time of writing them in the data warehouse, which entails better performances and simpler query formulation with no need for query rewriting. Schema-on-write approaches are based on the literature on (i) multidimensional design from NoSQL data [13,23] and (ii) NoSQL data warehouses, that aim at storing warehoused data in document/columnar/graph form by following design guidelines.

In this tutorial we distinguish between *mono-model* approaches, in which multidimensional data are stored in the data warehouse according to a single model (e.g., document-based), and *multi-model* approaches, in which a multi-model DBMS is used to grant higher storage flexibility.

3.1 Mono-Model Approaches

Several examples of schema-on-write approaches targeting a single NoSQL model can be found in the literature.

We start with the document-based model, for which some papers have proposed and compared different solutions to multidimensional design.Specifically, four solutions are proposed in [10]: (i) a *denormalized flat schema* (where a fact is stored using a single collection of documents including all its measures and levels with no nesting); (ii) a *deco schema* (denormalized like the previous one, but the measures and the levels of each dimension are stored in separate subdocuments); (iii) a *shattered schema* (where each dimension is stored in a separate collection of documents and connected to the fact documents using a reference, see Fig. 4 for an example); and (iv) a *hybrid schema* (like a shattered schema, but with all documents stored within a single collection). Based on experimental tests, it is argued that (i) the first two schemas require about 4 times the space required by the other two, which leads to significantly higher loading times; and (ii) denormalized flat schemas and shattered schemas tend to have better querying performances; however, there is not a single winner between these two since the execution times largely depend on the query features (mostly, on the number of joins they require). Similarly, two solutions are proposed in [8] and [28]: (i) a *simple schema* (where the fact and each dimension are stored in separate documents of the same collection, like in the hybrid schema mentioned above) and (ii) a *hierarchical schema* (like a simple schema, but using separate documents for each dimension hierarchy, much like the shattered schema mentioned above). The experimental comparison does not highlight significant differences in loading time and querying performance.

As to the graph-based model, in [26] two solutions are proposed. In the first one, the fact is stored in a graph node having measures as properties, and each level is stored in a node with its properties; the fact node points to the dimension nodes, which in turn point to the level nodes following the structure of the hierarchies (as also suggested in [7]). The second one is similar, except that the fact node points to a single node, which in turn points to each dimension node. A third solution is proposed in [27], where the fact node points to the dimension nodes, and each dimension node includes *all* the levels and properties of the corresponding hierarchy. Note that these three solutions are not experimentally compared in terms of efficiency.

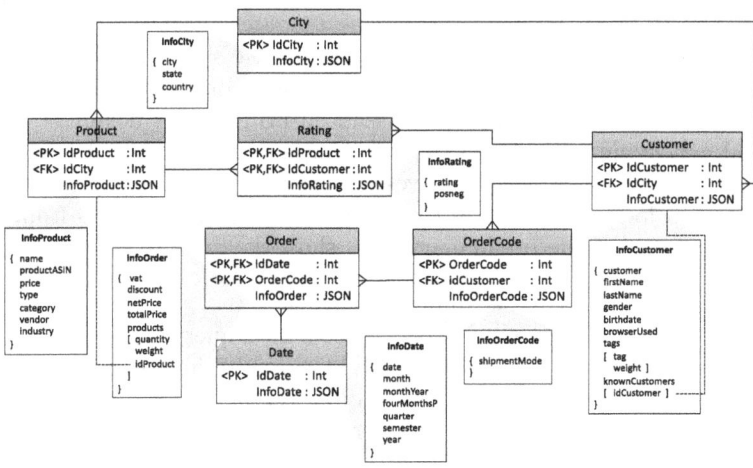

Fig. 4. A shattered schema

Finally, as to the column-based model, different strategies to arrange attributes into column-families (CFs) are proposed in [25]: (i) a *sameCF* schema, where all attributes are put in the same CF; (ii) a *CNSSB schema*, where each dimension is stored in a different CF; and (iii) a *factDate schema*, where some of the most-frequently used dimensions (in their example, the date dimension) are grouped together with fact data. Based on experimental tests, the authors conclude that the sameCF schema provides better performance for high-dimensional queries (three or four dimensions), while the CNSSB and factDate schemas are preferable for low-dimensional queries (one or two dimensions). In the same direction, in [6] the authors propose an approach that clusters in the same CFs attributes that are frequently used together in the workload queries.

3.2 Multi-model Approaches

A DBMS normally handles a specific data model (e.g., relational DBMSs, document-based DBMSs, etc.). When an application needs different types of data, the first possible solution is to integrate all data into a single DBMS; however, this means that some types of data cannot be stored and analyzed, and that querying performances may be unsatisfactory. The second solution is to use two or more DBMSs together (*polyglot persistence*); even in this case there are drawbacks, since technically managing more DBMSs is a challenge, the learning curve for developers is steep, performance optimization may be inadequate, and there is a risk of data inconsistency. To overcome these issues, *multi-model databases* (MMDBMSs, e.g., PostgreSQL and ArangoDB) natively support different data models under a single query language to grant performance, scalability, and fault tolerance, so as to reduce maintenance and data integration issues, speed up development, and eliminate migration problems.

As argued in [4], a *multi-model data warehouse* (MMDW) can store data according to the multidimensional model and, at the same time, let each of its elements be natively represented through the most appropriate model; among the benefits, reducing the cost for ETL and ensuring better flexibility, extensibility, and evolvability thanks to the use of schemaless models. However, in a multi-model setting, several alternatives emerge for the logical representation of dimensions and facts, and some of them may be better than others from one or more points of view.Some preliminary tests show that:

- Different dimensions can use different models.
- From the points of view of querying performance, query formulation conciseness, data storage, and complexity of ETL, a multidimensional implementation via the relational model is generally better than a document-based one, which in turn is better than a graph-based one.
- From the point of view of flexibility, extensibility, and evolvability, schemaless models (namely, document- and graph-based) are preferable to the relational one.

Figure 5 shows an optimal multi-model schema for the same multidimensional data of Fig. 4.

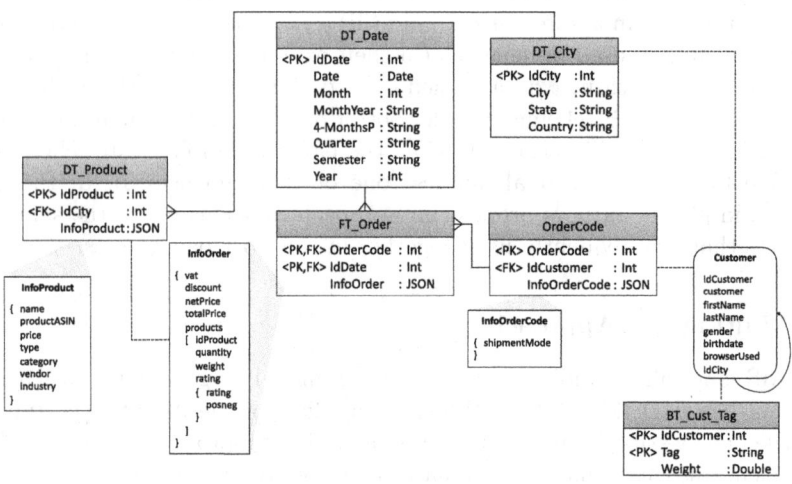

Fig. 5. A multi-model schema that mixes relational tables (e.g., DT_Date), documents (e.g., InfoProduct), and graphs (Customer)

4   Conclusion

Enabling OLAP queries over NoSQL data is getting more and more important today, but dealing with heterogeneity and schema variety intrinsic to NoSQL DBs is a challenge. In this tutorial we have discussed some directions for enabling

OLAP on schemaless NoSQL data, using either schema-on-read or schema-on-write approaches. Though several solutions have been proposed in the literature, their level of maturity is not comparable yet to the one reached by relational implementations. Among the relevant issues to be further investigated, we mention the following:

– increase the efficiency of the querying phase in schema-on-read approaches by paving the way to a more sophisticated optimization of query;
– develop techniques for online repairing of approximate functional dependencies present in schemaless data, so that the user can get correct analysis results without modifying the original data;
– extend the existing conceptual models to cope with schemaless data [20, 21];
– understand how to select and use materialized views in MMDWs, and what ad-hoc indexing strategies to adopt for them.

References

1. Abelló, A., et al.: Fusion cubes: towards self-service business intelligence. IJDWM **9**(2), 66–88 (2013)
2. Beheshti, S.-M.-R., Benatallah, B., Motahari-Nezhad, H.R., Allahbakhsh, M.: A framework and a language for on-line analytical processing on graphs. In: Wang, X.S., Cruz, I., Delis, A., Huang, G. (eds.) WISE 2012. LNCS, vol. 7651, pp. 213–227. Springer, Heidelberg (2012). https://doi.org/10.1007/978-3-642-35063-4_16
3. Bex, G.J., Gelade, W., Neven, F., Vansummeren, S.: Learning deterministic regular expressions for the inference of schemas from XML data. ACM TWEB **4**(4), 14 (2010)
4. Bimonte, S., Gallinucci, E., Marcel, P., Rizzi, S.: Data variety, come as you are in multi-model data warehouses. Inf. Syst. **104**, 101734 (2022)
5. Bohannon, P., Elnahrawy, E., Fan, W., Flaster, M.: Putting context into schema matching. In: Proceedings of VLDB, pp. 307–318 (2006)
6. Boussahoua, M., Boussaid, O., Bentayeb, F.: Logical schema for data warehouse on column-oriented NoSQL databases. In: Proceedings of DEXA, Lyon, France, pp. 247–256 (2017)
7. Castelltort, A., Laurent, A.: NoSQL graph-based OLAP analysis. In: Proceedings of KDIR, Rome, Italy, pp. 217–224 (2014)
8. Challal, Z., Bala, W., Mokeddem, H., Boukhalfa, K., Boussaid, O., Benkhelifa, E.: Document-oriented versus column-oriented data storage for social graph data warehouse. In: Proceedings of SNAMS, Granada, Spain, pp. 242–247 (2019)
9. Chen, C., Yan, X., Zhu, F., Han, J., Yu, P.S.: Graph OLAP: a multi-dimensional framework for graph data analysis. Knowl. Inf. Syst. **21**(1), 41–63 (2009)
10. Chevalier, M., Malki, M.E., Kopliku, A., Teste, O., Tournier, R.: Document-oriented models for data warehouses - NoSQL document-oriented for data warehouses. In: Proceedings of ICEIS, Rome, Italy, pp. 142–149 (2016)
11. Chouder, M.L., Rizzi, S., Chalal, R.: EXODuS: exploratory OLAP over document stores. Inf. Syst. **79**, 44–57 (2019)
12. Cuzzocrea, A., Bellatreche, L., Song, I.Y.: Data warehousing and OLAP over big data: current challenges and future research directions. In: Proceedings of DOLAP, pp. 67–70 (2013)

13. Dehdouh, K.: Building OLAP cubes from columnar NoSQL data warehouses. In: Proceedings of MEDI, Almería, Spain, pp. 166–179 (2016)
14. Dhamankar, R., Lee, Y., Doan, A., Halevy, A., Domingos, P.: iMAP: discovering complex semantic matches between database schemas. In: Proceedings of ICMD, pp. 383–394 (2004)
15. Gallinucci, E., Golfarelli, M., Rizzi, S.: Schema profiling of document-oriented databases. Inf. Syst. **75**, 13–25 (2018)
16. Gallinucci, E., Golfarelli, M., Rizzi, S.: Approximate OLAP of document-oriented databases: a variety-aware approach. Inf. Syst. **85**, 114–130 (2019)
17. Gartner Research: Market share: Database management systems, worldwide, 2019, April 2020. https://www.gartner.com/en/documents/3984279
18. Golfarelli, M., Rizzi, S.: Data Warehouse Design: Modern Principles and Methodologies. McGraw-Hill Inc., New York (2009)
19. Gómez, L.I., Kuijpers, B., Vaisman, A.A.: Online analytical processsing on graph data. Intell. Data Anal. **24**(3), 515–541 (2020)
20. Holubová, I., Contos, P., Svoboda, M.: Multi-model data modeling and representation: state of the art and research challenges. In: Proceedings of IDEAS, Montreal, QC, Canada, pp. 242–251 (2021)
21. Holubová, I., Svoboda, M., Lu, J.: Unified management of multi-model data - (vision paper). In: Proceedings of ER, Salvador, Brazil, pp. 439–447 (2019)
22. Izquierdo, J.L.C., Cabot, J.: Discovering implicit schemas in JSON data. In: Proceedings of ICWE, pp. 68–83 (2013)
23. Ouaret, Z., Chalal, R., Boussaid, O.: An overview of XML warehouse design approaches and techniques. IJICoT **2**(2/3), 140–170 (2013)
24. Ruiz, D.S., Morales, S.F., Molina, J.G.: Inferring versioned schemas from NoSQL databases and its applications. In: Proceedings of ER, pp. 467–480 (2015)
25. Scabora, L.C., Brito, J.J., Ciferri, R.R., de Aguiar Ciferri, C.D.: Physical data warehouse design on NoSQL databases - OLAP query processing over HBase. In: Proceedings of ICEIS, Rome, Italy, pp. 111–118 (2016)
26. Sellami, A., Nabli, A., Gargouri, F.: Transformation of data warehouse schema to NoSQL graph data base. In: Proceedings of ISDA, Vellore, India, pp. 410–420 (2018)
27. Sellami, A., Nabli, A., Gargouri, F.: Graph NoSQL data warehouse creation. In: Proceedings of iiWAS, Chiang Mai, Thailand, pp. 34–38 (2020)
28. Yangui, R., Nabli, A., Gargouri, F.: Automatic transformation of data warehouse schema to NoSQL data base: comparative study. In: Proceedings of KES, York, UK, pp. 255–264 (2016)

What's New in Temporal Databases?

Johann Gamper[(✉)] [iD], Matteo Ceccarello[iD], and Anton Dignös[iD]

Free University of Bozen-Bolzano, Dominikanerplatz/piazza Dominican 3,
39100 Bolzano, Italy
{johann.gamper,matteo.ceccarello,anton.dignoes}@unibz.it

Abstract. Temporal databases has been an active research area since
many decades, ranging from research work on query processing, most
dominantly on selection and join queries, to new directions in models
and semantics, such as for instance temporal probabilistic or streaming
data. At the same time more database vendors have been integrating
temporal features into their systems, most notably, the temporal features
of the SQL standard. In this paper, we summarize the latest research
developments as presented in 30 research papers over the last five years
in the context of temporal relational databases. Additionally, we also
describe the developments of industrial database systems and vendors.

Keyword: Temporal databases

1 Introduction

Temporal databases is an active research area since several decades, with a
renewed interest in recent years. The interest in temporal databases is driven by
a variety of old and new applications that require to store and process temporal data, such as versioning of web documents [21], management of normative
texts [27], air traffic monitoring and patient care [5], video surveillance [44], sales
analysis [41], financial market analysis [25], and data warehousing and analytics [48], to name a few.

In temporal databases every fact is associated with one or more timestamps [7]. The timestamps are typically formed by either a time period or a
set of time points, though other forms of timestamps exist, such as temporal elements. While time points are easier from a conceptual viewpoint, time periods
are practically more relevant and allow for efficient implementations. The timestamps can represent different aspects of time, most importantly *valid time* [32]
that indicates the validity of a fact in the real world (e.g., a contract that exists
over a given period of time) and *transaction time* [31] that indicates the time
when a tuple is/was stored in the database (e.g., a contract that was stored over

Supported by the Autonomous Province of Bozen-Bolzano with research call "Research
Südtirol/Alto Adige 2019" (project Enabling Industrial-Strength, Open-Source Temporal Query Processing – ISTeP).

© Springer Nature Switzerland AG 2022
S. Chiusano et al. (Eds.): ADBIS 2022, LNCS 13389, pp. 45–58, 2022.
https://doi.org/10.1007/978-3-031-15740-0_5

a given period of time and later on updated or deleted from the database). When both aspects of time are present in a relation, we have a bitemporal relation [30].

Figure 1 shows two temporal relations **Mgr** and **Pro**. The timestamps are half-open intervals and represent the tuples' valid time. Relation **Mgr** records managers of departments, where **Dept** is the name of the department, **MName** is the name of the manager and **T** is the time period for which the person manages the department. Relation **Pro** records projects running in departments, where **PName** is the name of the project, **PDept** is the department which runs the project, and **T** is the time period over which the project runs.

Mgr

Dept	MName	T
AI	Tom	[01/2022, 06/2022)
AI	Sue	[06/2022, 01/2024)
DB	Ann	[10/2021, 01/2023)

Pro

PName	PDept	T
ExTAI	AI	[04/2022, 05/2023)
AIHuM	AI	[01/2023, 01/2024)
TauDB	DB	[10/2021, 06/2022)

Fig. 1. A temporal database with two temporal relations.

The most widely used semantics for temporal databases is known as *sequence semantics*, where temporal queries are defined using the concept of *snapshot reducibility* [15,35,49]. Snapshot reducibility views a temporal database as a sequence of snapshots and constrains a temporal operator applied to a temporal relation to produce, at a time point t, the same result as the corresponding non-temporal operator applied to the snapshot at t, i.e., all input tuples that are valid at t. For instance, the result of a temporal count aggregation is defined "pointwise" by the result of a non-temporal count aggregation. The aim of temporal databases is to facilitate such kind of operations in queries that would otherwise result in long, error prone, and inefficient SQL queries [47].

Figure 2 reports the result of two temporal queries on our example database from Fig. 1. In particular, the result of the temporal join between the two relations to retrieve for each project the responsible manager is shown in Fig. 2a. The temporal join is performed by joining two temporal relations according to overlapping timestamps. That is, the result tuples are timestamped with the intersection of the overlapping time periods. In the literature the step of finding overlapping pairs of tuples is also referred to as overlap or interval join. This result of the temporal join is consistent with the traditional (non-temporal) join performed at each snapshot of the data. For instance, the snapshot at time point 04/2022 for relation **Pro** contains two projects: ExTAI from the AI department and TauDB from the DB department. The snapshot for relation **Mgr** at the same time point contains two managers: Tom for the AI department and Ann for the DB department. Performing a non-temporal join on these two snapshots gives the same result as the snapshot at time point 04/2022 for the relation in Fig. 2a, i.e., ExTAI is managed by Tom and TauDB is managed by Ann.

PName	PDept	MName	T
ExTAI	AI	Tom	[04/2022, 06/2022)
ExTAI	AI	Sue	[06/2022, 05/2023)
AIHuM	AI	Sue	[01/2023, 01/2024)
TauDB	DB	Ann	[10/2021, 06/2022)

PDept	Count	T
AI	1	[04/2022, 01/2023)
AI	2	[01/2023, 05/2023)
AI	1	[05/2023, 01/2024)
DB	1	[10/2021, 06/2022)

(a) Temporal Join   (b) Temporal Aggregation

Fig. 2. Example operations *temporal join* and *aggregation*.

Figure 2b shows the result of a temporal aggregation that counts, for each department, the number of projects stored in relation **Pro**. Also in this case the result is defined according to the snapshots, and in the result we have one project in the AI department from 04/2022 to 01/2023, because each snapshot within this period contains exactly one project for department AI.

Past research on temporal databases has been focusing on various aspects of managing and processing temporal data, most notably on data models, SQL-based query languages, and efficient evaluation algorithms for query processing. Due to the ubiquity of temporal data and the need for processing such data, more recently also industry caught up with the topic, resulting in several extensions of commercial and Open Source database systems (e.g., IBM DB2, Oracle, Teradata, and PostgreSQL) with various degrees of support for temporal data. Finally, the major extension in the SQL:2011 standard was the support for the representation of temporal data [7,34].

In this paper, we review the newest developments in temporal relational databases as presented in 30 research papers over the last five years, which extend the state of the art as described in [7]. More specifically, Sect. 2 provides an overview of the works on temporal query processing, mostly focusing on selection and join queries. Section 3 provides the works that focus on new research directions in the area of temporal data models and semantics, followed by an overview on the newest developments in industrial systems in Sect. 4. Finally, Sect. 5 concludes the paper and provides interesting topics for future work that received scant attention in the last years.

2   Query Processing of Primitive Operators

In this section, we focus on recent advancements in query processing, mainly from an algorithmic point of view.

2.1   Temporal Selection

A classic query involving intervals is the *overlap* query, which retrieves all tuples whose timestamp overlaps with the query period. Christodoulou et al. [14] introduce HINT, an index addressing this kind of problem. It partitions the timeline in a hierarchy of regular grids of geometrically increasing granularities. The addition of an auxiliary index of non-empty partitions allows to improve efficiency on skewed data.

A richer type of selection query are *range-duration* queries, first introduced by Behrend et al. [4]: matching intervals need to both overlap with the query and have a duration within a bound that is also defined in the query. Traditional index structures for intervals deal with only one aspect of intervals, and thus miss the opportunity to leverage the selectivity of queries on both dimensions.

In [4], the authors introduce PERIOD-INDEX⋆ to explicitly support range-duration queries. The index partitions the timeline in *buckets* that will contain any interval they intersect. Then, intervals in a bucket are further partitioned in *levels* based on their duration, with the minimum duration in each level increasing geometrically. Finally, each level is further partitioned in the time domain in order to efficiently retrieve intervals of a given duration based on their start time. This approach is adaptive to the distribution of start times, but assumes a Zipf-like distribution of durations.

Recently, an index deemed RD-INDEX supporting range-duration queries has been introduced by Ceccarello et al. [13]. This index partitions tuples in a two-dimensional grid according to their start time and their duration by taking into account the distribution of both dimensions. This allows to adapt to the distribution of the data, providing better performance than the state of the art. Experiments show that this index performs better than the state of th e art also on *mixed* workloads, where some queries constrain only the duration, some constrain only the position on the timeline, and some constrain both.

2.2 Temporal Joins

Binary Joins. Temporal binary joins are joins between two relations where the join predicate requires that the interval timestamps of the tuples in the two relations overlap. A specialized data structure, called the Overlap Interval Inverted Index, is proposed by Luo et al. [37] to efficiently compute binary interval joins. The index uses the end points of intervals as anchor points and approximates the nesting structure of intervals by establishing relationships between these anchor points. This information is then used to prune unnecessary comparisons.

Interval joins for in-memory data have also been studied by Bouros et al. [10]. The paper proposes optimizations of the forward scan algorithm [11] and develops a parallel version, where a thread is responsible of sweeping the timeline, and then forward scans are executed in parallel.

Another adaptation of the forward scan algorithm, specifically tailored to *skewed* data, is proposed by Hellings and Wu [28]. This algorithm enriches the forward scan algorithm with an auxiliary data structure, termed stab-forests, which allows to skip portions of the input relations that are provably not part of the result.

An approach that does not necessarily involve the development of ad-hoc index structures is presented by Dignös et al. [16]: overlap joins are rewritten as the union of two range-joins. This rewriting enables the computation of overlap joins using an efficient sort-merge based algorithm for range-joins that is on par with other state-of-the-art techniques and allows to efficiently implement overlap joins on widely available DBMS systems using B+-trees. Besides traditional overlap joins, this work also considers additional equality predicates in overlap joins as well as period boundaries that can have different interval definitions (e.g., closed or half-open) for different tuples.

Joins involving predicates on intervals can be extended in several directions. The most natural extension involves considering all of Allen's interval relations, as is done by Piatov et al. [46]. The paper takes the moves from [45], leveraging the *endpoint index* and the *gapless hash map* to efficiently process intervals in a cache-friendly way.

The *interval count semi-join* problem [9] requires instead to count for each interval of a relation R, the number of intervals in another relation S with which it overlaps. The paper extends the plane-sweep algorithm of [45] to solve this problem directly, without requiring a join followed by an aggregation step.

Finally, another extension is that of *band join* of intervals [8]. Specifically, the problem requires to join intervals that either overlap or whose smallest difference between endpoints is smaller than a parameter ϵ.

Multi-way Joins. In many cases, multiple temporal relations are to be joined. The traditional way of addressing this type of queries relies on finding the best sequence of binary joins. This approach has the drawback of potentially producing intermediate results which are much larger than the final output. In the past few years there has been a growing interest for multi-way equi-joins, following the development of the output optimal worst-case join [40], where the output of the join is computed by considering all involved relations at the same time.

Very recently, Hu et al. [29] developed an approach based on worst-case optimal join algorithms to deal with multi-way temporal joins. These algorithms are worst case optimal in the sense that, for a given query, one can bound the worst case output size based on the characteristics of the query: the algorithm will then run in time proportional to this worst case size. Furthermore, they introduce the problem of *durable* joins: only the intervals with duration longer than a given parameter τ are part of the output, allowing to ignore transient patterns.

The complexity of multi-way interval joins is studied by Khamis et al. [33]. Specifically, the paper provides a reduction of a multi-way interval join to a disjunction of multi-way equi-joins, and the corresponding backward reduction. This allows to both upper bound the complexity of multi-way interval joins and to state hardness results.

General intersection joins are the topic of the work by Tao and Yi [56]. Intersection joins consider overlaps between d-dimensional rectangles; for $d = 1$ the problem corresponds to overlap joins in temporal databases. The paper focuses on the *dynamic* variant of the problem, where one wants to update the solution as the relations involved in the join are modified. For binary joins of intervals, the paper provides an optimal data structure requiring $O(n)$ space and $O(\log n)$ amortized update time.

Multi-way temporal joins also arise in the context of finding temporal subgraphs, like k-cliques of overlapping intervals [57].

3   New Directions in Models and Semantics

In this section, we provide an overview about new research directions which go beyond traditional temporal databases and include new semantics, data models, and query types.

Semantics. Most works in temporal databases focus on duplicate free temporal relations, i.e., set semantics, where value-equivalent tuples are not allowed to overlap. The work by Dignös et al. [17] provides the first theoretical foundations for processing temporal data with multiset semantics under full relational algebra and aggregation. In particular, this paper defines multiset semantics by adopting a novel data model based on the concepts of K-relations and semirings, which satisfy the properties of snapshot reducibility. The authors show how the temporal operators over temporal relations with multiset semantics can be translated into standard SQL queries via a query rewriting approach.

Implementing *sequenced semantics* using standard relational algebra is the goal of [19]. To this end, the paper proposes to use *log-segmented timestamps* [18] rather than time intervals. Assuming that the timeline has $n = 2^k$ chronons, labels of $b \leq k$ bits can be univocally associated to pre-determined time intervals. Therefore, any arbitrary time interval can be encoded using a collection of at most k labels. The paper proposes to transform a temporal relation in a non-temporal relation featuring labels in place of temporal intervals, and where each tuple is replicated up to k times, depending on the temporal interval to which it is associated. This transformation allows to express temporal queries using standard (non-temporal) relational algebra, and thus allowing to implement sequenced semantics in standard DBMSs without modifications.

Temporal Probabilistic Databases. A *temporal probabilistic database* [42] is a database complying with both the possible world semantics [51] of probabilistic databases and the sequenced semantics of temporal databases. In summary, a temporal probabilistic database can be thought of a collection of probabilistic databases, one for each time instant. The query semantics then requires that the result of any operation at any time point t is equivalent to the result derived from the corresponding probabilistic operation applied to the probabilistic database at time t. Set operations in this model are investigated by Papaioannou et al. [42], whereas [43] studies the problems of outer and anti joins.

Streaming Data. Nowadays many applications have to deal with incoming streams of data, rather than static datasets to be stored and processed offline.

The work by Suzanne et al. [52, 53] considers the aggregation of spanning events (events with time periods) in the context of data streams. The work provides a framework that extends window aggregation over regular events with time points to spanning events with time periods. The framework supports a wide range of common window definitions for the aggregation, and it considers different ways how spanning events may be received in a data stream, e.g., an event may be received only at its end time, or an event is partially received first at its start time and later on completed at its end time. How window-slicing for spanning events can be performed to share computational costs among overlapping windows, is introduced in [54, 55].

The work by Grandi et al. [26] proposes a query language and a unified algebraic framework that integrates streaming, temporal, and standard relational data in an all-in-one approach. This framework provides an extended relational algebra for one-time queries with temporal and non-temporal semantics as well as continuous queries with different types of window expressions, together with a translation that allows the execution of continuous queries using traditional temporal operators.

Ongoing Databases. The paper by Mülle and Böhlen [39] studies the concept of "now" [2, 20] in temporal databases. While many approaches deal with time points declared as now by instantiating them to a given reference time (e.g., the current time), this solution provides a principled approach to deal with "now" during query processing, by keeping it uninstantiated and evaluating predicates and functions at all possible reference times. The result of a query is an ongoing relation that includes reference times. The authors introduce ongoing data types and their operations, a relational algebra for ongoing relations, and an implementation in PostgreSQL.

Historical What-If Queries. The work by Campbell et al. [12] introduces historical what-if queries that allow to determine the effect of hypothetical changes in the transactional history of a database. The approach exploits reenactment [3], a declarative replay technique for transactions, to simulate the evaluation of histories together with time travel [50] on transaction time to find the corresponding history of the data to apply the what-if scenario. The authors provide an optimization to apply historical changes only to the affected data together with an implementation as a middleware. While this approach does not focus on explicitly timestamped data, it exploits the transactional history of database systems.

Temporal Keyword Search. The work by Gao et al. [24] studies the problem of evaluating keyword queries with temporal predicates in temporal databases. The work shows how multiple interpretations and their corresponding SQL queries including temporal joins can be generated from the temporal predicates in the keyword search. The work in [23] shows how temporal aggregation and span temporal aggregation [22] can be employed in temporal keyword search in order to allow users to query statistical information over time.

Data Warehouse. Ahmed et al. [1] show how to generalize and extend the multidimensional model used in data warehouses with temporal features and temporal online analytical processing (OLAP) operators. The work introduces a multidimensional model that is capable of representing traditional (non-temporal) and time-varying data independently with consistency constraints. The authors provide a mapping from the temporal model into the relational model and show how the temporal OLAP operators can be answered using standard SQL.

The work by Mahlknecht et al. [38] proposes different logical models how temporal data can be represented in a data warehouse to support efficient aggregations over time. The models differ in the way how data with time periods is stored: as the set of all time points in a time period, as the start and end time points of a time period, or as a combination of the two. The different models may or may not facilitate different aggregation operators over time that are frequently used in data warehouses. The authors show the queries in standard SQL and provide and experimental evaluation for the different models and aggregation operators on ETL performance and query time.

4 Systems

Database vendors have been gradually enhancing their database systems with support for temporal features, particularly with respect to the temporal features in the SQL:2011 standard [34]. After IBM DB2, Oracle DBMS, Teradata, and MS SQL Server, other database vendors have been following in the implementation of temporal features. In this review we focus on the new additions of the last five years and refer the reader to [6,7] for a more exhaustive study on the temporal features offered before.

MariaDB as of version 10.3.4 (Jan 2018) supports system-versioned tables[1] from the SQL:2011 standard, which provide integrated transaction time support. As of version 10.4.3 (Feb 2019), the database added also support for application-time period tables[2] (i.e., valid time relations). Since both temporal dimensions can be combined, MariaDB can also represent bitemporal tables. As of version 10.5.3 (May 2020), temporal uniqueness (WITHOUT OVERLAPS) was added, which can be used in the declaration of a table schema[3]. This feature allows to enforce temporal primary key constraints.

In a similar fashion, the in-memory, column-oriented relational database system SAP HANA introduced system-versioned tables as of version 2.0 SPS03[4]

---

[1] https://mariadb.com/kb/en/mariadb-1034-release-notes/.

[2] https://mariadb.com/kb/en/mariadb-1043-release-notes/.

[3] https://mariadb.com/kb/en/mariadb-1053-release-notes/.

[4] https://help.sap.com/doc/d25e2e530606453c9866c695298423b3/2.0.03/en-US/
Whats_New_SAP_HANA_Platform_Release_Notes_en.pdf.

(Oct 2018), and application-time period tables as of version 2.0 SPS04[5] (Oct 2019).

The Open Source database system PostgreSQL followed a different route from the SQL:2011 standard. It offers support for temporal features through the build-in range types (period datatype) with associated operators and functions since version 9.2 (Sep 2012). As of version 14.0[6] (Sep 2021) PostgreSQL added support for multiranges[7] as a new data type, which are ordered lists of ranges with associated operators and functions.

The work by Lu et al. [36] provides a prototype built-in temporal implementation in Tencent's distributed database management system. The work integrates the features of the SQL:2011 standard into the system. It employs query rewriting in the parser to map queries on valid time into non-temporal queries. For transaction time several optimizations are proposed: a lazy migration strategy from the current to the history table that exploits the database management systems storage claiming procedure; a key/value store based approach to maintain only changed data instead of copies between current and history tables; and an optimized operator that retrieves current and historical data.

Other systems, such as CockroachDB and Snowflake, support time travel functionalities within a given retention time period. These systems allow to query and restore historical states of the data (if available). However, unlike in the SQL:2011 standard, versions and timestamps of the data are implicit and cannot be accessed.

5 Conclusion and Future Directions

In this paper, we reviewed new contributions in the field of temporal relational databases from the last five years. As a result, we survey 30 papers that span different areas in temporal relational databases: query processing with selection queries and joins, and new directions with topics such as improved temporal semantics, temporal probabilistic databases, streaming data, and more. Finally, we also summarized the newest developments with regards to temporal features in commercial and Open Source database systems.

While we have noticed that join algorithms have received most attention in the last few years, we also identified several topics that did not receive attention at all or are underrepresented. One such topic that requires deeper investigations is cost or cardinality estimation for temporal query operators. This is particularly important for temporal joins. As of today most query optimizers use heuristics or constants for the selectivity estimation of joins in the presence of inequalities. More precise cost estimation algorithms and their tight integration into query optimizers would be helpful to further improve the efficiency of query processing.

---

[5] https://help.sap.com/doc/d25e2e530606453c9866c695298423b3/2.0.04/en-US/ Whats_New_SAP_HANA_Platform_Release_Notes_en.pdf.

[6] https://www.postgresql.org/docs/release/14.0/.

[7] https://www.postgresql.org/docs/current/functions-range.html.

Secondly, more research work on SQL extensions for temporal operators is needed, which is not covered in the SQL:2011 standard. While there exists some past research on this aspect, none of the proposed extensions received wide acceptance. The availability of a standard for the easy formulation of temporal queries in SQL may also help industry with the integration of temporal operators in their DBMS, in a similar fashion as the SQL:2011 standard pushed the development of temporal features.

Another direction for future research is concerned with query processing of bitemporal operators, which consider both valid time and transaction time. While the SQL:2011 standard allows to define and represent bitemporal tables, there exists only one work in the last five years that considers the computation of joins on more than one time dimension; all other works only focus on data that has either a valid time or a transaction time.

References

1. Ahmed, W., Zimányi, E., Vaisman, A.A., Wrembel, R.: A temporal multidimensional model and OLAP operators. Int. J. Data Warehous. Min. **16**(4), 112–143 (2020). https://doi.org/10.4018/IJDWM.2020100107
2. Anselma, L., Piovesan, L., Sattar, A., Stantic, B., Terenziani, P.: A comprehensive approach to 'now' in temporal relational databases: Semantics and representation. IEEE Trans. Knowl. Data Eng. **28**(10), 2538–2551 (2016). https://doi.org/10.1109/TKDE.2016.2588490
3. Arab, B.S., Gawlick, D., Krishnaswamy, V., Radhakrishnan, V., Glavic, B.: Reenactment for read-committed snapshot isolation. In: Proceedings of the 25th ACM International Conference on Information and Knowledge Management, CIKM 2016, Indianapolis, IN, USA, 24–28 October 2016, pp. 841–850. ACM (2016). https://doi.org/10.1145/2983323.2983825
4. Behrend, A., Dignös, A., Gamper, J., Schmiegelt, P., Voigt, H., Rottmann, M., Kahl, K.: Period index: A learned 2d hash index for range and duration queries. In: Proceedings of the 16th International Symposium on Spatial and Temporal Databases, SSTD 2019, Vienna, Austria, 19–21 August 2019. pp. 100–109. ACM (2019). https://doi.org/10.1145/3340964.3340965
5. Behrend, A., Schmiegelt, P., Xie, J., Fehling, R., Ghoneimy, A., Liu, Z.H., Chan, E., Gawlick, D.: Temporal state management for supporting the real-time analysis of clinical data. In: Bassiliades, N., Ivanovic, M., Kon-Popovska, M., Manolopoulos, Y., Palpanas, T., Trajcevski, G., Vakali, A. (eds.) New Trends in Database and Information Systems II. AISC, vol. 312, pp. 159–170. Springer, Cham (2015). https://doi.org/10.1007/978-3-319-10518-5_13
6. Böhlen, M.H., Dignös, A., Gamper, J., Jensen, C.S.: Database technology for processing temporal data (invited paper). In: 25th International Symposium on Temporal Representation and Reasoning, TIME 2018, Warsaw, Poland, October 15–17, 2018. LIPIcs, vol. 120, pp. 2:1–2:7. Schloss Dagstuhl - Leibniz-Zentrum für Informatik (2018). https://doi.org/10.4230/LIPIcs.TIME.2018.2
7. Böhlen, M.H., Dignös, A., Gamper, J., Jensen, C.S.: Temporal data management – an overview. In: Zimányi, E. (ed.) eBISS 2017. LNBIP, vol. 324, pp. 51–83. Springer, Cham (2018). https://doi.org/10.1007/978-3-319-96655-7_3

8. Bouros, P., Lampropoulos, K., Tsitsigkos, D., Mamoulis, N., Terrovitis, M.: Band joins for interval data. In: Proceedings of the 23rd International Conference on Extending Database Technology, EDBT 2020, Copenhagen, Denmark, March 30–April 02, 2020, pp. 443–446. OpenProceedings.org (2020). https://doi.org/10.5441/002/edbt.2020.53

9. Bouros, P., Mamoulis, N.: Interval count semi-joins. In: Proceedings of the 21st International Conference on Extending Database Technology, EDBT 2018, Vienna, Austria, 26–29 March 2018, pp. 425–428. OpenProceedings.org (2018). https://doi.org/10.5441/002/edbt.2018.38

10. Bouros, P., Mamoulis, N., Tsitsigkos, D., Terrovitis, M.: In-memory interval joins. The VLDB J. **30**(4), 667–691 (2021). https://doi.org/10.1007/s00778-020-00639-0

11. Brinkhoff, T., Kriegel, H., Seeger, B.: Efficient processing of spatial joins using r-trees. In: Proceedings of the 1993 ACM SIGMOD International Conference on Management of Data, Washington, DC, USA, 26–28 May 1993, pp. 237–246. ACM Press (1993). https://doi.org/10.1145/170035.170075

12. Campbell, F.S., Arab, B.S., Glavic, B.: Efficient answering of historical what-if queries. In: SIGMOD 2022: International Conference on Management of Data, Philadelphia, PA, USA, 12–17 June 2022, pp. 1556–1569. ACM (2022). https://doi.org/10.1145/3514221.3526138

13. Ceccarello, M., Dignös, A., Gamper, J., Khnaisser, C.: Indexing temporal relations for range-duration queries. CoRR abs/2206.07428 (2022). https://doi.org/10.48550/arXiv.2206.07428

14. Christodoulou, G., Bouros, P., Mamoulis, N.: HINT: A hierarchical index for intervals in main memory. In: SIGMOD 2022: International Conference on Management of Data, Philadelphia, PA, USA, 12–17 June 2022, pp. 1257–1270. ACM (2022). https://doi.org/10.1145/3514221.3517873

15. Dignös, A., Böhlen, M.H., Gamper, J., Jensen, C.S.: Extending the kernel of a relational DBMS with comprehensive support for sequenced temporal queries. ACM Trans. Database Syst. **41**(4), 26:1–26:46 (2016). https://doi.org/10.1145/2967608

16. Dignös, A., Böhlen, M.H., Gamper, J., Jensen, C.S., Moser, P.: Leveraging range joins for the computation of overlap joins. The VLDB J. **31**(1), 75–99 (2021). https://doi.org/10.1007/s00778-021-00692-3

17. Dignös, A., Glavic, B., Niu, X., Gamper, J., Böhlen, M.H.: Snapshot semantics for temporal multiset relations. Proc. VLDB Endow. **12**(6), 639–652 (2019). https://doi.org/10.14778/3311880.3311882

18. Dyreson, C.E.: Using CouchDB to compute temporal aggregates. In: 18th IEEE International Conference on High Performance Computing and Communications; 14th IEEE International Conference on Smart City; 2nd IEEE International Conference on Data Science and Systems, HPCC/SmartCity/DSS 2016, Sydney, Australia, 12–14 December 2016, pp. 1131–1138. IEEE Computer Society (2016). https://doi.org/10.1109/HPCC-SmartCity-DSS.2016.0159

19. Dyreson, C.E., Ahsan, M.A.M.: Achieving a sequenced, relational query language with log-segmented timestamps. In: 28th International Symposium on Temporal Representation and Reasoning, TIME 2021, 27–29 September 2021, Klagenfurt, Austria. LIPIcs, vol. 206, pp. 14:1–14:13. Schloss Dagstuhl - Leibniz-Zentrum für Informatik (2021). https://doi.org/10.4230/LIPIcs.TIME.2021.14

20. Dyreson, C.E., Jensen, C.S., Snodgrass, R.T.: Now in temporal databases. In: Liu, L., Özsu, M.T. (eds.) Encyclopedia of Database Systems, Second Edition. Springer, New York (2018). https://doi.org/10.1007/978-1-4614-8265-9_248

21. Dyreson, C.E., Lin, H., Wang, Y.: Managing versions of web documents in a transaction-time web server. In: Proceedings of the 13th International Conference on World Wide Web, WWW 2004, New York, NY, USA, 17–20 May 2004, pp. 422–432. ACM (2004). https://doi.org/10.1145/988672.988730

22. Gamper, J., Böhlen, M.H., Jensen, C.S.: Temporal aggregation. In: Liu, L., Özsu, M.T. (eds.) Encyclopedia of Database Systems, Second Edition. Springer, New York (2018). https://doi.org/10.1007/978-1-4614-8265-9_386

23. Gao, Q., Lee, M.L., Ling, T.W.: Temporal keyword search with aggregates and group-by. In: Ghose, A., Horkoff, J., Silva Souza, V.E., Parsons, J., Evermann, J. (eds.) ER 2021. LNCS, vol. 13011, pp. 160–175. Springer, Cham (2021). https://doi.org/10.1007/978-3-030-89022-3_14

24. Gao, Q., Lee, M.L., Ling, T.W., Dobbie, G., Zeng, Z.: Analyzing temporal keyword queries for interactive search over temporal databases. In: Hartmann, S., Ma, H., Hameurlain, A., Pernul, G., Wagner, R.R. (eds.) DEXA 2018. LNCS, vol. 11029, pp. 355–371. Springer, Cham (2018). https://doi.org/10.1007/978-3-319-98809-2_22

25. Grandi, F., Mandreoli, F., Martoglia, R., Penzo, W.: A relational algebra for streaming tables living in a temporal database world. In: 24th International Symposium on Temporal Representation and Reasoning, TIME 2017, 16–18 October 2017, Mons, Belgium. LIPIcs, vol. 90, pp. 15:1–15:17. Schloss Dagstuhl - Leibniz-Zentrum für Informatik (2017). https://doi.org/10.4230/LIPIcs.TIME.2017.15

26. Grandi, F., Mandreoli, F., Martoglia, R., Penzo, W.: Unleashing the power of querying streaming data in a temporal database world: A relational algebra approach. Inf. Syst. **103**, 101872 (2022). https://doi.org/10.1016/j.is.2021.101872

27. Grandi, F., Mandreoli, F., Tiberio, P.: Temporal modelling and management of normative documents in XML format. Data Knowl. Eng. **54**(3), 327–354 (2005). https://doi.org/10.1016/j.datak.2004.11.002

28. Hellings, J., Wu, Y.: Stab-forests: Dynamic data structures for efficient temporal query processing. In: 27th International Symposium on Temporal Representation and Reasoning, TIME 2020, 23–25 September 2020, Bozen-Bolzano, Italy. LIPIcs, vol. 178, pp. 18:1–18:19. Schloss Dagstuhl - Leibniz-Zentrum für Informatik (2020). https://doi.org/10.4230/LIPIcs.TIME.2020.18

29. Hu, X., Sintos, S., Gao, J., Agarwal, P.K., Yang, J.: Computing complex temporal join queries efficiently. In: SIGMOD 2022: International Conference on Management of Data, Philadelphia, PA, USA, 12–17 June 2022, pp. 2076–2090. ACM (2022). https://doi.org/10.1145/3514221.3517893

30. Jensen, C.S., Snodgrass, R.T.: Bitemporal relation. In: Liu, L., Özsu, M.T. (eds.) Encyclopedia of Database Systems, pp. 243–244. Springer, New York (2009). https://doi.org/10.1007/978-0-387-39940-9_1409

31. Jensen, C.S., Snodgrass, R.T.: Transaction time. In: Liu, L., Özsu, M.T. (eds.) Encyclopedia of Database Systems, Second Edition, pp. 4200-4201. Springer, New York (2018). https://doi.org/10.1007/978-1-4614-8265-9_1064

32. Jensen, C.S., Snodgrass, R.T.: Valid time. In: Liu, L., Özsu, M.T. (eds.) Encyclopedia of Database Systems, Second Edition, pp. 4359–4360. Springer, New York (2018). https://doi.org/10.1007/978-1-4614-8265-9_1066

33. Khamis, M.A., Chichirim, G., Kormpa, A., Olteanu, D.: The complexity of Boolean conjunctive queries with intersection joins. In: PODS 2022: International Conference on Management of Data, Philadelphia, PA, USA, 12–17 June 2022, pp. 53–65. ACM (2022). https://doi.org/10.1145/3517804.3524156

34. Kulkarni, K.G., Michels, J.: Temporal features in SQL: 2011. SIGMOD Rec. **41**(3), 34–43 (2012). https://doi.org/10.1145/2380776.2380786

35. Lorentzos, N.A., Mitsopoulos, Y.G.: SQL extension for interval data. IEEE Trans. Knowl. Data Eng. **9**(3), 480–499 (1997). https://doi.org/10.1109/69.599935
36. Lu, W., Zhao, Z., Wang, X., Li, H., Zhang, Z., Shui, Z., Ye, S., Pan, A., Du, X.: A lightweight and efficient temporal database management system in TDSQL. Proc. VLDB Endow. **12**(12), 2035–2046 (2019). https://doi.org/10.14778/3352063.3352122
37. Luo, J.-Z., Shi, S.-F., Yang, G., Wang, H.-Z., Li, J.-Z.: O2iJoin: an efficient index-based algorithm for overlap interval join. J. Comput. Sci. Technol. **33**(5), 1023–1038 (2018). https://doi.org/10.1007/s11390-018-1872-x
38. Mahlknecht, G., Dignös, A., Kozmina, N.: Modeling and querying facts with period timestamps in data warehouses. Int. J. Appl. Math. Comput. Sci. **29**(1), 31–49 (2019). https://doi.org/10.2478/amcs-2019-0003
39. Mülle, Y., Böhlen, M.H.: Query results over ongoing databases that remain valid as time passes by. In: 36th IEEE International Conference on Data Engineering, ICDE 2020, Dallas, TX, USA, 20–24 April 2020, pp. 1429–1440. IEEE (2020). https://doi.org/10.1109/ICDE48307.2020.00127
40. Ngo, H.Q., Ré, C., Rudra, A.: Skew strikes back: new developments in the theory of join algorithms. SIGMOD Rec. **42**(4), 5–16 (2013). https://doi.org/10.1145/2590989.2590991
41. Papaioannou, K., Böhlen, M.H.: Temprora: top-k temporal-probabilistic results analysis. In: 32nd IEEE International Conference on Data Engineering, ICDE 2016, Helsinki, Finland, 16–20 May 2016, pp. 1382–1385. IEEE Computer Society (2016). https://doi.org/10.1109/ICDE.2016.7498350
42. Papaioannou, K., Theobald, M., Böhlen, M.H.: Supporting set operations in temporal-probabilistic databases. In: 34th IEEE International Conference on Data Engineering, ICDE 2018, Paris, France, 16–19 April 2018, pp. 1180–1191. IEEE Computer Society (2018). https://doi.org/10.1109/ICDE.2018.00109
43. Papaioannou, K., Theobald, M., Böhlen, M.H.: Outer and anti joins in temporal-probabilistic databases. In: 35th IEEE International Conference on Data Engineering, ICDE 2019, Macao, China, 8–11 April 2019, pp. 1742–1745. IEEE (2019). https://doi.org/10.1109/ICDE.2019.00187
44. Persia, F., Bettini, F., Helmer, S.: An interactive framework for video surveillance event detection and modeling. In: Proceedings of the 2017 ACM on Conference on Information and Knowledge Management, CIKM 2017, Singapore, 06–10 November 2017, pp. 2515–2518. ACM (2017). https://doi.org/10.1145/3132847.3133164
45. Piatov, D., Helmer, S., Dignös, A.: An interval join optimized for modern hardware. In: 32nd IEEE International Conference on Data Engineering, ICDE 2016, Helsinki, Finland, 16–20 May 2016, pp. 1098–1109. IEEE Computer Society (2016). https://doi.org/10.1109/ICDE.2016.7498316
46. Piatov, D., Helmer, S., Dignös, A., Persia, F.: Cache-efficient sweeping-based interval joins for extended Allen relation predicates. The VLDB J. **30**(3), 379–402 (2021). https://doi.org/10.1007/s00778-020-00650-5
47. Snodgrass, R.T.: Developing Time-Oriented Database Applications in SQL. Morgan Kaufmann (1999)
48. Snodgrass, R.T.: A case study of temporal data. Teradata Corporation (2010)
49. Soo, M.D., Jensen, C.S., Snodgrass, R.T.: An algebra for TSQL2. In: The TSQL2 Temporal Query Language, chap. 27, pp. 501–544. Kluwer (1995)
50. Stonebraker, M.: The design of the POSTGRES storage system. In: VLDB 1987, Proceedings of 13th International Conference on Very Large Data Bases, September 1–4, 1987, Brighton, England, pp. 289–300. Morgan Kaufmann (1987)

51. Suciu, D., Olteanu, D., Ré, C., Koch, C.: Probabilistic Databases. Synthesis Lectures on Data Management. Morgan & Claypool Publishers (2011). https://doi.org/10.2200/S00362ED1V01Y201105DTM016

52. Suzanne, Aurélie, Raschia, Guillaume, Martinez, José: Temporal aggregation of spanning event stream: a general framework. In: Hartmann, Sven, Küng, Josef, Kotsis, Gabriele, Tjoa, A Min, Khalil, Ismail (eds.) DEXA 2020. LNCS, vol. 12392, pp. 385–395. Springer, Cham (2020). https://doi.org/10.1007/978-3-030-59051-2_26

53. Suzanne, A., Raschia, G., Martinez, J., Jaouen, R., Hervé, F.: Temporal aggregation of spanning event stream: an extended framework to handle the many stream models. Trans. Large Scale Data Knowl. Centered Syst. **49**, 1–32 (2021). https://doi.org/10.1007/978-3-662-64148-4_1

54. Suzanne, A., Raschia, G., Martinez, J., Tassetti, D.: Window-slicing techniques extended to spanning-event streams. In: 27th International Symposium on Temporal Representation and Reasoning, TIME 2020, 23–25 September 2020, Bozen-Bolzano, Italy. LIPIcs, vol. 178, pp. 10:1–10:14. Schloss Dagstuhl - Leibniz-Zentrum für Informatik (2020). https://doi.org/10.4230/LIPIcs.TIME.2020.10

55. Suzanne, A., Raschia, G., Martinez, J., Tassetti, D.: Slicing techniques for temporal aggregation in spanning event streams. Inf. Comput. **281**, 104807 (2021). https://doi.org/10.1016/j.ic.2021.104807

56. Tao, Y., Yi, K.: Intersection joins under updates. J. Comput. Syst. Sci. **124**, 41–64 (2022). https://doi.org/10.1016/j.jcss.2021.09.004

57. Zhu, K., Fletcher, G.H.L., Yakovets, N., Papapetrou, O., Wu, Y.: Scalable temporal clique enumeration. In: Proceedings of the 16th International Symposium on Spatial and Temporal Databases, SSTD 2019, Vienna, Austria, 19–21 August 2019, pp. 120–129. ACM (2019). https://doi.org/10.1145/3340964.3340987

Graph Processing

An Algebra for Path Manipulation in Graph Databases

Roberto García[1,3]($\boxtimes$) and Renzo Angles[2,3]

[1] Engineering Systems Doctoral Program, Faculty of Engineering,
Universidad de Talca, 3340000 Curicó, Chile
roberto.garcia@utalca.cl

[2] Department of Computer Science, Faculty of Engineering, Universidad de Talca,
3340000 Curicó, Chile
rangles@utalca.cl

[3] Millennium Institute for Foundational Research on Data (IMFD), Santiago, Chile

Abstract. A key characteristic of current graph query languages is their support for path queries. Although a path query looks for paths in a graph database, current graph query languages are restricted to return just the source and target nodes connected by each solution path. Therefore, the user is not able to manipulate the elements (nodes and edges) of the resulting paths. In order to overcome such restriction, this paper presents an algebra for path manipulation. Inspired by the relational algebra, we defined the operators of selection, projection, node-based join, edge-based join, node-based cartesian product, edge-based cartesian product, union, intersection and difference. These operators are closed under sets of paths, i.e. the input and the output are sets of paths. We study the algebraic properties of the operators and describe use cases that justify the usefulness of the algebra.

Keywords: Path manipulation · Path operations · Path algebra · Paths · Graph databases

1 Introduction

In the last years, there has been an increasing interest around the development of technologies based on graphs and their use in many application domains [9]. Among such technologies, several graph database systems have been developed to facilitate the tasks of storing, manipulating and querying graphs.

A key feature of any graph database system is its query language. The research around graph query languages has been very intensive, in particular around two main features: graph pattern matching and path queries. The aim of a path query is to obtain all the paths that connect two nodes in a graph (recall that a path is a sequence of nodes and edges connecting two nodes). A popular way to express a path query is a triple of the form (s, r, p) where s is the source node, p the target node, and r is a regular path expression that defines the sequence of edges that a resulting path must satisfy [7].

© Springer Nature Switzerland AG 2022
S. Chiusano et al. (Eds.): ADBIS 2022, LNCS 13389, pp. 61–74, 2022.
https://doi.org/10.1007/978-3-031-15740-0_6

Most of the current practical graph query languages support path queries, however the functionalities for manipulating the resulting paths is very restricted. In the case of SPARQL [13], the standard query language for RDF, a path query just returns the pairs of nodes satisfying a path expression, so the user is not able to access specific elements (i.e. nodes and edges) of the resulting paths. In Cypher [1], the query language of Neo4j, the paths returned by a path query can be assigned to a variable, so the elements of each path can be accessed by using specific functions; unfortunately, the facilities provided by such functions is reduced. In PGQL [19] and OrientDB [3], each solution path is returned as a text. In GSQL [2], Gremlin [4] and TypeQL (GraQL) [5] the result of a path query is a set of objects, so the resulting paths can be processed by using a programming language. G-Core [6] defines a data model which allows storing paths, and the query language supports regular path queries; however, there exists no operators for path manipulation.

The notion of path appears in several application domains, and path manipulation operations can be related to specific use cases. For example, consider route planning as an application domain where cities and roads in a country are modeled as a graph. If it is the case that the country is divided into regions, we could have a database with multiple graphs where each graph contains information about a region. Now, consider that we want to obtain the shortest path between city A which is located in region R_1, and city B which is located in region R_2. Current query languages are not able to answer such query as the scope of a path query is a single graph. Moreover, optimization techniques like path indexing and query rewriting do not work due to data isolation.

Our hypothesis is that the problems mentioned above can be solved by extending current query languages with operations allowing the manipulation of paths (e.g. specifically, operations over collections of paths). Hence, the general objective of this article is to develop the foundations of query languages supporting path manipulation. Specifically, we define a set of operators that allow the manipulation of paths (a.k.a. a path algebra closed under sets of paths) and study some algebraic properties of such operators. Additionally, we present examples showing that essential queries of graph analysis can be expressed by using path manipulation operations.

The rest of the article is organized as follows: Sect. 2 presents the related work; Sect. 3 introduces a data model where paths are first class citizens; Sect. 4 defines the syntax and semantics of the path algebra; Sect. 5 presents basic rewriting rules among the operators; Sect. 6 describes use cases for the designed algebra; and Sect. 7 discusses issues and further research.

2   Related Work

In the current literature we can find some related studies with path operations. In [12], Gondran defines a general algebraic structure for path operations, including sum and multiplication operators. The aim of the algebra is to simplify the representation of complex path problems. In [16], Manger defines the operators of join and product. In this case, the aim is to provide a simpler and more computationally efficient path algebra. In [18], Naudziunas and Griffin define a specific domain language for path algebra specification. In [20], Rodriguez and Neubauer present some path operations whose evaluation is based on automatons.

Path manipulation operations are also used with related data models. In [10], Frasincar et al. present an algebra for querying XML data. This algebra is based on the relational algebra, and is used for query rewriting and optimization. The same idea was applied for RDF query optimization [11]. Stuckenschmidt et al. [21] shown how to use indexes for optimizing the use of paths in a distributed RDF path query. In this case, paths are processed by using indexed sub paths which represent local minimum paths, and they can be concatenated by using a join operator in order to reconstruct the original path.

On the other hand, multiple academic and commercial graph query languages support path queries (e.g. Cypher [1], Gremlin [4], G-Core [6], GSQL [2], OrientDB [3], PGQL [19] and TypeQL [5]). These languages support different types of path queries, however, they are very restricted with respect to the result of a path query, and its subsequent manipulation. For example, the paths are returned by using a textual representation (OrientDB and PGQL), or by using a object-based representation (Cypher, Gremlin, GSQL and TypeQL). G-Core allows to store the resulting paths in the database, but they cannot be manipulated. Languages such as G-Core and PGQL introduces the notion of "path template" as a way to define reusable path expression.

Cypher provides procedures and functions which allow path manipulation (via its APOC Library). Among them: *Create* allows to create a path from a start node and a list of relationships; *Combine* allows to combine two paths satisfying that the last node of the former is the same as the start node of the latter; *Slice* allows to split a path in sub-paths with a desired length and from a desired node; and *Elements* allows to convert a path into a list of nodes and edges. The characteristics and expressiveness of these operations have not been studied.

In conclusion, current systems and query languages have a limited support for path manipulation.

3   Data Model

In this section we introduce a graph data model which allows the manipulation of paths. The proposed model is based on the "Path Property Graph model" introduced by the G-CORE query language.

In general terms, a graph (database) will be constituted by nodes, edges and paths, where each of them has an identifier and a label. In comparison with the graphs supported by current graph database systems, our model considers paths as first class citizens. It implies that a query language using our model will be able to return nodes, edges and paths.

Let **O** be an infinite set of object identifiers and **L** be an infinite set of labels. Given a set $X = \{a_1, \ldots, a_n\}$, we use $\mathrm{FLIST}(X) = [a_1, \ldots, a_n]$ to denote a sequence of elements of X.

Definition 1. *A graph is a tuple* $G = (N, E, P, \rho, \delta, \lambda)$ *where:*

1. $N \subset O$ *is a finite set of node identifiers;*
2. $E \subset O$ *is a finite set of edge identifiers;*
3. $P \subset O$ *is a finite set of path identifiers;*
4. N, E *and* P *are disjoint sets;*
5. $\rho : E \rightarrow (N \times N)$ *is a total function;*
6. $\delta : P \rightarrow FLIST(N \cup E)$ *is a total function satisfying that, for each* $p \in P$, *we have* $\delta(p) = [n_1, e_1, n_2, e_2, \ldots, e_x, n_{x+1}]$, *where* $x \geq 0$, $n_i \in N$ *where* $1 \leq i \leq x + 1$, $e_j \in E$ *where* $1 \leq j \leq i$, *and* $\rho(e_j) = (n_j, n_{j+1})$ *or* $\rho(e_j) = (n_{j+1}, n_j)$ *for each* $e_j \in E$ *where* $1 \leq j \leq x$;
7. $\lambda : (N \cup E \cup P) \rightarrow L$ *is a total function.*

Function ρ defines the pairs of nodes connected by each edge. Given an edge e, if $\rho(e) = (n_1, n_2)$ then n_1 is the *source node* of e and n_2 is the *target node*. Function δ defines the sequence of nodes and edges related to each path identifier. Note that δ allows paths containing a single node. Finally, function λ defines a single label for each object (i.e. nodes, edges and paths) in the graph.

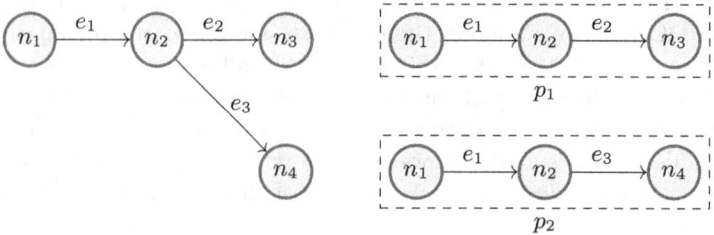

Fig. 1. Graphical representation of a graph with paths. Nodes are represented as circles (persons), edges as arrows (friend), and paths as dashed squares (potential friend).

Example 1 (Sample graph). Figure 1 shows the graphical representation of a sample graph whose formal definition is given as follows. Let $G = (N, E, P, \rho, \delta, \lambda)$ be a graph where $N = \{n_1, n_2, n_3, n_4\}$ is the set of node identifiers; $E = \{e_1, e_2, e_3\}$ is the set of edge identifiers; $P = \{p_1, p_2\}$ is the set of path identifiers; the edges are defined by $\rho(e_1) = (n_1, n_2)$, $\rho(e_2) = (n_2, n_3)$ and $\rho(e_3) = (n_2, n_4)$; the paths are defined by $\delta(p_1) = [n_1, e_1, n_2, e_2, n_3]$ and $\delta(p_2) = [n_1, e_1, n_2, e_3, n_4]$; and, the labels are defined by $\lambda(n_1) = \lambda(n_2) = \lambda(n_3) = \lambda(n_4) = $ "person", $\lambda(e_1) = \lambda(e_2) = \lambda(e_3) = $ "friend", $\lambda(p_1) = \lambda(p_2) = $ "potential friend".

It is good to mention that for this issue, for reasons of simplicity, the use of properties in the graph elements will not be considered for this definition.

4   Path Algebra

This section defines a set of operators (or functions) to manipulate paths and collections of paths. First, we defined operators to extract elements in a path, and operators to obtain specific properties of a path.

4.1   Operations over Paths

Given an edge e, we have that $Source(e)$ returns the source node identifier of e, and $Target(e)$ returns the target node identifier of e. Given the path $\delta(p_1) = [n_1, e_1, n_2, e_2, n_3]$ from the Fig. 1, we have that:

- $First(p)$: returns the identifier of the first node occurring in path p, e.g. $First(p_1) = n_1$;
- $Last(p)$: returns the identifier of the last node occurring in path p, e.g. $Last(p_1) = n_3$;
- $Node(p, i)$: returns the identifier of the node occurring in the position i of the path p, e.g. $Node(p_1, 2) = n_2$;
- $Edge(p, j)$: returns the identifier of the edge occurring in the position j of the path p, e.g. $Edge(p_1, 1) = e_1$;
- $SubPath(p, i, j)$: returns a subpath p' from p where $First(p') = Node(p, i)$ and $Last(p') = Node(p, j)$, e.g. $SubPath(p_1, 2, 3) = [n_2, e_2, n_3]$;
- $Length(p)$: returns the length (number of edges) of the path p, e.g. $Length(p_1) = 2$;
- $LeftSubPath(p, j)$: returns a subpath p' from the path p where $First(p') = First(p)$ and $Last(p') = Node(p, j)$, e.g. $LeftSubPath(p_1, 2) = [n_1, e_1, n_2]$;
- $RightSubPath(p, i)$: returns a subpath p' from the path p where $Last(p') = Last(p)$ and $First(p') = Node(p, i)$, e.g. $RightSubPath(p_1, 2) = [n_2, e_2, n_3]$;
- $Label(p, o)$: returns the label of an object (node or edge) o from the path p, e.g. $Label(p_1, Node(p_1, 2)) = \text{"person"}$.

Let p_1 and p_2 be paths in a graph G. We will say that p_1 is equal to p_2, denoted $p_1 = p_2$, if they meet the following conditions:

1. p_1 and p_2 have the same length, i.e. $Length(p_1) = Length(p_2)$; and
2. p_1 and p_2 have the same sequence of nodes and edges (identifiers).

$$\bigwedge_{i=1}^{n=Length(p_1)} Obj(p_1, i) = Obj(p_2, i)$$

Based on the above definition, a path p_1 and p_2 are equals only and only if they have the same length and the same sequence of nodes and edges. On the other hand, we will use $p_1 \neq p_2$ to denote that p_1 is not equal to p_2.

We say that p_1 is node-linkable with p_2, denoted $p_1 \triangleright_N p_2$, if $Last(p_1) = First(p_2)$. Assuming that $p_1 \triangleright_N p_2$, the *node-based natural concatenation* of p_1 and p_2, denoted $p_1 \circ_N p_2$, returns a path p such that $SubPath(p, 1, Length(p_1) +$

1) $= p_1$ and $SubPath(p, Length(p_1) + 1, Length(p_1) + Length(p_2)) = p_2$. Additionally, the *node-based cross concatenation* of p_1 and p_2, denoted $p_1 *_N p_2$, returns a set of paths S where a path $p = p' \circ_N p''$ is in S if p' is a *leftsubpath* of p_1 denoted $p' \sqsubseteq p_1$, p'' is a *rightsubpath* of p_2 denoted $p'' \sqsupseteq p_2$, and $p' \rhd_N p''$.

Given a pair of paths $p_1 = [n_1^1, e_1^1, ..., e_{i-1}^1, n_i^1]$ and $p_2 = [n_1^2, e_1^2, ..., e_{j-1}^2, n_j^2]$ in a graph G, we say that p_1 is edge-linkable with p_2, denoted $p_1 \rhd_E p_2$, if it exists an edge e in G such that $Last(p_1) = Source(e)$ and $First(p_2) = Target(e)$. Assuming that $p_1 \rhd_E p_2$, the *edge-based natural concatenation* of p_1 and p_2, denoted $p_1 \circ_E p_2$, returns a path p such that $SubPath(p, 1, Length(p_1) + 1) = p_1$ and $SubPath(p, Length(p_1) + 2, Length(p_1) + Length(p_2) + 1) = p_2$. Furthermore, the result of $p_1 \circ_E p_2$ will be the sequence $[n_1^1, \ldots, n_i^1, e, n_1^2, \ldots, n_j^2]$. Additionally, the *edge-based cross concatenation* of p_1 and p_2, denoted $p_1 *_E p_2$, returns a set of paths S where a path $p = p' \circ_E p''$ is in S if p' is a *leftsubpath* of p_1 denoted $p' \sqsubseteq p_1$, p'' is a *rightsubpath* of p_2 denoted $p'' \sqsupseteq p_2$, and $p' \rhd_E p''$.

Let p be a path, v be a value, k is an integer, $id \in \mathbf{O}$ and $\odot \in \{=, <, >, \leq, \geq\}$. A *selection condition* is defined as follows: a simple selection condition is any of the expressions $First(p) = id$, $Last(p) = id$, $Node(p, i) = id$, $Edge(p, j) = id$, $Length(p) \odot k$ and $Label(p, o) = v$; if c_1 and c_2 are selection conditions then $(c_1 \wedge c_2)$, $(c_1 \vee c_2)$ and $\neg(c_1)$ are complex selection conditions.

The evaluation of a selection c over a path p, denoted $ev(c, p)$, returns *true* or *false*. If p is a simple selection condition then $ev(c, p)$ is *true* if the equivalence applies, and *false* otherwise. The evaluation of a complex selection condition is defined by the following Table 1:

Table 1. Evaluation of complex selection conditions. c_1 and c_2 are selection conditions.

c_1	c_2	$(c_1 \wedge c_2)$	$(c_1 \vee c_2)$	$\neg(c_1)$
true	true	true	true	false
true	false	false	true	
false	true	false	true	true
false	false	false	false	

4.2 Operations over Sets of Paths

Definition 2 (Path Algebra). *Let S and S' be sets of paths, c be a filter expression, i and j are integers. We define a path algebra integrated by the following operators:*

- *Selection:*

$$\sigma_c(S) = \{p \in S \mid ev(p, c) = true\}$$

- *Projection:*

$$\pi(S, i, j) = \{p \in S \mid SubPath(p, i, j)\}$$

- *Node-based Join:*

$$S_1 \bowtie_N S_2 = \{p_1 \circ_N p_2 \mid p_1 \in S_1, p_2 \in S_2, p_1 \triangleright_N p_2\}$$

- *Edge-based Join*

$$S_1 \bowtie_E S_2 = \{p_1 \circ_E p_2 \mid p_1 \in S_1, p_2 \in S_2, p_1 \triangleright_E p_2\}$$

- *Node-based Cartesian Product:*

$$S_1 \times_N S_2 = \{p_1' \circ_N p_2' \mid p_1' \sqsubseteq p_1 \in S_1, p_2' \sqsupseteq p_2 \in S_2, p_1' \triangleright_N p_2'\}$$

- *Edge-based Cartesian product*

$$S_1 \times_E S_2 = \{p_1' \circ_E p_2' \mid p_1' \sqsubseteq p_1 \in S_1, p_2' \sqsupseteq p_2 \in S_2, p_1' \triangleright_E p_2'\}$$

- *Union*

$$S_1 \cup S_2 = \{p \mid p \in S_1 \ or \ p \in S_2\}$$

- *Intersection:*

$$S_1 \cap S_2 = \{p \mid p \in S_1 \ and \ p \in S_2\}$$

- *Difference*

$$S_1 \backslash S_2 = \{p_1 \in S_1 \mid \nexists \ p_2 \in S_2 \ satisfying \ that \ p_1 = p_2\}$$

It is worth mentioning that every operation defined above, its result is a set of paths, i.e. our algebra is closed under sets of paths.

4.3   Examples of Queries

Next we present examples of queries using the operators of the algebra. The examples are based in the routes representations of the Fig. 2, this figure presents a set of cities, towns and beaches, connected by different roads. Additionally in the Table 2 is presented a set of paths that represent different bus and train routes.

- **Selection:**
  Query: I want to get the paths from c_4 of the bus bus_3.

$$\sigma_{first(p)=c_4}(bus_3) = \{p \in bus_3 \mid first(p) = c_4\}$$

Result:

$$result_1 = \{[c_4, r_5, b_2], [c_4, r_3, t_2]\}$$

Table 2. Routes of different bus lines, represented as paths of the graph in Fig. 2. This representation is based on the textual representation of a path.

bus_1	bus_2	bus_3	$train_1$
$\{[c_1, h_1, c_2, h_2, c_3, h_3, c_4],$ $[c_4, h_3, c_3, h_2, c_2, h_1, c_1]\}$	$\{[c_3, r_4, b_1], [b_1, r_4, c_3],$ $[c_3, r_2, t_1], [t_1, r_2, c_3]\}$	$\{[c_4, r_5, b_2], [b_2, r_5, c_4],$ $[c_4, r_3, t_2], [t_2, r_3, c_4]\}$	$\{[c_5, r_6, b_3],$ $[b_3, r_6, c_5]\}$

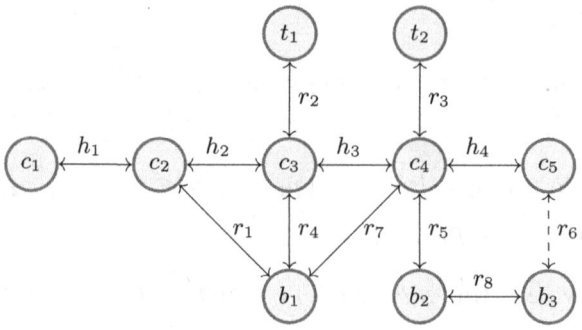

Fig. 2. Graphical representation of a routes between cities. Nodes are represented as circles (cities c, towns t and beach b) and edges as arrows (highway h and road r).

- **Projection:**
  Query: I want to know which beaches or towns I can get to from c_4 by the bus_3.

$$\pi(\sigma_{first(p)=c_4}(bus_3), 2, 2) = \{p \in \sigma_{first(p)=c_4}(bus_3) \mid SubPath(p, 2, 2)\}$$

  Result:

$$result_2 = \{[b_2], [t_2]\}$$

- **Node-based Join:**
  Query: I want to go from the city c_1 to the beach b_2, using the bus_1 and bus_3 correspondingly.

$$\sigma_{first(p)=c_1}(bus_1) \bowtie_N \sigma_{last(p)=b_2}(bus_3)$$

  Result:

$$result_3 = \{[c_1, h_1, c_2, h_2, c_3, h_3, c_4, r_5, b_2]\}$$

- **Node-based Cartesian product:**
  Query: I want to go from the city c_1 through c_3 and see what is nearby, using first the bus_1 and then the bus_2.

$$\sigma_{first(p)=c_1}(bus_1) \times_N \sigma_{first(p)=c_3}(bus_2)$$

  Result:

$$result_4 = \{[c_1, h_1, c_2, h_2, c_3, r_4, b_1], [c_1, h_1, c_2, h_2, c_3, r_2, t_1]\}$$

- **Edge-based Join:**
  Query: I want to create new routes based on the routes of bus_3 and $train_1$.

$$bus_3 \bowtie_E train_1$$

  Result:

$$result_5 = \{[b_2, r_5, c_4, h_4, c_5, r_6, b_3], [t_2, r_3, c_4, h_4, c_5, r_6, b_3],$$
$$[c_4, r_5, b_2, r_8, b_3, r_6, c_5]\}$$

- **Edge-based Cartesian product:**
  Query: I want to create new routes based on the routes of bus_1 and $train_1$, and the existing road in the Fig. 2.

$$bus_1 \times_E train_1$$

Result:

$$result_6 = \{[c_1, h_1, c_2, h_2, c_3, h_3, c_4, h_4, c_5, r_6, b_3], [c_1, h_1, c_2, h_2, c_3, h_3, c_4, h_4, c_5],$$
$$[c_4, h_4, c_5, r_6, b_3], [c_4, h_4, c_5]\}$$

- **Union:**
  Query: I want to create a new bus_x using the routes from the bus_2 starting in t_1 and bus_3 starting in c_4.

$$(\sigma_{first(p)=t_1}(bus_2)) \cup (\sigma_{first(p)=c_4}(bus_3))$$

Result:
$$result_7 = \{[t_1, r_2, c_3], [c_4, r_3, t_2]\}$$

- **Intersection:**
  Query: I want to get the common routes between the $result_1$ and $result_7$.

$$result_1 \cap result_7$$

Result:
$$result_8 = \{[c_4, r_3, t_2]\}$$

- **Difference:**
  Query: I want to get the routes that differ between the $result_1$ and $result_8$.

$$result_1 \backslash result_8$$

Result:
$$result_9 = \{[c_4, r_5, b_2]\}$$

5 Properties

In this section we study the algebraic properties of the path operators defined above.

We will say that two sets of paths S_1 and S_2 are equivalent, denoted by $S_1 \equiv S_2$, if $[\![S_1]\!]_G = [\![S_2]\!]_G$ for each path p from S evaluated in G. The next lemma allows us to define the next properties: Compositionality, commutativity, associativity and distributivity.

Lemma 1. *The union and intersection operators are associative and commutative.*

$$(S_1 \cup S_2) \equiv (S_2 \cup S_1); \; S_1 \cup (S_2 \cup S_3) \equiv (S_1 \cup S_2) \cup S_3; \; (S_1 \cap S_2) \equiv (S_2 \cap S_1);$$
$$S_1 \cap (S_2 \cap S_3) \equiv (S_1 \cap S_2) \cap S_3$$

Lemma 2. *The Join and cartesian product operators are associative and not commutative. The Difference operator is not associative and not commutative. Furthermore, these operators are distributive over the union based in a non-repeated values semantics.*

- $(S_1 \bowtie_N S_2) \not\equiv (S_2 \bowtie_N S_1)$
- $S_1 \bowtie_N (S_2 \bowtie_N S_3) \equiv (S_1 \bowtie_N S_2) \bowtie_N S_3$
- $S_1 \bowtie_N (S_2 \cup S_3) \equiv (S_1 \bowtie_N S_2) \cup (S_1 \bowtie_N S_3)$
- $(S_1 \times_N S_2) \not\equiv (S_2 \times_N S_1)$
- $S_1 \times_N (S_2 \times_N S_3) \equiv (S_1 \times_N S_2) \times_N S_3$
- $S_1 \times_N (S_2 \cup S_3) \equiv (S_1 \times_N S_2) \cup (S_1 \times_N S_3)$
- $(S_1 \backslash S_2) \not\equiv (S_2 \backslash S_1)$
- $S_1 \backslash (S_2 \backslash S_3) \not\equiv (S_1 \backslash S_2) \backslash S_3$
- $S_1 \backslash (S_2 \cup S_3) \equiv (S_1 \backslash S_2) \cup (S_1 \backslash S_3)$
- $(S_1 \bowtie_E S_2) \not\equiv (S_2 \bowtie_E S_1)$
- $S_1 \bowtie_E (S_2 \bowtie_E S_3) \equiv (S_1 \bowtie_E S_2) \bowtie_E S_3$
- $S_1 \bowtie_E (S_2 \cup S_3) \equiv (S_1 \bowtie_E S_2) \cup (S_1 \bowtie_E S_3)$
- $(S_1 \times_E S_2) \not\equiv (S_2 \times_E S_1)$
- $S_1 \times_E (S_2 \times_E S_3) \equiv (S_1 \times_E S_2) \times_E S_3$
- $S_1 \times_E (S_2 \cup S_3) \equiv (S_1 \times_E S_2) \cup (S_1 \times_E S_3)$

Note that the join and cartesian product operators (for both, node-based and edge-based) are not commutative due to the concatenation operations defined in Sect. 4, which are defined based on the order in which the sets are operated, that is, if there are two paths p_1 and p_2, $p_1 \circ_N p_2 \not\equiv p_2 \circ_N p_1$, $p_1 *_N p_2 \not\equiv p_2 *_N p_1$, $p_1 \circ_E p_2 \not\equiv p_2 \circ_E p_1$ and $p_1 *_E p_2 \not\equiv p_2 *_E p_1$ depending on the case.

Lemma 3. *The node-based join, node-based cartesian product, edge-based join, edge-based cartesian product, difference, union, and intersection operators are compositional, since each of these always returns a new set of paths P.*

6   Use Cases

This section presents application domains where the proposed path operations can be useful.

6.1   Transportation

A transportation network can be naturally abstracted as a graph [14], where a node represents a point of interest (POI), and the an edge represents a road between POIs. A path in a transportation network represents a way to go from a point A to a point B.

If we have two sets of path respectively, S_1 and S_2, representing the paths of two people to some points in a map, we can perform many path operations with the goal of retrieving some important information like comparisons or other useful data. Consider the following path operations that can be used to perform basic operations in a transportation network: *Select*: to filter the paths connecting two points according to a given selection condition; *Project*: to project sub-routes from the set of routes; *Node-based Join*: extend routes and create new paths to explore; *Edge-based Join*: explore a link between two paths; *Intersect*: get the common routes between two person; *Difference*: get the different routes from one person.

Based on the operations described above, we can formulate some specific cases that involve the use of more than one operation:

- Found two routes for two persons, from a specific point to another, like going from the mall (A) to the university (B), without them crossing the others path. In this case, the *select, project* and *difference* operations will be useful, because we can get the routes from A to B, get the middle path and get different sub routes.
- If we think that a country is divided into regions, and each region is represented by a graph that contains the cities and their connections, the operation *Select* would be useful to find subpaths within regions and the finding the desired path by joining with a *Node-based Join* or *Edge-based Join* depending the case.

6.2   Proteins

Some of the path operations can be used in protein analysis. A protein can be represented as a graph where the nodes represent amino acids and ligands, and the relationships among them (e.g. precedence, distance or gap) can be represented as edges.

One useful task to perform with path operations in the context of proteins is the search of common chains between two proteins. If we see each protein as a set of paths, where each path represents a specific chain in a protein, we can perform some path operations in it in order to obtain basic information: *Select*: To search specific chains in a protein, or specific amino acids in it; *Project*: To project specific objects of the chains; *Intersect*: To compare two proteins and retrieve the common chains.

Another useful task to perform with our operations on proteins is to look for the interaction of ligands with protein subchains. For this case, we can use the *Edge-based Cartesian Product*, and *Select* which allows us, given two sets of paths, one representing the amino acid subchains and the other the ligands to interact, to generate the different combinations, and filter by a desired length or some specific condition.

The search for common chains between proteins or the ligands interaction with amino-acids, allows discovering similarities between them, with the aim of analyzing them and discovering new drugs [15,17] and applying this knowledge in the treatment of diseases [8].

7   Discussion

In this section we discuss three issues related to path manipulation which are not covered in this paper but deserve further research: labels and repeated results; notions of compatibility; and path queries across multiple graphs.

7.1   Labels and Repeated Results

Labels are fundamental to express path queries, as they are used to describe a regular path expression. In our case, labels are just used by the *Select* operation to filter paths occurring in a set of paths. Next we discussion the use of labels in other operators.

In the *Project* operation, labels can be used to project sub-paths (occurring in a set of paths) by using a regular path expression. We have not included this feature because there are multiple interpretations for such projection (e.g. semantics for repeated results). Moreover, depending on the given semantics, the operation could be complex in the sense of the number of paths that can be returned.

The *Join* of paths based on labels looks possible at the first time, however it implies some issues (e.g. create a new node join the paths). Similar problems arise with the *cartesian product* operation.

In the case of the set-based operations (*Union*, *Intersection* and *Difference*), the use of labels can cause a loss of paths because we are comparing labels and not identifiers. Specifically, two path can have the same structure, have the same length, have the same order of elements (nodes, edges labels), and can be "compatible" under a label comparison, but they are not the same path as their components are different.

7.2   Notions of Compatibility

In Sect. 4 we define a notion of equality with the objective of evaluating two paths, and such notion was used to define set-based operations (Union, Difference and Intersection). Such notion of compatibility is based on identifiers, i.e. two paths are considered equal if their sequence of identifiers is the same. A label-based definition of compatibility implies a comparison of the labels.

Another option is to compare source and target nodes of a path, in this case, we have a pair of paths p_1 and p_2 that will be compatibles if $First(p_1) = First(p_2)$ and $Last(p_1) = Last(p_2)$. This option allows us to obtain paths with the same origin and destination, regardless of their length or their intermediate sub-path sequence.

Finding an option to define the compatibility of two paths is not an easy task, due to the complexity of defining the correct semantics depending of what we want to evaluate and return.

7.3 Paths Queries Across Multiple Graphs

An interesting use-case for the path algebra presented here is the evaluation of path queries over multiple graphs. For example, suppose that the graph shown in Fig. 2 is divided in three sub-graphs: the graph G_1 which includes the nodes $\{c_1, c_2, c_3, t_1, b_1\}$ and the edges $\{h_1, h_2, r_1, r_2, r_4\}$; the graph G_2 which includes the nodes $\{c_4, c_5, t_2, b_2, b_3\}$ and the edges $\{h_4, r_3, r_5, r_6, r_8\}$; and the graph G_3 which includes the nodes $\{c_3, c_4, b_1\}$ and the edges $\{h_3, r_7\}$. Based on our example, G_1 and G_2 can represent two regions in a country, and G_3 represents the border between both regions. Now, suppose we want to obtain the paths between city c_1 and beach b_3. Note that this query cannot be answered by using a single path query (e.g. an RPQ expression) as c_1 and b_3 are in different graphs.

The above query can be solved by using the following algebra expression:

$$(\sigma_{first(p)=c_1}(G_1) \bowtie_N \sigma_{true}(G_3)) \bowtie_N \sigma_{last(p)=b_3}(G_2)$$

First, we *Select* the paths starting in c_1 to any other node in $region_1$, and storing them in pr_1; second, we *Select* the paths from any node in $region_2$ to b_3, and storing them in pr_2; third, we *Select* the connecting paths occurring in the "connection graph", and storing them in *cpaths*. Then, we can use the *Node-based Join* operation, to connect the sets of paths pr_1, pr_2 and *cpaths*. It results in a set containing the paths between c_1 and b_3.

Now suppose that the user requires the shortest paths between c_1 and b_3. In this case, we can extend the path algebra with operations that allow to obtain the set of shortest paths, e.g. *Shortest(S)* where S is a set of paths. This type of query can be also solved by introducing operator for grouping and aggregation over sets of paths.

Acknowledgements. R. García was supported by CONICYT PFCHA/BECA DE DOCTORADO NACIONAL/2019 under Grant 21192157. R. Angles was supported by ANID FONDECYT Chile through grant 1221727.

References

1. Cypher Query Language. https://neo4j.com/developer/cypher-query-language/
2. GSQL Documentation. https://docs.tigergraph.com/gsql-ref/current/intro/intro
3. OrientDB Community Edition. https://orientdb.org/
4. TinkerPop Documentation. https://tinkerpop.apache.org/docs/current
5. TypeQL Queries: Vaticle. https://docs.vaticle.com/docs/query/overview
6. Angles, R., et al.: G-CORE: a core for future graph query languages. In: Proceedings of the 2018 International Conference on Management of Data, pp. 1421–1432. ACM (2018). https://doi.org/10.1145/3183713.3190654
7. Angles, R., Arenas, M., Barceló, P., Hogan, A., Reutter, J., Vrgoč, D.: Foundations of modern query languages for graph databases. ACM Comput. Surv. **50**(5), 1–40 (2017). https://doi.org/10.1145/3104031
8. Cassandri, M., et al.: Zinc-finger proteins in health and disease. Cell Death Discov. **3**(1), 17071 (2017). https://doi.org/10.1038/cddiscovery.2017.71

9. Foulds, L.R.: Graph Theory Applications. Springer, New York (1992). https://doi. org/10.1007/978-1-4612-0933-1

10. Frasincar, F., Houben, G.J., Pau, C.: XAL: an algebra for XML query optimization. Aust. Comput. Sci. Commun. **24**(2), 49–56 (2002)

11. Frasincar, F., Houben, G.J., Vdovjak, R., Barna, P.: RAL: an algebra for querying RDF. WWW **7**(1), 83–109 (2004). https://doi.org/10.1023/B:WWWJ. 0000015866.43076.06

12. Gondran, M.: Path algebra and algorithms. In: Roy, B. (ed.) Combinatorial Programming: Methods and Applications. ASIC, vol. 19, pp. 137–148. Springer, Dordrecht (1975). https://doi.org/10.1007/978-94-011-7557-9_6

13. Harris, S., Seaborne, A.: SPARQL 1.1 Query Language (W3C Recommendation), 21 March 2013. https://www.w3.org/TR/sparql11-query/

14. Hegeman, T., Iosup, A.: Survey of graph analysis applications (2018). https://doi. org/10.48550/ARXIV.1807.00382

15. Konc, J., Janežič, D.: Binding site comparison for function prediction and pharmaceutical discovery, April 2014. https://doi.org/10.1016/j.sbi.2013.11.012

16. Manger, R.: A new path algebra for finding paths in graphs. In: 26th International Conference on Information Technology Interfaces, vol. 1, pp. 657–662 (2004)

17. Mavromoustakos, T., et al.: Strategies in the rational drug design. Curr. Med. Chem. **18**(17), 2517–2530 (2011). https://doi.org/10.2174/092986711795933731

18. Naudziunas, V., Griffin, T.G.: A domain-specific language for the specification of path algebras. In: Proceedings of the First Workshop on ATE, Wrocław, Poland, 31 July 2011, vol. 760, pp. 46–57. CEUR-WS (2011). https://ceur-ws.org/Vol-760/ paper7.pdf

19. van Rest, O., Hong, S., Kim, J., Meng, X., Chafi, H.: PGQL: a property graph query language. In: GRADES 2016. ACM, New York (2016). https://doi.org/10. 1145/2960414.2960421

20. Rodriguez, M.A., Neubauer, P.: A path algebra for multi-relational graphs. In: IEEE 27th International Conference on Data Engineering Workshops, pp. 128–131 (2011). https://doi.org/10.1109/ICDEW.2011.5767613

21. Stuckenschmidt, H., Vdovjak, R., Broekstra, J., Houben, G.J.: Towards distributed processing of RDF path queries. Int. J. Web Eng. Technol. **2**(2/3), 207–230 (2005). https://doi.org/10.1504/IJWET.2005.008484

Road Network Graph Representation for Traffic Analysis and Routing

Chiara Bachechi$^{(\boxtimes)}$ ⓘ and Laura Po ⓘ

"Enzo Ferrari" Engineering Department, University of Modena and Reggio Emilia,
Modena, Italy
{chiara.bachechi,laura.po}@unimore.it

Abstract. The road network is the infrastructure along which the mobility of users and goods takes place; the analysis of these networks in terms of spatial and graph theoretical approaches can provide insights to understand urban mobility, improve daily commuting, and reflect on new, more sustainable, scenarios. This paper presents an open-source framework to analyze the road network and investigate the relationship between its topology and traffic conditions. Open-source geographical data are stored in a graph database containing roads, junctions, and Points of Interest (POI), allowing importing of traffic data. The framework includes routing algorithms to obtain the optimal path based on different aspects such as distance, traffic volume, and the number of traversed junctions; furthermore, it allows simulating road closures to observe how they affect road viability. The framework was tested in the use case of the city of Modena (Italy) providing promising results.

Keywords: Road network · Graph database · Routing · Road traffic · Smart city · Network modelling

1 Introduction

At least once in a lifetime, everyone has been stuck in traffic. In recent decades, researchers have studied the topology and characteristics of the road network to understand, plan and optimize traffic management, improve commuting time and reduce emissions. In particular, when it comes to road networks that are complex non-planar spatial networks, weighted multi-digraphs play a massive role in the analysis and understanding of their properties. Two main approaches to road traffic modeling are suggested in the literature: a primal approach described in [26] and a dual approach presented and exploited in [14,25]. The primal approach generates a graph where each road intersection (junction) is a node and lanes or road sections connect the junctions with relationships: a *primal graph*. The primal graph is necessary for routing purposes and provides an in-depth representation of the road network. The dual approach instead is inverse: it considers roads as nodes, and the connections between them are created when they meet at a junction. The obtained graph is a *dual graph*. Most studies adopt the primal

© Springer Nature Switzerland AG 2022
S. Chiusano et al. (Eds.): ADBIS 2022, LNCS 13389, pp. 75–89, 2022.
https://doi.org/10.1007/978-3-031-15740-0_7

approach because, in the dual graph, the information regarding the geo-location and the geometry of the roads may be lost. However, both representations can be useful to unveil different aspects of the road network topology: the dual graph can help to study the connections between roads and easily determine the ones where traffic flows converge. For this reason, we realized an open-source framework available on git repository[1] that allows directly generating a graph instance from Open Street Map (OSM)[2] data composed of two graph representations: a primal and a dual graph. The two representations are generated together and added to the same graph database, keeping a connection between the node representation of the road in the *dual graph* and its geometry memorized in the *primal graph*. This combined approach allows exploiting the advantages of a simplified representation of the relations between roads without losing the complexity of their geometry construction.

The obtained graph model also includes the Points Of Interest (POI) in the geographical area and supports the integration of mobility information, such as traffic volume data. Moreover, the graph can be changed by closing or opening roads to investigate alternative routes in the different road closure scenarios. The analysis of the road network can be performed by applying centrality and community detection algorithms. This paper highlights the importance of integrating traffic information directly into the graph and taking into account the distance, the volume of traffic and the number of nodes crossed when evaluating the shortest path. For this reason, three different routing methods have been compared. Since the framework imports data from OpenStreetMap, it enables the easy generation of a digital twin of the road network of any city, area, or region; also, routing can be performed in very few steps. We strongly believe that this solution will simplify the modeling, analysis, and improvement of road networks in different cities. We present a use case analysis conducted on the road network of the city of Modena (Italy) that shows how traffic information can be generated, integrated, and exploited to gain additional insight into the traffic conditions of the urban area.

The rest of the paper is organized as follows. Related work is presented in Sect. 2. The generation of the graph model is described in Sect. 3. Section 4 explains how traffic data can be generated; while, Sect. 5 describes the methodology adopted to integrate traffic data in the existing road graph model. Furthermore, Sect. 6 presents three routing approaches and the management of road closures. In Sect. 7, the results of the analysis performed in the city of Modena are presented. Finally, conclusions are discussed in Sect. 8.

2   Related Work

There are many examples of the profitable use of graphs for representing road networks [1,13,20,29]. In [21], several modelling solutions that can be employed to represent and simplify a road network are presented. A data-driven graph

---

[1] https://github.com/ChiaraBachechi/roadRouting/.

[2] https://wiki.openstreetmap.org/.

model generated from traffic sensors observations based on the primal approach is described in [16]. They model the road network as an undirected graph weighted according to the spatial correlation among adjacent traffic intersections; however, to adopt this solution, traffic sensors are needed at every road intersection. In [15], a tool that allows studying efficiency in a road network is described. They employed a PostgreSQL database with pgRouting[3] and PostGIS extension to convert OSM ways into geometric features. Graph algorithms, however, are generally performing better when executed on a graph database as discussed in [7,22]. For this reason, we decide to employ the Neo4j graph data platform[4] for our framework. In [5], a library to convert the OSM road network into a road graph is presented. This tool allows downloading and analyzing the road network as a graph. We rely on this library for the generation of the initial *primal graph*, but we enriched the obtained graph with properties and relations extracted from traffic data, POI nodes and their connections, additional properties related to the status of the street, and we generate a *dual graph*. Moreover, our developed framework allows closing and opening streets and generating different routes according to the new road status. Several studies discussed the use of centrality measures to explain traffic flows in urban areas [11,12] investigating the correlation among different centrality measures and the corresponding simulated or real traffic flows. The studies are mainly focused on the primal approach, we further investigate this correlation in this paper considering also the *dual graph*. Pathfinding algorithms have been widely studied to determine the shortest path between a source and a destination. In [30], the labels associated with each road are taken into account to find label-constrained paths, employing index-based techniques and decomposing the road network into a tree-like structure. Moreover, in [27], the authors suggest that the users are looking for the fastest path rather than the shortest path; thus, they include traffic influence factors when evaluating the shortest path and develop 'Trafforithm' a traffic-aware shortest path algorithm.

3   Road Network Graph Modelling

To ensure that this methodology can be easily applied to any urban environment, OSM is the main open data source. OSM provides very high-resolution data regarding road networks collected through the collaboration of a big community of users worldwide. The topological data structure has two main elements: nodes and ways. Nodes represent map features without a size that can be approximated as points. While, ways are lists of nodes representing polylines and polygons. The more intuitive conversion of this structure in a graph model is to generate graph nodes from OSM nodes and relationships from ways that connect them. As a result, a *primal graph* is obtained. This graph can be used for routing purpose since each node maintains a reference to the real point in space where the junction is located. However, the number of nodes and relationships is very high, while the density is very low. A simplified representation of the road network

---

[3] https://pgrouting.org/.
[4] https://neo4j.com/.

can be generated by the dual approach where the graph nodes are roads (each identified by their OSM identifier). The resulting *dual graph* is a modified version of the named street graph (described in [8]) that assumes as street names their OSM identifiers. The *dual graph*'s analysis highlights the interactions between roads. These two representations of the road network are regenerated in the same graph database instance maintaining a connection between them to exploit the advantages of both approaches.

3.1 Primal Graph

The *primal graph* represent point-to-point relationships between junctions. The OSMnx Python package [5] was employed to construct the graph database instance directly from OSM data. Given a point and a radius, the data in the circular area are converted into a 'graphml' format; then, a Neo4j instance is generated through a query in Neo4j proprietary language: Cypher. To do that we employed the Awesome Procedure On Cypher (APOC)[5] library. Since our framework is devoted to vehicular traffic, we decide to consider only the roads where vehicles can travel. We modify the structure of the graph obtained directly with OSMnx: we changed the label of OSM nodes to 'Node', and we insert a spatial property containing the node's location (latitude and longitude). Moreover, relationships between the junction nodes are called 'ROUTE' and contain information about the type of highway, the street name, the OSM identifier, the length of the road segment, and the status. The status property has also been added and is 'active' when the road is open to vehicular traffic, or 'closed' if not. A relationship is generated for each travel direction.

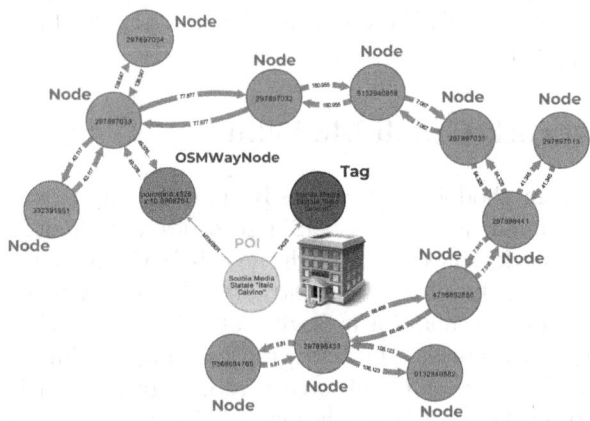

Fig. 1. An example of the primal graph structure.

For routing purposes, also POIs are of key importance as sources and targets of vehicles' routes. For this reason, the Overpass API[6] is queried to get data

---

[5] https://github.com/neo4j-contrib/neo4j-apoc-procedures.

[6] http://overpass-api.de/.

Fig. 2. Comparison between the paths evaluated with the three approaches.

regarding amenities (e.g. restaurants, bars, pubs, and schools) and insert POI in the already existing graph as new nodes. The OSM's flexible structure enables the association of a variable number and different type of tags to each element (way or node) that describes its properties. Thus, the POI nodes are labelled as 'PointOfInterest' with a connected 'Tag' node containing all the additional properties of the POI retrieved from OSM. Each POI is connected to a node labelled as 'OSMWayNode' with a 'MEMBER' relationship and then connected to the node of the road network nearest to the exact geographic position of the POI. When the POI is represented as a point (OSM node), the 'OSMWayNode' is a single node. While, when the POI has a complex geometry (e.g. polygon) and is represented as a way in OSM, it is represented as a collection of connected 'OSMWayNodes'. Figure 1 shows an example of a POI node connected to: a Tag node with additional information, and an OSMWayNode connected to the other nodes of the road network. The Gray relationships are of 'ROUTE' type and the value displayed is the distance in meters. Exploiting the framework for our use case, we generate a *primal graph* for the city of Modena containing 19,607 junction nodes, 841 POIs nodes, and 33,571 'ROUTE' relationships. The average un-directed degree of nodes is 3.41: the incoming and outgoing average degree are very similar and around 1.71. The graph has a very low density as expected for a road network (8×10^{-5}).

3.2 Dual Graph

The *primal graph* is a point-to-point oriented graph where the relationship that connects the nodes are segments of road lanes and to represent a single road you may need more than one relationship. For example, in Fig. 3, all the nodes and relationships highlighted in orange in the *primal graph* correspond to the same road. For this reason, we decide to generate a simplified version of the graph that is derived from the *primal graph*, and thus we will refer to it as *dual graph*. In this graph, each road is represented by a node 'RoadOsm', and the relationships 'CONNECTED' are generated between connected roads. Two roads are connected if there is a junction that allows driving from the source road to the target road. Thus, each 'CONNECTED' relationship is associated with a junction 'Node' in the *primal graph*. However, for each node in the *primal graph* there could be several 'CONNECTED' relationships in the *dual graph*. This version of the graph is distant from the real street map since the nodes do not have a spatial reference. For example, in Fig. 3, the node highlighted in light blue in the *primal graph* corresponds to the three relationships in the same color in the *dual graph*. This alternative representation can be useful to better understand the relationship between roads and identify the ones that are more important in the road network. Our framework automatically generate this simplified *dual graph* as an additional layer over the *primal graph*. Moreover, the graph contains only the roads in the 'active' status: open to vehicular circulation. In this way, the user can generate alternative *dual graphs* in different scenarios of road closures and study the different relationships' contexts that will emerge. The *dual graph* generated for the city of Modena contains 4421 nodes and 15037 relationships, the average undirected degree of nodes is 6.8 (the incoming and the outgoing average degrees are very similar and around 3.4), and a decimal order higher density compared with the *primal graph* (7×10^{-4}).

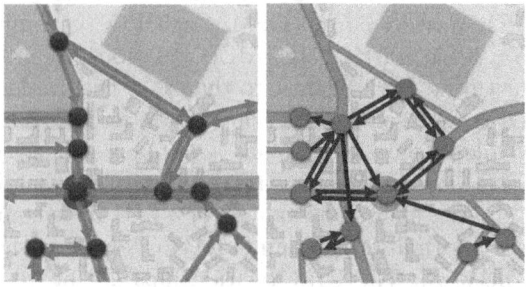

Fig. 3. Example of a *primal graph* (on the left) and the *dual graph* (on the right). (Color figure online)

4   Generation of Traffic Data

Traffic information can be generated in many ways from traffic sensors observations (e.g. induction loop detectors, cameras, Bluetooth sensors), GPS routes, and open data sources (e.g. Open Transport data[7]), OD matrices or through simulations. Each method can provide different data for each road lane such as vehicle count, speed, type of vehicles, traffic flow (Veh/hour), etc. We are interested in integrating the annual average daily traffic volume (AADT) for each road lane in the road network. AADT is a traffic volume metric defined as the

---

[7] http://opentransportmap.info/.

average daily traffic volume at a given road lane over a full 365 days/year [10]. In the following, we described the solution adopted to generate traffic data and calculate AADT in our use case.

About 400 induction loop detectors are placed under the surface of the street in the urban area of the city of Modena; these sensors provide a value of vehicle count and average speed every minute. Traffic flow data have been collected from November 2018 to April 2021, cleaned (as described in [4]), and used to feed a micro-simulation traffic model: SUMO (Simulation for Urban MObility) [17]. The traffic sensor data of each day are the input of the SUMO model that simulates the traffic in all the main streets of the urban area. This simulation allows predicting the traffic flow in all the road segments of the road network, and it is better described in [2,3,24]. Considering all the simulations of the year 2019, for each road r, AADT was evaluated as:

$$AADT_r = \frac{\sum_{i=0}^{N} \sum_{j=0}^{24} flow_{r,j,i}}{N}$$

where N is the number of simulated days in the given year, and $flow_{r,j,i}$ is the observed traffic flow (veh/hour) in the road lane r for the jth hour of the ith day of the year. The traffic data generated by the traffic model for the city of Modena are visualized in a dashboard[8] and available as open data[9]. Moreover, the CSV formatted file used in our use case is available for testing in the git repository.

5   Traffic Data Integration

The framework allows integrating traffic data directly in the graph, through a python script. This traffic data need to be formatted as a CSV file containing the OSM id of the starting OSM node, and the ending OSM node between which the traffic is measured. A new relationship named 'AADT' is inserted in the *primal graph* between the source and the target node; then, the traffic volume is added as a property of this new relation. A new relationship is needed because the *primal graph* and the road network the traffic data refers to can be different. In particular, in our use case, traffic data are lane-based; thus, in the same direction there could be more lanes and the *primal graph* has a single ROUTE relation. Additionally, a new property is added to the 'ROUTE' relation: the average traffic volume in each direction. Moreover, traffic data may not be complete: they can cover only a reduced part of the total roads. Where traffic data are not provided, in order to exploit as much as possible the available traffic information, two main approximations are adopted:

– the traffic volume is evaluated as the average traffic volume of all the observed relationships in the same direction between 1 to 5-order neighbors,

---

[8] https://trafair.eu/trafficflow/annual-average-traffic-volume.
[9] https://dati.emilia-romagna.it/dataset?organization=comune-di-modena& tags=features&res_format=WFS&_res_format_limit=0&page=2.

– if there are no neighboring traffic relationships, the traffic volume is evaluated as the average AADT of all the roads of the same type (e.g. highway, primary, secondary, and residential).

Moreover, in the *dual graph*, a new 'traffic' property is added to each road node. Since longer roads are supposed to have a higher traffic volume, the traffic property is evaluated as the ratio between the average AADT and the total distance of all the ROUTE relationships that correspond to the given road.

Fig. 4. Comparison between the shortest paths before and after via Wiligelmo closure.

6   Routing

Detecting the optimal route between two points in a network based on different traffic conditions is of key importance for traffic management and analysis. Moreover, routing algorithms can help in generating realistic random routes to feed simulation models. Pathfinding algorithms find the best path between two nodes in a graph, comparing all the possible paths based on their cost. The cost can be evaluated in very different ways: summing the values of a property of the relation that associates the two nodes (used as weight), or counting the number of traversed nodes in the path. We explored three main approaches for shortest path evaluation based on: the number of traversed nodes, the distance, and the traffic volume. The shortest path based on the number of traversed nodes was evaluated with the fast algorithm provided by Neo4j. This solution is unweighted and only needs to search for the path; thus, the implementation is based on the fast bidirectional breadth-first search algorithm [23]. However, when considering the 'distance' attribute that corresponds to the length of the road segment that connects the two nodes, we need to employ a more complex algorithm: the A*

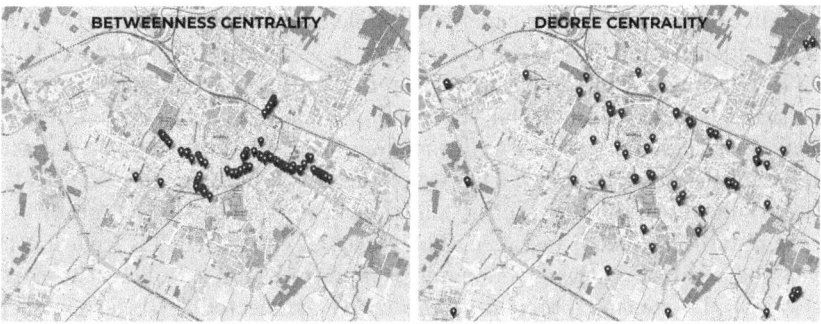

Fig. 5. Map of the 100 junctions with the highest betweenness centrality (on the left) and the highest degree centrality (on the right) in the city of Modena.

informed search algorithm that uses a heuristic function. This heuristic function is the Haversine distance between two geo-located points on the earth sphere [18]. For this reason, each node contains information regarding its position (the coordinates). Moreover, the distance between the nodes in a road network can be significantly different from the real distance to travel; thus, the A* algorithm use as weight the length of the road segment. Finally, if traffic data are available, the optimal path that considers also the amount of traffic between each node can be evaluated. In a first attempt, we try to employ the A* algorithm considering the traffic volume as weight; however, since there are several rural roads outside the urban area where traffic volume is low, the obtained path was very long and not realistic. For this reason, we decide to define a new relationship property based on both distance and traffic volume. The value of this property is estimated as:

$$w_x = 0.5 * \frac{d_x - min_d}{max_d - min_d} + 0.5 * \frac{AADT_x - min_{AADT}}{max_{AADT} - min_{AADT}}$$

where d_x is the road length corresponding to the 'ROUTE' relation and $AADT_x$ its average traffic volume, min_d and max_d are the minimum and maximum distance values computed on all the road network, and min_{AADT} and max_{AADT} the minimum and maximum AADT values on all the road network. As a result, the weight of each relationship is the equal-weighted sum of the normalized values of distance and AADT. This solution allows finding the optimal path, considering both the traffic conditions and the length of the resulting path. In Fig. 2, the three paths obtained applying the routing procedure between the same source and the same target node but with different approaches are displayed. We can observe that the three resulting paths are slightly different.

6.1 Managing Road Closures

For an event or the presence of maintenance work, a street may needs to be closed for a short period. In this case, the traffic manager need to know an alternative path that avoids traveling through certain streets. To enable this functionality, our street graph can be dynamically modified, setting the status of a road (identified by its road name) to 'close' or 'open' to traffic. All the 'ROUTE' relations in our road network are characterized by the 'status' property automatically set to 'open' when the *primal graph* is generated; however, the user can easily change this status by running our 'ChangeStreetStatus' script (available in the git repository). This script takes as input the name of the street (e.g. Via Wiligelmo), the new status, and the parameters needed to establish the connection with the Neo4j instance. The status of each 'ROUTE' relation that involves the road with the given name is updated and, to allow the correct execution of the routing procedure, the road's connected POIs are detached and new relationships are established with the other roads in 100 m from each POI. When the road is re-opened, the original relationships with the POIs are re-established.

Figure 4 compares the shortest paths obtained considering the traffic volume when the road via Wiligelmo is open or closed. Our routing algorithm finds the best path that avoids passing through the closed street.

7   Road Network Analysis

Now that we have converted the road network into a graph, and we have integrated traffic information, we can employ several graph-based algorithms to investigate the graph structure and its relation to traffic. For doing this, we rely on the Graph Data Science library of Neo4j. In order to identify the junction that, given the graph topology, are involved in the majority of the shortest paths, we employed the Betweenness Centrality (BC) algorithm on the *primal graph*. The shortest paths connect two points passing through the minimum possible number of relations. The BC evaluates the shortest paths between all the couples of nodes in the network and then associates to each node a score [9]. This score depends on the number of the shortest paths crossing the node and is evaluated as:

$$score(n) = \sum_{s,t \epsilon N} \frac{sp(s,t|n)}{path(s,t)}$$

where n is the actual node, N is the ensemble of all nodes in the graph, $sp(s,t|n)$ is the number of the shortest paths between s and t that passes through n, and $path(s,t)$ the total number of the shortest paths between s and t. In the city of Modena, we obtained a BC score between 0 and 8×10^7, with an average score of about 3×10^6. Figure 5 displays the first 100 junctions with the highest score. Moreover, we investigate if the BC score of the junction is correlated with the traffic in the incoming and outgoing roads by evaluating the correlation between the BC score and the sum of all the traffic volumes of the roads. Since the Person's and Spearman's coefficients were both lower than 0.2 (0.12 and 0.15 respectively), as already discussed and demonstrated in [11], we prove that, also in our use case, the BC score is not significantly correlated with the traffic volume. Then, we try to test the Harmonic Centrality (HC) algorithm:

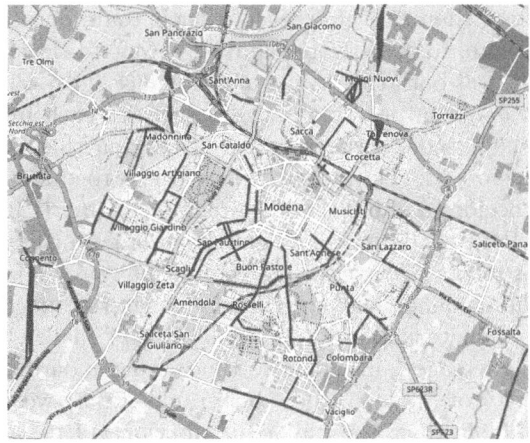

Fig. 6. PR results for the city of Modena.

a modified version of Closeness Centrality that works better with unconnected graphs [19]. The HC score depends on the average shortest path between the node and all the other nodes in the graph. The average shortest path is evaluated summing the inverse of the distances between the given node and all the others. HC algorithm has been applied to the *primal graph* of the city of Modena obtaining a Spearman's correlation coefficient significantly higher (0.61). Similarly, we evaluated the HC score of the *dual graph* and compared it with the sum of all the traffic volumes along each road. We obtain a Spearman's correlation coefficient of 0.45. Thus, there is a relation between HC score and traffic volume. If we want to employ centrality to find the most congested junction, however, the best solution is to exploit the presence of traffic data in the properties of our graph and use the Degree Centrality (DC) algorithm. The DC algorithm is a weighted algorithm; thus, it can be weighted by the ratio between AADT and distance. In this way, the degree of each junction is evaluated as the sum of this ratio in all of the incoming routes. In the city of Modena, we obtained a DC between 0 and 8×10^3, with an average degree of 142. The Pearson's and Spearman's correlation coefficients with the traffic flow are obviously very high (0.95) because the traffic has been used as weight. In Fig. 5, the most important nodes evaluated with BC and DC are compared. We can observe that the result is very different since it conveys different information. The BC assumes that: the drivers are always choosing the shortest path, all the positions in the city have the same importance, and the same number of vehicles are driving through them. The DC, instead, considers the AADT traffic and shows the nodes with the highest incoming traffic: the most congested junctions. The application of centrality algorithms to the *primal graph* is an interesting solution to transfer the traffic information from roads to junctions. As a matter of fact, not only the roads with an high traffic volume are affected by slowdowns; all the vehicles driving through the roads incoming in a congested junction will have a longer traversal times than expected, even if their route does not involve congested roads. Therefore, to investigate the roads prone to slowdowns, we set up a methodology that uses both the *primal* and the *dual graph*. Firstly, we employ DC on the *primal graph* in order to assign a score to each junction based on traffic. Then, the new 'score' property of the junction that connects the two roads is transferred in each relationship between two road sections in the *dual graph*. Finally, we decide to apply the PageRank algorithm (PR) [6] to the *dual graph*. We employ PR considering as weight the 'score' property evaluated with DC. In this way, a road can have a high rank if many roads are connected to it, or if some of these connections are through congested junctions. As can be seen in Fig. 6, the framework allows to automatically visualize the roads with a PR higher than the average rank plus two times the standard deviation. In the *dual graph* of the city of Modena, the roads' ranks are between 0.15 and 7.28, with an average of 0.95; thus, the rank's distribution is left-skewed with a low number of roads with a high page rank. The displayed road sections are 201. These are only some examples of the insights that can be obtained through the *primal graph* and *dual graph* analysis supported by our framework.

8   Conclusion

This paper presented an open-source framework to perform an analysis of the road network, investigate the relation between topology and traffic conditions, and exploit routing algorithms to obtain the optimal path based on different aspects such as distance, traffic volume, number of traversed junctions.

The proposed representation of road networks as the combination of *primal graph* and *dual graph* allows users to apply graph algorithms on cascade on both levels, offering the opportunity to analyze the relationship between the topology of the road network and the traffic distribution. The framework can be efficiently employed by users that are not aware of the graph theory or do not know how a graph database works; moreover, it can provide a good starting point for knowledgeable users that want to conduct deep analytic by applying graph data science and machine learning algorithms. To the best of our knowledge, there are no available open-source frameworks that allow generating both primal and dual graphs and integrating traffic data.

We also explained how we evaluate traffic in our use case and how we manage to evaluate the AADT on the roads where traffic data were missing. Thus, a user that has information about the traffic between a reduced number of nodes can employ our framework to estimate the traffic in the rest of the road network. In future work, the framework can be employed to study the robustness of the road network to different road closure scenarios [28] or compare the road network graphs of different cities considering their traffic.

References

1. Ahmadzai, F., Rao, K., Ulfat, S.: Assessment and modelling of urban road networks using Integrated Graph of Natural Road Network (a GIS-based approach). J. Urban Manag. **8**(1), 109–125 (2019). https://doi.org/10.1016/j.jum.2018.11.001. https://www.sciencedirect.com/science/article/pii/S2226585618301341
2. Bachechi, C., Po, L.: Implementing an urban dynamic traffic model. In: IEEE/WIC/ACM International Conference on Web Intelligence, WI 2019, Thessaloniki, Greece, 14–17 October 2019. ACM (2019)
3. Bachechi, C., Po, L.: Traffic analysis in a smart city. In: Web4City, International IEEE/WIC/ACM Smart City Workshop: Web for Smart Cities, Thessaloniki, Greece, 14–17 October 2019 (2019)
4. Bachechi, C., Rollo, F., Po, L.: Detection and classification of sensor anomalies for simulating urban traffic scenarios. Clust. Comput. **25**, 2793–2817 (2021). https://doi.org/10.1007/s10586-021-03445-7. https://link.springer.com/article/10.1007/s10586-021-03445-7#citeas
5. Boeing, G.: OSMnx: new methods for acquiring, constructing, analyzing, and visualizing complex street networks. Comput. Environ. Urban Syst. **65**, 126–139 (2017). https://doi.org/10.1016/j.compenvurbsys.2017.05.004. https://www.sciencedirect.com/science/article/pii/S0198971516303970
6. Brin, S., Page, L.: The anatomy of a large-scale hypertextual web search engine. Comput. Netw. ISDN Syst. **30**(1), 107–117 (1998). https://doi.org/10.1016/S0169-7552(98)00110-X. https://www.sciencedirect.com/science/article/pii/S016975529800110X. Proceedings of the Seventh International World Wide Web Conference
7. Chen, J., Song, Q., Zhao, C., Li, Z.: Graph database and relational database performance comparison on a transportation network. In: Singh, M., Gupta, P.K., Tyagi, V., Flusser, J., Ören, T., Valentino, G. (eds.) ICACDS 2020. CCIS, vol. 1244, pp. 407–418. Springer, Singapore (2020). https://doi.org/10.1007/978-981-15-6634-9_37
8. Claramunt, C., Winter, S.: Structural salience of elements of the city. Environ. Plann. B. Plann. Des. **34**, 1030–1050 (2007). https://doi.org/10.1068/b32099
9. Freeman, L.: A set of measures of centrality based on betweenness. Sociometry **40**, 35–41 (1977). https://doi.org/10.2307/3033543
10. Fu, M., Kelly, J., Clinch, J.P.: Estimating annual average daily traffic and transport emissions for a national road network: a bottom-up methodology for both nationally-aggregated and spatially-disaggregated results. J. Transp. Geogr. **58**, 186–195 (2017)
11. Gao, S., Wang, Y., Gao, Y., Liu, Y.: Understanding urban traffic-flow characteristics: a rethinking of betweenness centrality. Environ. Plann. B. Plann. Des. **40**, 135–153 (2013). https://doi.org/10.1068/b38141
12. Jayasinghe, A., Sano, K., Nishiuchi, H.: Explaining traffic flow patterns using centrality measures. Int. J. Traffic Transp. Eng. **5**, 134–149 (2015). https://doi.org/10.7708/ijtte.2015.5(2).05
13. Jiang, B., Zhao, S., Yin, J.: Self-organized natural roads for predicting traffic flow: a sensitivity study. J. Stat. Mech. Theory Exp. **2008** (2008). https://doi.org/10.1088/1742-5468/2008/07/P07008

14. Jorge, A.A.S., Rossato, M., Bacelar, R.B., Santos, L.B.L.: A unified graph model for line and segment maps. In: Proceedings of the 10th International Space Syntax Symposium, pp. 146:1–146:11 (2015). https://www.sss10.bartlett.ucl.ac.uk/wp-content/uploads/2015/07/SSS10Proceedings146.pdf

15. Jorge, A.A.S., Rossato, M., Bacelar, R.B., Santos, L.B.L.: GIS4Graph: a tool for analyzing (geo) graphs applied to study efficiency in a street network. In: GEOINFO (2017)

16. Liu, T., Jiang, A., Miao, X., Tang, Y., Zhu, Y., Kwan, H.K.: Graph-based dynamic modeling and traffic prediction of urban road network. IEEE Sens. J. **21**(24), 28118–28130 (2021). https://doi.org/10.1109/JSEN.2021.3124818

17. López, P.Á., et al.: Microscopic traffic simulation using SUMO. In: 21st International Conference on Intelligent Transportation Systems, ITSC 2018, Maui, HI, USA, 4–7 November 2018, pp. 2575–2582. IEEE (2018)

18. Mahmoud, H., Akkari, N.: Shortest path calculation: a comparative study for location-based recommender system. In: 2016 World Symposium on Computer Applications Research (WSCAR), pp. 1–5 (2016). https://doi.org/10.1109/WSCAR.2016.16

19. Marchiori, M., Latora, V.: Harmony in the small-world. Phys. A Stat. Mech. Appl. **285**, 539–546 (2000). https://doi.org/10.1016/S0378-4371(00)00311-3

20. Marshall, S.: Line structure representation for road network analysis. J. Transp. Land Use **9**(1) (2015). https://doi.org/10.5198/jtlu.2015.744. https://www.jtlu.org/index.php/jtlu/article/view/744

21. Marshall, S., Gil, J., Kropf, K., Tomko, M., Figueiredo, L.: Street network studies: from networks to models and their representations. Netw. Spat. Econ. **18**, 1–15 (2018). https://doi.org/10.1007/s11067-018-9427-9

22. Miler, M., Odobašić, D., Medak, D.: The shortest path algorithm performance comparison in graph and relational database on a transportation network. Promet-Traffic Transp. **26**, 75–82 (2014). https://doi.org/10.7307/ptt.v26i1.1268

23. Needham, M., Hodler, A.: Graph Algorithms: Practical Examples in Apache Spark and Neo4j. O'Reilly Media (2019). https://books.google.it/books?id=UwIevgEACAAJ

24. Po, L., Rollo, F., Bachechi, C., Corni, A.: From sensors data to urban traffic flow analysis. In: 2019 IEEE International Smart Cities Conference, ISC2 2019, Casablanca, Morocco, 14–17 October 2019, pp. 478–485. IEEE (2019)

25. Porta, S., Crucitti, P., Latora, V.: The network analysis of urban streets: a dual approach. Phys. A Stat. Mech. Appl. **369**(2), 853–866 (2006). https://doi.org/10.1016/j.physa.2005.12.063. https://www.sciencedirect.com/science/article/pii/S0378437106001282

26. Porta, S., Crucitti, P., Latora, V.: The network analysis of urban streets: a primal approach. Environ. Plann. B. Plann. Des. **33**(5), 705–725 (2006). https://doi.org/10.1068/b32045

27. Qi, L., Schneider, M.: Trafforithm - a traffic-aware shortest path algorithm in real road networks with traffic influence factors. In: Proceedings of the 1st International Conference on Geographical Information Systems Theory, Applications and Management - GISTAM, pp. 105–112. INSTICC, SciTePress (2015). https://doi.org/10.5220/0005350701050112

28. Sohouenou, P.Y., Christidis, P., Christodoulou, A., Neves, L.A., Presti, D.L.: Using a random road graph model to understand road networks robustness to link failures. Int. J. Crit. Infrastruct. Prot. **29**, 100353 (2020). https://doi.org/10.1016/j.ijcip.2020.100353. https://www.sciencedirect.com/science/article/pii/S1874548220300172

29. Wang, G.M., Li, Y.Q., Xu, M.: Integrating the management and design of urban road network to alleviate tide traffic*. In: 2019 IEEE Intelligent Transportation Systems Conference (ITSC), pp. 708–713 (2019). https://doi.org/10.1109/ITSC.2019.8917083
30. Zhang, J., Yuan, L., Li, W., Qin, L., Zhang, Y.: Efficient label-constrained shortest path queries on road networks: a tree decomposition approach. Proc. VLDB Endow. **15**(3), 686–698 (2021). https://doi.org/10.14778/3494124.3494148

Parallel Discovery of Top-k Weighted Motifs in Large Graphs

Nikolaos Koutounidis and Apostolos N. Papadopoulos[✉][iD]

School of Informatics, Aristotle University of Thessaloniki, Thessaloniki, Greece
{koutounidis,papadopo}@csd.auth.gr

Abstract. The enumeration of all cliques in a graph or finding the largest clique are important problems that unfortunately are computationally intensive. Another alternative is to select only the most important motifs (e.g., small subgraphs, or patterns), where significance is expressed by means of a function that quantifies the importance of the subgraph. Given a weighted graph $G(V, E, w())$, where V is the set of nodes and E is the set of edges and $w()$ is a function that returns the weight of an edge e we are looking for the efficient computation of the top-k weighted triangles (and also higher-order cliques, e.g., 4-cliques, 5-cliques, etc.). More specifically, the proposed methodology is based on a parallel algorithm which is efficient and scalable and exploits the multi-threading capabilities of modern multi-core processors. Initially we present a solution for the discovery of top-k triangles, which are the simplest non-trivial cliques and then we generalize our solution for the discovery of top-k cliques of higher order. Performance evaluation results based on real-life networks show that the proposed algorithmic technique is significantly more efficient than the centralized one and also it is scalable showing very good speedups by increasing the number of cores.

Keywords: Graph mining · Graph motifs · Parallel algorithms

1 Introduction

Graph mining is an established and rapidly evolving research area with important and useful contributions. Graph mining is the process of extracting potentially useful patterns (and therefore knowledge) from graph-based data. In its simplest form a graph G consists of a set V of nodes or vertices and a set E of links or edges connecting pairs of nodes. The interpretation of the edge between two nodes depends on the application. For example, in a protein-protein interaction network, nodes represent proteins and an edge between nodes $u, v \in V$ denotes that the associated proteins interact with each other for a particular function. As another example, consider a graph where nodes represent authors and an edge between two nodes denotes that the associated researchers are co-authors in at least one scientific article. There are numerous applications that require the management and mining of graph data.

© Springer Nature Switzerland AG 2022
S. Chiusano et al. (Eds.): ADBIS 2022, LNCS 13389, pp. 90–103, 2022.
https://doi.org/10.1007/978-3-031-15740-0_8

In many cases, the nodes or the edges of the graph are annotated with additional information. For example, each edge e may carry a weight $w(e)$ denoting the importance of this edge. In a weighted graph, the significance of an edge is quantified by the weight. Likewise, the significance of a subgraph is quantified by a function that takes into account the edges of the subgraph and returns a real number representing the weight of the subgraph.

The existence of edge weights results in a difference on the importance of certain subgraphs. For example, in an unweighted graph all triangles (i.e., 3-cliques) are of the same importance. However, in a weighted graph, triangles formed by heavy-weighted edges are considered more important than triangles formed by light-weighted edges. The same rationale applies in the case of higher-order cliques.

In general, small graph patterns are very useful. For example, triangles have been used in other more complex knowledge-discovery tasks of graph data, such as community detection [10] and dense subgraph discovery [15]. It is not a coincidence that there exists a huge body of research focusing on the efficient discovery of triangles in large graphs [1,7,11,13]. A triangle is composed of three nodes u, v, z connected to each other. Essentially, a triangle corresponds to the relationship: "a friend of a friend is my friend".

In the same line, higher-order cliques have been also used as a basis for more advanced knowledge discovery. Clique percolation is a widely used technique for the analysis of overlapping communities in large graphs [9]. Moreover the concept of clique conductance has been used for community detection in graphs [8].

Taking into account the importance of triangles and cliques, the work in [6] proposes an algorithmic technique to detect the top-k triangles in a weighted graph. It is expected, that in a weighted graph some subgraphs are more important than others, depending on the accumulated weight of each subgraph. We focus on clique-based motifs, starting with triangles and generalizing for higher-order cliques. However, for large graphs it is expected that performance will degrade and this is expected to be more intense for higher-order cliques. To alleviate this performance issue, in this work, we propose a parallel algorithm for the discovery of top-k triangles and higher-order cliques. The proposed technique exploits the multi-core architecture of modern processors and shows satisfactory speedups with respect to the number of cores being used. Moreover, the proposed algorithm is extended towards the discovery of top-k cliques, after establishing a theoretical foundation of how cliques can be detected based on triangles. Performance evaluation results demonstrate that the proposed parallel algorithm is efficient, scalable and can be used in very large graphs containing millions of nodes and billions of edges.

The rest of the paper is organized as follows. In Sect. 2 we present related work in the area. Section 3 presents fundamental concepts whereas Sect. 4 describes in detail the proposed parallel algorithm. Performance evaluation results are presented and discussed in Sect. 5 whereas Sect. 6 concludes our work and present briefly future research directions.

2   Related Research

The literature is rich in algorithmic techniques for counting and listing all triangles and other small cliques in large graphs. For example, for a graph with m edges the best known algorithm to count all triangles requires $\mathcal{O}(m^{3/2})$. In fact, this matches the upper bound of the number of triangles that can be formed using m edges. The first algorithm achieving this bound has been reported in [5]. Later other alternatives were proposed, depending on weather we iterate over the set of nodes or the set of edges [7,11].

In addition to the basic techniques, more advanced algorithms have been reported focusing in different setting. Triangle discovery in disk-resident graphs has been covered in [4]. The algorithm performs multiple passes over the graph, since it cannot be loaded in main memory. Also, parallel MapReduce algorithms have been proposed for counting or listing the triangles of a single large graph, which is distributed across the nodes of a cluster [12]. Moreover, the triangle counting problem has been also addressed in the streaming model of computation, where the graph is seen as a stream of edges. In this setting, only a small part of the graph can be accommodated in main memory and therefore techniques in this category are approximate. More specifically, in [2] the authors propose algorithms for triangle counting that provide the final answer with error guarantees. The more edges we consume, the more accurate the result becomes.

In addition to the research works focusing on the efficient computation of triangles, others use the concept of the triangle to achieve more effective knowledge discovery. In [3] the concept of k-truss is introduced, which is based on the number of triangles formed by each graph node. The number of triangles per node quantifies how well the node is connected with respect to its 1-hop neighborhood. This concept has been used to detect subgraphs were the number of triangles per node exeed k [14,16]. Moreover, triangles have been used by [13] towards detecting dense subgraphs in large graphs. Also, in [10] the authors analyze the performance of a novel community detection mechanism which is based on the number of triangles formed by a set of nodes.

Recently, an algorithmic technique has been proposed in [6] that computes the top-k triangles based on a scoring function. This function takes into account the weights of the edges. In fact, this was the first work to use weighting graphs for triangle counting and listing. Motivated by that work, in the present work we propose a parallel algorithm for top-k triangle discovery, exploiting the parallelism capabilities of modern processors. Moreover, we extend this idea to top-k cliques as well.

3   Background

Let $G(V, E, w)$ denote an undirected weighted graph, where each edge $e = (u, v)$ is annotated with a weight $w(e)$ (or $w(u, v)$) representing the *strength* of the specific edge. Without loss of generality, we assume that higher weight values are preferable. Given a graph G as input, we need to discover the k best (i.e., the k heaviest) triangles in the graph, based on a scoring function applied on triangles. More specifically, the score of a triangle formed by nodes u, v and x ($u < v < x$) is defined as the generalized p-mean as follows:

$$score(u, v, x) = \left(\frac{1}{3}(w(u, v)^p + w(v, x)^p + w(u, x)^p) \right)^{1/p} \tag{1}$$

where $w(u, v)$ corresponds to the weight of the edge joining nodes u and v, and p is an integer constant.

The brute-force algorithm to solve the problem discovers all possible triangles, computes the score of each one and, finally, selects the k best. However, since the number of triangles in a graph with m edges is in $\mathcal{O}(m^{3/2})$, even using a priority queue to accommodate the best k triangles, leads to a complexity of $\mathcal{O}(\log k \cdot m^{3/2})$. Therefore, since we are interested in a small subset of the triangles (the value of k is usually very small in comparison to the total number of triangles in the graph) a more selective algorithm is required in order to avoid the enumeration of the complete set of triangles.

An algorithm that performs better than brute-force has been proposed in [6]. The algorithm works by first sorting the edges of G in a non-ascending weight order and computes the triangles that a specific edge is participating in. As long as the algorithm has not yet found k triangles with a score that exceeds a specific threshold t, it continues by reading the next edge and enumerating all triangles that this edge closes. The threshold t represents the score of the heaviest triangle that may exist in the graph but we have not enumerated it yet.

In order to compute an appropriate threshold value, the algorithm maintains three different lists of edges, L, H and S. Initially, all edges are inserted in the L list and once an edge is examined, it is moved to the H list. Once an edge in the H list, it is examined again and it is moved to the S list. We know that we have enumerated all the triangles that the edges in the S list are part of. The threshold t is computed as follows:

$$t = h_w + 2 \cdot l_w \tag{2}$$

where h_w and l_w are the edge weights that the h counter is pointing to, in the array with all the edges in descending weight order and the weight of the edge that the l counter points to in the aforementioned array, respectively.

The algorithm in each iteration enumerates all triangles that are formed by at least one edge in the H or S list, meaning that the remaining possible heaviest triangles can only have one edge in the H list and two edges in the L list. Therefore, r is the weight of the heaviest edge in the H list and two times the weight of the heaviest edge in the L list. When the algorithm has found k triangles that have a higher score than t, it is clear that it has found the k heaviest triangles in the graph.

To clarify the way the centralized algorithm works, we provide a simple example demonstrating the course of the algorithm. The algorithm is applied on the small graph shown in Fig. 1. Moreover, we assume that $k = 2$, i.e., we require the two best triangles of the graph.

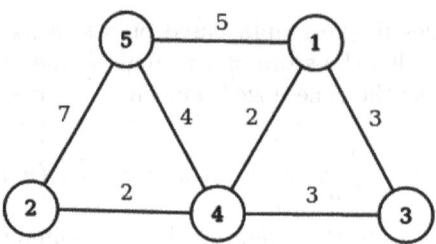

Fig. 1. A simple graph used in the example.

Because in the algorithm two different searches are executed and one of them is computationally more expensive, we use w in order to favor the less expensive search. w is the power to which we raise the weight of the edge that the l counter points to before evaluating the clause $l_w^w > h_w$ where l and h are the edges the corresponding counters point to. The value of w is affected by the distribution of the weights of the graph to be examined. For the following example the value of w will be set to 1.5.

Iteration 1. During the first iteration, all the edges are inserted into L and the H and S lists are empty. No triangles have been enumerated up to now. The variables have the following values: $l = 0$, $h = 0$, $L = [e_{2,5}, e_{1,5}, e_{4,5}, e_{1,3}, e_{3,4}, e_{1,4}, e_{2,4}]$, $H = [\ ]$, $S = [\ ]$, $T = [\ ]$. The edge $e_{2,5}$ is examined first. Since l and h are equal we search for triangles formed by $e_{2,5}$ and at least one edge in the HS list. $e_{2,5}$ currently does not participate in any triangles with at least one edge in the HS list, since that list is empty. $e_{2,5}$ is moved to H.

Iteration 2. $l = 1$, $h = 0$, $L = [e_{1,5}, e_{4,5}, e_{1,3}, e_{3,4}, e_{1,4}, e_{2,4}]$, $H = [e_{2,5}]$, $S = [\ ]$, $T = [\ ]$, $t = 17$. Since $l_w^{1.5} > h_w$ we search for triangles that are formed by the edge $e_{1,5}$ and at least one edge in the HS list. The edge $e_{1,5}$ does not participate in any triangles with at least one edge in list HS, thus the edge $e_{1,5}$ is moved to list H.

Iteration 3. $l = 2$, $h = 0$, $L = [e_{4,5}, e_{1,3}, e_{3,4}, e_{1,4}, e_{2,4}]$, $H = [e_{2,5}, e_{1,5}]$, $S = [\ ]$, $T = [\ ]$, $t = 13$. Since $l_w^{1.5} > h_w$ we search for triangles that are formed by the edge $e_{4,5}$ and at least one edge in the HS list. The edge $e_{4,5}$ does participate in two triangles with at least one edge in the list HS, the triangles (1, 4, 5) with weight 11 and the triangle (2, 4, 5) with weight 13. These triangles are stored in the list T. The edge $e_{4,5}$ is moved to H.

Iteration 4. $l = 3$, $h = 0$, $L = [e_{1,3}, e_{3,4}, e_{1,4}, e_{2,4}]$, $H = [e_{2,5}, e_{1,5}, e_{4,5}]$, $S = [\ ]$, $T = [(1,4,5), (2,4,5)]$, $t = 13$. Since $l_w^{1.5} < h_w$ we will search for triangles that are formed by the $e_{2,5}$ and two edges in list L. Edge $e_{2,5}$ does not participate in any triangles with two edges in list L. The edge $e_{2,5}$ is moved to S and continue.

Iteration 5. $l = 3$, $h = 1$, $\boldsymbol{L} = [e_{3,4}, e_{1,4}, e_{2,4}]$, $\boldsymbol{H} = [e_{2,5}, e_{1,5}, e_{4,5}, e_{1,3}]$, $\boldsymbol{S} = [\,]$, $\boldsymbol{T} = [145, 245]$, $t = 11$. Two triangles have been enumerated and they are above or equal to the threshold so the two heaviest triangles have been determined.

4   The Proposed Approach

In this section, we explain in detail the proposed parallel algorithm to compute the top-k heaviest triangles (and cliques in general) in a potentially large graph. The main characteristic of the centralized algorithm proposed in [6] is that in each iteration, the centralized algorithm processes only one edge and then proceeds by detecting triangles. Moreover, the centralized algorithm moves each examined edge from the light list L to the heavy list H or from heavy list H to the super heavy list S. This transfer happens before the enumeration of the triangles that the examined edge closes.

In contrast, the parallel algorithm increases the number of examined edges in each iteration and also applies a principled strategy in order for the available cores to work in parallel, boosting efficiency. In addition, moving edges from one list to another is performed in batches rather than in a one-by-one manner. The changes applied do not have an impact on the correctness of the algorithm. The pseudocode of the parallel algorithm is given in Algorithms 1 and 2.

4.1   Correctness

It must be proven that all relevant triangles will be detected. More specifically, it must be shown that moving the edges from the L list to the H list preemptively, will not affect the ability of the algorithm to detect the correct answer.

Only the first part of the algorithm (the body of the if statement) is moving an edge from one list to the other, therefor only the search of triangles that consist of that edge and one edge in the $H \cup S$ list and one edge in the L list can be affected. It is evident that the search for all triangles that consist of that edge and two edges in the $H \cup S$ list will not be affected, since the contents of the $H \cup S$ list after the move operation will always be a super set of the contents before the move operation.

The above statements are true when multiple edges are moved before searching for triangles that these edges are closing. Therefore, the only search that can be affected is the search for triangles that consist of the examined edge, one edge in the list $H \cup S$ and one edge in the list L. When the single-thread algorithm is moving an edge, the L list will always be the same list after the operation minus the examined edge and as a result, triangle detection is not affected. However, the parallel algorithm removes n edges before triangle detection and the L list will be a subset of itself after the move operation. We identify the following cases.

Algorithm 1. Parallel Algorithm for k heaviest triangles

Input: Weighted graph $G = (V, E, w)$, number of triangles k, parameter $\alpha\rho$, number of edges to read n, number of edges to read for the "expensive" search m, number of threads th.

1: Sort E in decreasing order of weight
2: Set threshold $t = \infty$, triangle set $T = \emptyset$
3: Set partitions $HS = \emptyset$, $L = E$
4: Set edge pointers $h = l = -1$
     // We take the convention that $E_{-1} = \infty$
5: **while** there are $< k$ triangles above weight t in T **do**
6:     Set edges list $ef = []$
7:     **if** $w_{l+1}^{\alpha\rho} > w_{h+1}$ **then**
8:         $i=0$.
9:         **while** $i < n$ **do**
10:             Move E_{l+1} from L to HS.
11:             add $El + 1$ to ef
12:         **end while**
13:         $Y =$ find triangles with one edge in ef list and 2 edges in HS using th
         threads using Algorithm 2
14:         $Z =$ find triangles with one edge in ef 1 edge from L and 1 edge from HS
         using th threads using Algorithm 2
15:         $T = T \cup (Y \cup Z)$
16:         $l = l + n$
17:     **else**
18:         **while** $i < th * m$ **do**
19:             add $Eh + 1$ to ef
20:             $i = i + 1$
21:         **end while**
22:         $Y =$ find triangles with one edge in ef list and 2 edges in L using th threads
         using Algorithm 2
23:         $T = T \cup Y$
24:         $h = h + th * m$
25:     **end if**
26:     Update threshold $t = w_h^p + 2w_l^p$
27: **end while**

In the first case, a triangle has one edge in the examined edge set (the edges we examine simultaneously) and two edges in list L that does not belong to the examined edge set. This triangle will not be detected during the enumeration but it is also not affected by the multiple edges that move from the L list to the $H \cup S$ list, since each edge is still in the L list except the examined edge.

In the second case, a triangle has one edge in the examined set and two edges in the $H \cup S$ list. In this case, there will be no impact either as the triangle will be enumerated and also the list that its edges are in has not changed by the move operation.

In the third case, a triangle has two edges in the examined set and one edge in the $H \cup S$ list. In this case, the search for triangles that consist of the examined

---

Algorithm 2. Parallel triangle detection

Input: List of edges to examine E, List L of light edges, List H of heavy edges, number of threads th, integer search mode.

1: Set priority queue $T = [\,]$
2: Split E to th sub-lists
3: Create th new threads and pass one sub list to each, the L and H list and priority queue T.
4: In each thread find triangles that the examined edge is part of, the examined edges are in the sublist for each thread
5: Join all threads
6: Return T

---

edge, one edge in list L list one in the list $H \cup S$ will fail to find the triangle for either the first or the second examined edge. Although the other search, the one that searches for triangles that consist of the examined edge and two edges in the list $H \cup S$ will find the triangle for both the examined edges. Since the two searches are happening before any other move operation may take place and before the threshold is reevaluated, we can perceive them as an atomic search and if one of them can find the triangle we can consider the impact as resolved.

In the fourth case, a triangle has three edges in the examined set. In this case, the single-thread algorithm would fail to find the triangle for the first edge but it would succeed finding it after the examination of the second edge. The parallel algorithm would move all the edges from list L to the list $H \cup S$ before executing any search and as a result, it would find the triangle three times, during the search for triangles which consists of the examined edge and two edges of the list $H \cup S$. Thus, the triangle will be detected by the parallel algorithm and this algorithm will find it one more time than the single-thread algorithm would.

In the fifth case, a triangle has one edge in the examined set, one edge in L and one edge in $H \cup S$. In that case, the search for triangles that consist of the examined edge and one edge in list L and one edge in list $H \cup S$ will detect the triangle, exactly as the single-thread algorithm would do.

In the sixth and final case, a triangle has two edges in the examined set and one edge in list L. In this case, the single-thread algorithm would fail to find the triangle when it would examine the first edge as the other two edges would be in list L, but it would successfully enumerate the triangle when examining the second edge as. The parallel algorithm would enumerate the triangle during the examination of both edges in the examined set.

With these six cases we cover all the possible distribution of the edges of a triangle in the examined set and the L and HS list. For all the case the multi-thread algorithm despite moving all the edges in the examined set from the L list to the HS list is able to find the triangles. The drawback is that for some cases the multi-thread algorithm will enumerate a triangle one additional time which is wasted computational time, the searching of triangles is happening in the threads and the fact that it is happening in parallel should even out the time lost in enumerating the triangles one excessive time.

4.2   Extension to 4-Cliques

In this section, we provide an extension of the algorithm in order to enumerate the k heaviest 4-cliques. Also, two theorems are provided to support the correctness of the algorithms.

Theorem 1. *Two $(k-1)$-cliques which have $k-2$ common vertices and their remaining non-common vertices are connected, are forming a k-clique, for $k >= 4$.*

Proof. One k-clique has k vertices and all of them are connected with each other. Every subgraph of a clique is also a clique of a smaller size. Hence, the $k-2$ common vertices of the two $(k-1)$-cliques also form a clique of size $k-2$. The non-common vertices of the two $(k-1)$-cliques are fully connected with the rest $k-2$ common vertices. And since those two non-common vertices are also connected we have k vertices (the $k-2$ common vertices and the 2 non-common vertices) fully connected with each other hence a clique of size k.

Theorem 2. *A k-clique consists of k $k-1$-cliques that have $k-2$ common vertices with each other and the non-common vertices are connected, for $k >= 4$.*

Proof. Let's assume that there exists a clique of size k and that it consists of a clique c of size $k-1$ and one vertex connected with all the vertices of the $k-1$-clique named u but it does not form a second or subsequent clique with these vertices. That is not possible as the vertex u will form exactly $k-1$ new $k-1$-cliques with the vertices of the c $k-1$-clique as those vertices already are connected with each other and we have stated that the new vertex has to connect with each one of them. Also the $k-1$ newly created $k-1$-cliques will have exactly $k-2$ common vertices with each other and with the initial $k-1$-clique c as the new vertex will need $k-2$ vertices to form each one of the $k-1$ new $k-1$-cliques and those vertices already are part of the initial $k-1$-clique c. And since the k new $k-1$-cliques will have each $k-2$ vertices from a pool of $k-1$ vertices, they can have only 1 different vertex but the new vertex is common for all $k-1$ new $(k-1)$-cliques and therefor they have $k-2$ common vertices.

In the sequel, we provide details of how to compute all k-cliques that a specific edge participates in. First, we describe the 4-clique case and then we extend to k-cliques for $k > 4$. To enumerate all the 4-cliques for Theorem 1 we have to find triangles with one common edge (two common vertices) and their remaining vertices should be connected to each other. The first step is to enumerate all the triangles the examined edge is part of. We have to maintain a data structure for all the triangles we have enumerate in a previous step. In that data structure we store one vertex for each triangle, this vertex is the one that is not part of the examined edge. Then we are searching for connected vertices in this data structure $O(n^2)$ where n is the number of the triangles we enumerated during the previous step. The data structure should be able to handle queries if two edges

---

Algorithm 3. Find all 4-cliques an edge is part of

Input: HashMap with all edges of a graph Edges, edge e.

1: Enumerate all triangles that e is part of and store all the vertices that are part of a triangles but they are not part of e in a list L
2: **for** v in L **do**
3:     **for** $u \neq v$ in L **do**
4:         **if** u connected to v **then**
5:             Edge e and vertices u and v form a 4-clique
6:         **end if**
7:     **end for**
8: **end for**

---

are connected in $O(1)$ time. From Theorem 2 we have found all the 4-cliques one edge is part of since we have checked all the triangles with two common vertices.

For cliques where $k >= 5$ we have to maintain a data structure with buckets, one bucket for each $k - 2$-clique computed during a previous step. While we enumerate the $k - 1$-cliques during a previous step of the algorithm we have to store the vertex that belongs to the $k - 1$-clique but not to the $k - 2$-clique to that $k - 2$-clique's corresponding bucket. From Theorem 1 we can find the $k - cliques$ one edge is part of by finding vertices that are connected and belong to the same bucket. This requires $O(m * n^2)$ time where m is the number of $k - 2$-cliques enumerated for this edge and n is number of vertices inside the bucket with the most vertices. For this we will again need a data structure that can answer if two vertices are connected in $O(1)$ time. From Theorem 2 we have enumerated all the k-cliques that an edge is part of (Fig. 2).

Algorithm 4. Find all k-cliques an edge is part of

Input: HashMap with all edges of a graph Edges, edge e.

1: Enumerate all triangles that e is part of and store all the vertices that are part of a triangles but they are not part of e in a list L.
2: Enumerate all $k - 1$-cliques and store each vertex that is part of the $k - 1$-clique but not part of one of the $k - 2$-cliques that constitute the $k - 1$-clique in a hashmap H where the key is the $k - 2$-clique the vertex is not part of and the key is a list with vertices, add the vertex to that list.
3: **for** c in H's keys **do**
4:     **for** v in $H[c]$ **do**
5:         **for** $u \neq v$ in $H[c]$ **do**
6:             **if** u connected to v **then**
7:                 clique c and vertices u and v form a k-clique
8:             **end if**
9:         **end for**
10:     **end for**
11: **end for**

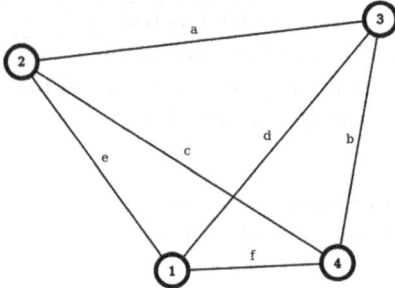

Fig. 2. An example of 4-clique.

With Algorithms 3 and 4 we can enumerate all the k-cliques a specific edge is part of. If we apply these two algorithms to the previous parallel algorithm, we can extend it and find the heaviest k-cliques in a given graph. Algorithm 3 detects all 4-cliques for a given edge only if all the triangles that this edge is part of have been found in a previous step. Algorithm 1 does not find all the triangles that on edge is part in the same step. It can find the triangles that consist of the examined edge and at least one edge in the $H \cup S$ list and in a next step it can find the triangles that consist of the examined edge and two edges in the L list. With this limitation in mind, a new way to compute the threshold in each iteration is required and proposed in the sequel.

Theorem 3. *The weight of the heaviest 4-cliques not yet enumerated in the graph is up to $3w_h + 3w_l$.*

Proof. Omitted due to lack of available space.

Utilizing Algorithm 3 and the triangles enumerated previously, the k heaviest 4-cliques can be found. This centralized approach can be converted to parallel if we take into consideration the proof of the correctness of Algorithm 1.

5   Performance Evaluation

In this section, we evaluate the performance of the algorithms. In the experiments we compare the centralized, Dynamic Heavy Light (DHL), against the parallel algorithm[1]. For the performance evaluation, four real-life networks have been used (available at http://snap.stanford.edu/), shown in Table 1. Using these networks, weights have been generated for the edges by using two types of distributions: normal and power-law, as shown in Table 2. This enables the evaluation of the algorithm for different values of the weights. For those experiments, the graphs described in Table 2 were used.

---

[1] The source code is freely available at https://github.com/nikkout/FindCliques.

Table 1. Real-life networks used in experimental evaluation.

Network	Nodes	Edges	Triangles
com-LiveJournal	3,997,962	34,681,189	177,820,130
com-Orkut	3,072,441	117,185,083	627,584,181

Table 2. Synthetically generated weighted networks.

Network	Based on	Distribution
Graph 1	com-LiveJournal	Normal
Graph 2	com-LiveJournal	Power-Law
Graph 3	com-Orkut	Normal
Graph 4	com-Orkut	Power-Law

Next, we show the time the two algorithms required in order to find the k heaviest triangles of the aforementioned four graphs. The number k of heaviest triangles requested for each graph, is 1000, 100,000 and 1,000,000. The parallel algorithm is executed with 4 threads (Fig. 3). We observe that the parallel algorithm outperforms the centralized one(DHL). Also, it is evident that the distribution of the weights is affecting the performance. As a result, the parallel algorithm has a greater performance improvement over the centralized one when the weights are following a normal distribution. This happens because in this

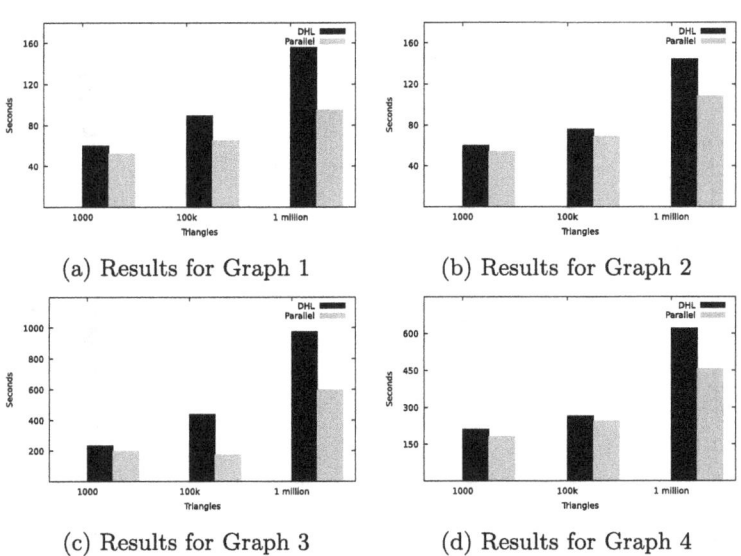

(a) Results for Graph 1   (b) Results for Graph 2

(c) Results for Graph 3   (d) Results for Graph 4

Fig. 3. Comparison of centralized and parallel algorithms.

Table 3. Comparison of centralized and parallel algorithms.

Algo and Graph	1k triangles	100k triangles	1m triangles
DHL 1	1 m 0.059 s	1 m 29.361 s	2 m 35.73 s
Parallel 1	52.570 s	1 m 5.306 s	1 m 34.940 s
DHL 2	59.876 s	1 m 15.845 s	2 m 22.266 s
Parallel 2	54.295 s	1 m 9.198 s	1 m 48.445 s
DHL 3	3 m 51.11 s	7 m 18.155 s	15 m 73.77 s
Parallel 3	3 m 18.322 s	5 m 22.469 s	9 m 58.563 s
DHL 4	3 m 30.851 s	4 m 24.355 s	10 m 20.921 s
Parallel 4	3 m 2.742 s	4 m 4.292 s	7 m 36.709 s

case the threshold that the two algorithms are sharing converges slower. Table 3 shows the runtimes for all the experiments discussed in this section.

Next, we report the runtime of the parallel algorithm to detect the k heaviest 4-cliques when different number of threads is being used. For these experiments the first four graphs of Table 2 have been used. As we observe in Fig. 4a, when more threads are used, the time required for the enumeration of the k heaviest 4-cliques is reduced significantly. The distribution of the weights affects the gain of the algorithm when more threads are being used. This happens because the power-law distribution helps the algorithm to converge faster and this reduces the margin that the additional threads can improve performance.

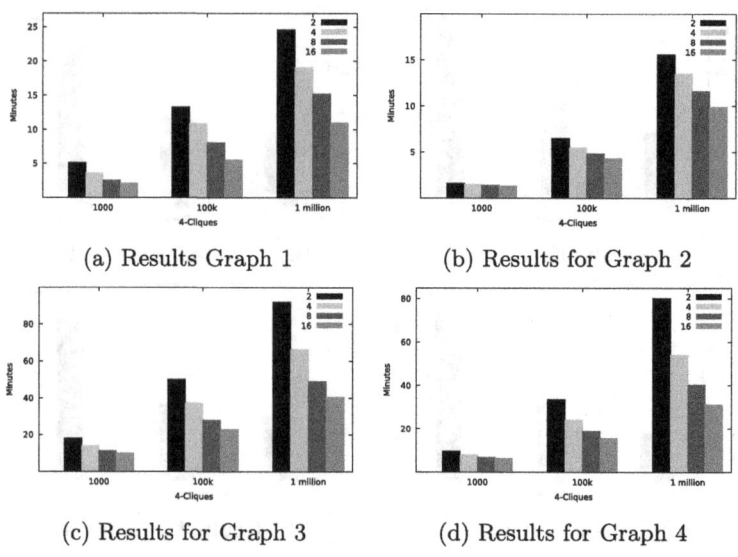

(a) Results Graph 1   (b) Results for Graph 2

(c) Results for Graph 3   (d) Results for Graph 4

Fig. 4. Scalability results. Runtime vs number of cores and number k.

6   Conclusions

This paper proposes a parallel algorithm for the discovery of important motifs in large graphs, where importance is quantified by the weight of the motifs. More specifically, we develop a solution which validated both theoretically and experimentally which manages to detect heavy motifs efficiently and with satisfactory scalability by increasing the number of CPU cores being used. Performance evaluation results are also offered, that are based on real-life networks using different distributions for the edge weights. An interesting future work direction is to design a distributed version of the parallel algorithm to work on a cluster.

References

1. Becchetti, L., Boldi, P., Castillo, C., Gionis, A.: Efficient semi-streaming algorithms for local triangle counting in massive graphs. In: Proceedings of ACM SIGKDD, pp. 16–24 (2008)
2. Buriol, L.S., Frahling, G., Leonardi, S., Marchetti-Spaccamela, A., Sohler, C.: Counting triangles in data streams. In: Proceedings of the Twenty-Fifth ACM SIGACT-SIGMOD-SIGART Symposium on Principles of Database Systems, Chicago, Illinois, USA, 26–28 June 2006, pp. 253–262 (2006)
3. Cohen, J.: Trusses: cohesive subgraphs for social network analysis, December 2008
4. Hu, X., Tao, Y., Chung, C.W.: Massive graph triangulation. In: Proceedings of SIGMOD, pp. 325–336 (2013)
5. Itai, A., Rodeh, M.: Finding a minimum circuit in a graph. In: Proceedings of STOC, pp. 1–10 (1977)
6. Kumar, R., Liu, P., Charikar, M., Benson, A.R.: Retrieving top weighted triangles in graphs. CoRR abs/1910.00692 (2019). https://arxiv.org/abs/1910.00692
7. Latapy, M.: Main-memory triangle computations for very large (sparse (power-law)) graphs. Theor. Comput. Sci. **407**(1–3), 458–473 (2008)
8. Lu, Z., Wahlström, J., Nehorai, A.: Community detection in complex networks via clique conductance. Sci. Rep. **8**, 1–16 (2018)
9. Palla, G., Derényi, I., Farkas, I., Vicsek, T.: Uncovering the overlapping community structure of complex networks in nature and society. Nature **435**(7043), 814–818 (2005)
10. Prat-Pérez, A., Dominguez-Sal, D., Brunat, J.M., Larriba-Pey, J.L.: Shaping communities out of triangles. In: CIKM 2012, pp. 1677–1681. Association for Computing Machinery, New York (2012)
11. Schank, T., Wagner, D.: Finding, counting and listing all triangles in large graphs, an experimental study. In: Nikoletseas, S.E. (ed.) WEA 2005. LNCS, vol. 3503, pp. 606–609. Springer, Heidelberg (2005). https://doi.org/10.1007/11427186_54
12. Suri, S., Vassilvitskii, S.: Counting triangles and the curse of the last reducer. In: Proceedings of WWW, pp. 607–614 (2011)
13. Tsourakakis, C.E.: A novel approach to finding near-cliques: the triangle-densest subgraph problem. CoRR abs/1405.1477 (2014)
14. Wang, J., Cheng, J.: Truss decomposition in massive networks. PVLDB **5**(9), 812–823 (2012)
15. Wang, N., Zhang, J., Tan, K.L., Tung, A.K.H.: On triangulation-based dense neighborhood graph discovery. Proc. VLDB **4**(2), 58–68 (2010)
16. Zhao, F., Tung, A.K.H.: Large scale cohesive subgraphs discovery for social network visual analysis. PVLDB **6**(2), 85–96 (2012)

A Data Quality Framework
for Graph-Based Virtual Data Integration
Systems

Yalei Li, Sergi Nadal$^{(\boxtimes)}$ ⓘ, and Oscar Romero ⓘ

Universitat Politècnica de Catalunya (BarcelonaTech), Barcelona, Spain
yalei.li@estudiantat.upc.edu, {snadal,oromero}@essi.upc.edu

Abstract. Data Quality (DQ) plays a critical role in data integration. Up to now, DQ has mostly been addressed from a single database perspective. Popular DQ frameworks rely on Integrity Constraints (IC) to enforce valid application semantics, which lead to the Denial Constraint (DC) formalism which models a broad range of ICs in real-world applications. Yet, current approaches are rather monolithic, considering a single database and do not suit data integration scenarios. In this paper, we address DQ for data integration systems. Specifically, we extend virtual data integration systems to elicit DCs from disparate data sources to be integrated, using DC-related state-of-the-art, and propagate them to the integrated schema (global DCs). Then, we propose a method to manage global DCs and identify (i) minimal DCs and (ii) potential clashes between them.

Keywords: Data Quality · Data integration · Denial constraints

1 Introduction

We are nowadays witnessing an unprecedented growth in the volume of data that organizations are collecting as part of their decision making processes. With the proliferation of large-scale repositories of heterogeneous data, such as *data lakes* or open-data related initiatives, the ability to perform cross-analysis with high data quality deems a competitive advantage. Indeed, data quality is essential for the decision making process, where poor data quality can lead to wrong decision making, poor model performance, and operational instability [9,13,15]. Yet, in such large-scale data repositories, there coexists data generated by different providers who independently maintain them adhering to their own business rules and needs. Hence, the presence of missing, erroneous, out-of-date, or conflicting data is the norm rather than the exception [3,19].

Data integration systems, which have the main objective of providing an integrated view over an evolving and heterogeneous set of data sources [8], have mostly addressed data quality aspects from a *warehousing* perspective as part of their Extract-Transform-Load (ETL) processes [11]. This is, quality rules and

© Springer Nature Switzerland AG 2022
S. Chiusano et al. (Eds.): ADBIS 2022, LNCS 13389, pp. 104–117, 2022.
https://doi.org/10.1007/978-3-031-15740-0_9

constraints are enforced when materializing source data into the target schema. Yet, in those scenarios that require fresh query results, which are implemented via *virtual* integration systems that rewrite queries posed over the global schema in terms of queries over the data sources leveraging declarative mappings, the management of data quality remains a challenge [2, 21]. Indeed, the kind of mappings adopted in this settings, represented as logical expressions, focus on specifying relations between source and target schemata but not how quality in the target schema must be enforced [12].

The state of the art on data quality management is focused on the automatic derivation of quality rules. This is, from a particular database instance, a set of rules are inferred via *rule mining* techniques and then implemented to detect errors (i.e., violations) [1]. To that end, the formalism of *denial constraints* has been widely adopted, as it is expressive enough to represent most data dependencies found in the literature such as key dependencies, functional dependencies, or order dependencies [5]. Succinctly, denial constraints are first-order formulae that express that a set of predicates cannot be all true for any combination of tuples in a relation, which are expressed as relationships between pairs of tuples of that relation. Despite the wide success of such model, which has given the rise to systems for denial constraint discovery (e.g., FastDC [6], Hydra [4], DynFD [20], or DCfinder [17]), or error detection and data repairing (e.g., Llunatic [7], HoloClean [18], or HoloDetect [10]), to the best of our knowledge they have not been studied in the context of virtual data integration systems.

In order to overcome the previously identified gap (i.e., manage data quality via denial constraints in a virtual data integration system), we present an approach that leverages related work on denial constraint discovery in order to synthesize rules from different data sources and manage them at the global (i.e., integration) level. To that end, our approach builds and extends a graph-based data integration system, which enables expressive visual query paradigms to non-expert users [16]. Precisely, we perform a bottom-up approach propagating the rules discovered at the sources to the target graph. In terms of DQ management, we take advantage of the techniques based on DC to express DQ rules. Two main phases are identified. The first step in this process is to elicit DCs at the source level and then propagate DCs to the target schema. Global rules consolidation is enabled in the integration graph, where users can actively manage and verify the rules even before propagating them to the integrated system.

Contributions. We summarize our contributions as follows:

- We define a data quality management framework that identifies quality rules (as denial constraints) per source, model them into a graph-based representation, and incrementally propagate them into the integrated schema.
- We globally conciliate rules automatically, which facilitates the identification of data cleaning tasks in the form of User Defined Functions.
- We extend a query rewriting algorithm to consider global DCs and enforce them over the underlying data sources.

Outline. The paper is structured as follows. Sections 2 and 3 discuss related work and introduce background concepts. Section 4, presents our approach, while Sect. 6 validates it. Section 7, concludes the paper and outlines future work.

2   Related Work

In this section, we review the state of the art in discovery and management of DCs. As previously stated, our objective is to leverage and benefit of from such methods on a graph-based data integration system. Thus, we narrow the scope to these projects that are openly available and guarantee reproducibility.

FASTDC [6]. It defines the syntax and fundamental semantics for DCs, and derives sound and complete inference rules. The implication testing algorithm checks whether a DC is implied by a set of DCs linearly, which effectively reduces the number of DCs in the output. The main DC discovery algorithm first builds the predicate space by comparing every tuple pair in the instance set, which contains all the possible predicates that can be formed into DCs. To overcome overfitting, FASTDC introduces an approximation parameter in A-FASTDC to allow flexibility in DCs satisfiability requirement. A DC stays valid if the percentage of violations on a instance over the total number of tuple pairs is below a given approximation threshold.

HYDRA [4]. In the spirit of FASTDC, aims to address the quadratic complexity in predicate evaluations and accelerate the DCs generation from the evidence sets. HYDRA devises a sampling technique to quickly approximate the DCs by processing only a small fraction of all tuples, providing adaptability to scale up with the number of tuples. It proposes to first samples tuple pairs to build an initial set of DCs for a dataset. The algorithm corrects the tuple pair samples from its focused sampling process and determines the complete evidence set to avoid the expensive comparison of all tuple pairs in FASTDC.

DCFINDER [17]. It follows FASTDC's approach with improvement on building evidence sets. DCFINDER generates a predicate space from the input database, and builds a data structure from the data records. It utilizes attribute value indexing to avoid the expensive tuple pair comparison of FASTDC. DCFINDER introduces predicate selectivity to drive efficiency even further to avoid the unnecessarily large number of logical operations when generating evidence sets. In an approach comparable to FASTDC, DCFINDER uses the DFS procedure to discover all minimal DCs based on evidence set coverage of DC candidates.

ADCMiner [14]. Focuses on mining approximate DCs, which discovers constraints in inconsistent databases and obtains more general and less contrived constraints. ADCMiner defines a novel approximation function that does not assume any specific definition of an approximate DC but takes the semantics as input. The function consists of two properties called *monotonicity* and *indifference*. ADCMiner generates all minimal ADCs if the approximation score is under a given threshold. Unlike AFASTDC, ADCMiner reduces the process of finding ADCs by avoiding the post-process after detecting valid exact constraints. In general, the algorithm involves four parts: a predicate space generator, a sampler, an evidence set constructor, and an enumeration algorithm. Similar to HYDRA, ADCMiner includes a sampling process to reduce the running time significantly.

3 Preliminaries

In this section, we introduce the running example, which will be used to illustrate our approach, and later, discuss the formal background, which is also exemplified.

3.1 Running Example

We consider a (simplified) data integration scenario on the finance domain, related to organizations and their performance in the stock market. Table 1 presents exemplary data generated from three independent data sources. D_1 (see Table 1a) provides information about companies and their standard industrial classification of economic activities (SIC). D_2 (see Table 1b) yields contextual information related to the history about companies (i.e., founded year and founder/s). Finally, D_3 (see Table 1c) maintains information about the stock prices per company and date. In all cases, we consider the stock symbol to be the attribute used to join the different data sources.

Table 1. Three independent datasets providing information about companies, their history and stock prices

(a) D_1 – SIC

Symb	Comp	SIC
AAPL	Apple	3571
PYPL	Paypal	7389
V	Visa Inc	7389
GOOGL	Google	3571
...	...	...

(b) D_2 – History

S	N	Y	F
GOOGL	Alphabet Inc.	2015	LP&SB
F	Ford Motor C.	1903	HF
APPLE	Apple	1976	SJ&SW&RW
...	...	...	...

(c) D_3 – Stock

Code	Date	Price
AAPL	20220406	171.28
MMM	01/01/2001	100
V	20220406	220.86
...	...	...

As shown in the exemplary data, there exist data quality issues when considering each dataset individually. D_2's attribute names are coded and non-descriptive, while D_3 presents dates encoded in different formats and contains erroneous data (i.e., *MMM* is not a valid stock symbol). Note, however, additional data quality problems arise when considering their integration. If, as expected from the domain, we consider the stock symbol to be a company's primary key, then we can assume the existence of a functional dependency stating the symbol determines the company name. This, however, does not hold in the running example, where the symbol *GOOGL* has associated different names in different data sources. To manage such kind of situations (i.e., quality problems at both the local and the global level), the remainder of this section is devoted to present the formal background that our approach will build upon.

3.2   Formal Background

3.2.1   Graph-Based Virtual Data Integration

Here, we present the core components of our graph-based virtual data integration system. We refer the reader to [16] for further details on how queries are processed over such constructs.

Relations and Wrappers. A schema R is composed of a finite nonempty set of relational symbols $\{r_1, \ldots, r_m\}$, where each r_i has a fixed arity n_i. Let A be a set of attribute names, then each $r_i \in R$ is associated to a tuple of attributes denoted by $att(r_i)$. Let D be a set of values, a tuple t in r_i is a function $t : att(r_i) \to D$. For any relation r_i, $tuples(r_i)$ denotes the set of all possible tuples for r_i. We define the set of wrappers W as those elements in R that contain a function $exec(w)$ that returns a set of tuples $T \subseteq tuples(w)$. In practice, wrappers can be implemented via any language as long as there exists a mapping function from their data model to *first normal form* (1NF).

Global Graph. The global graph $\mathcal{G} = \langle V_{\mathcal{G}}, E_{\mathcal{G}} \rangle$ is an unweighted, directed, connected graph with no self loops. The vertex set $V_{\mathcal{G}}$ is partitioned into two disjoint sets C and F, respectively concepts and features. The set F itself is further partitioned into two disjoint subsets F_{id} and F_{id}^-, consisting of *id* features and *non-id* features, respectively. Next, labels in $E_{\mathcal{G}}$ contain the analyst's domain $\mathcal{L}$ as well as the set of *semantic annotations* $\mathcal{A}$. Semantic annotations are system specific labels and have a special treatment (e.g., `hasFeature`). Hence, we formalize the edge set $E_{\mathcal{G}}$ as the union of the sets $(C \times \mathcal{L} \times C)$ and $(C \times \{\text{hasFeature}\} \times F)$, the former assigning labels in $\mathcal{L}$ between concepts and the latter linking concepts and their features.

Source Graph. A source graph $\mathcal{S}$ is analogous to $\mathcal{G}$. However, here the vertex set $V_{\mathcal{S}}$ is composed of $(W \cup A)$, respectively the set of wrappers and attributes from the previous definition (note that $\mathcal{S}$ is a graph-based representation of the wrappers and their attributes). We use $wrap(\mathcal{S})$ to denote the set of wrappers in $V_{\mathcal{S}}$. Here, we introduce the semantic annotation `hasAttribute`, meant to connect a wrapper with its attributes. Thus, in $\mathcal{S}$ the edge set $E_{\mathcal{S}}$ is composed of $(W \times \{\text{hasAttribute}\} \times A)$.

Schema Mappings. A LAV schema mapping for a wrapper w is a pair $\mathcal{M}(w) = \langle \mathcal{F}, \gamma \rangle$, where $\mathcal{F}$ is an injective function $\mathcal{F} : att(w) \to F$; and γ is a subgraph of $\mathcal{G}$. Consequently, we define the functions $\mathcal{F}(w)$ and $\gamma(w)$ respectively denoting, for w, the mapping from attributes to features $\mathcal{F}$ and the subgraph γ. Recall that we encode mappings as part of the graph, precisely $\mathcal{M}$ contains $\mathcal{F}$ and φ. Thus, to encode $\mathcal{F}$ we extend the set of semantic annotations $\mathcal{A}$ with the `sameAs` label, linking attributes in $\mathcal{S}$ to features in $\mathcal{G}$.

Example 1. Figure 1 depicts the complete integration graph based on the running example depicted in Sect. 3.1.

3.2.2   Data Quality Management

Denial Constraints. A *predicate* P is a comparison unit in the form $v_1 \phi v_2$ or $v_1 \phi c$ where v_1, v_2 are values, respectively from the tuples t_x, t_y, ϕ is a comparison operator and c is a constant.

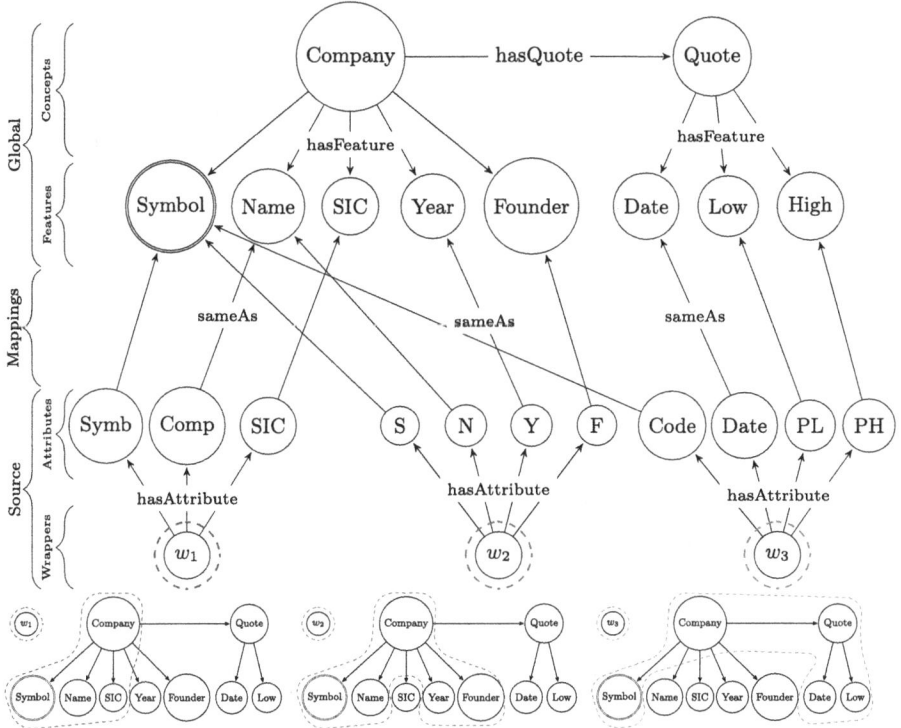

Fig. 1. An example integration graph. Doubly circled features denote IDs. The bottom colored graphs represent mappings (i.e., subgraphs of $\mathcal{G}$) for each wrapper dashed with the same color. There, some features have been omitted for clarity.

Definition 1 (Denial constraint). *A DC φ over a set of tuples T is an expression of the form $\forall t_x, t_y \in T, \neg(P_1 \wedge ... \wedge P_m)$ where φ is satisfied by T if and only if for any pair $t_x, t_y \in T$, at lease one of the predicates $P_1, ..., P_m$ is false.*

$r \models \varphi$ denotes a valid DC φ over a set of tuples T. This is, all predicates cannot be true for any tuple pair, otherwise, there is a violation. $\varphi.Pred$ denotes the set of predicates in φ. Then, we say a DC φ_1 is *minimal* if there does not exist a φ_2 such that $r \models \varphi_1, r \models \varphi_2$, and $\varphi_2.Pred \subset \varphi_1.Pred$.

Ranking DCs. In order to reduce the search space of valid DCs, a scoring function is defined to rank them. The *interestingness score* of each DC is calculated based on its succinctness and support from data. *Succinctness* models how overfitting a constraint rule is. This definition follows Occam's razor principle, where the competing hypothesis making fewer assumptions is preferred [6].

Definition 2 (Succinctness). *The succinctness of a DC φ, denoted $Succ(\varphi)$, is the minimal possible length of a DC divided by its own length $Len(\varphi)$. This is defined as $Succ(\varphi) = Min(\{Len(\phi)|\forall\phi\}) / Len(\varphi)$.*

Coverage determines the interestingness of a DC and measures its statistical significance. By definition, a valid DC φ needs to violate at least one predicate from the evidence. The higher number of satisfied predicates from the evidence, the more support it gives to φ. A pair of tuples satisfying k predicates is a k-*evidence (kE)*. In the best case, the maximum k for a tuple pair in a DC φ is equal to $|\varphi.Pred| - 1$, otherwise it violates φ. A weight parameter is introduced to reflect a higher score to high values of k, from 0 to 1.

Definition 3 (Coverage). *A k-evidence (kE) for φ is a tuple pair $\langle t_x, t_y \rangle$, where k is the number of predicates in φ that are satisfied by $\langle t_x, t_y \rangle$ and $k \leq |\varphi.Pres| - 1$. The weight for a kE for φ is $w(k) = (k + 1) \,/\, |\varphi.Pres|$. The Coverage($\varphi$) is then defined as:*

$$Coverage(\varphi) = \frac{\sum_{k=0}^{|\varphi.Pred|-1} |kE| * w(k)}{\sum_{k=0}^{|\varphi.Pred|-1} |kE|}$$

4   Managing Data Quality in Virtual Data Integration

In this section, we present our proposed system structure and algorithms in detail. As depicted in Fig. 2, rectangles state the solutions we propose in the following sections. We first generate local DCs for each wrapper (i.e., source), and then propagate the DCs to the global graph. There, we establish algorithms to resolve the potential conflicts. Precisely, we address (a) minimal DCs maintenance to prune the redundancy of DCs at the global level; and (b) potential conflicts between DCs derived from contradictory predicates.

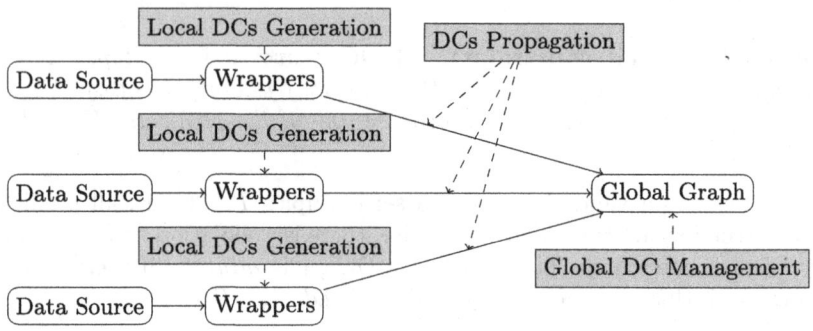

Fig. 2. Overview of the proposed solution process.

4.1   DC Generation and Graph-Based Representation

For each wrapper, we first utilize DCFINDER to produce DC rules at the local level and model them into its equivalent graph representation. For the sake of simplicity and ease of presentation, we narrow the kind of DCs we deal with to those comparing attributes from the same wrapper, and thus no cross-attribute predicates are involved. Hence, DCFINDER generates DCs from each wrapper in the form of: $\{t.w.A_i \; \phi \; t.w.A_j \wedge ... \wedge t.w.A_x \; \phi \; t.w.A_y\}$. Additionally, due to the fact that DCFINDER tends to generate a large amount of DCs, we keep the interestingness score for each DC and filter top-k DCs based on the ranking.

Next, we describe how we model DC rules in an integration graph. To that end, we extend the vertex set of the source graph $\mathcal{S}$ with the set D of DCs. A DC $d \in D$ can be connected to a wrapper w via the semantic annotation hasDC. For each DC, we identify its predicates (via hasPred), which must be connected to two attributes from the same wrapper via hasAtt1 and hasAtt2, and to an operator via hasOp. Additionally, we encode as nodes of $\mathcal{S}$ the confidence values of DCs, and link them via hasScore. Such model is likewise for $\mathcal{G}$, however here we consider the concept of *global DC*, which identifies a DC that must hold for all tuples at the global level (i.e., those generated from any of the wrappers via rewriting in the integration graph). To guarantee traceability and maintenance of the framework, DCs at the source level are connected to DCs at the global level via the sameAs semantic annotation.

Example 2. Consider a local DC expressed over w_3 stating that the high price of a stock quote should always be greater than the low one at any given date, expressed as the predicate: $\{t.PL \geq t.PH\}$. Figure 3, depicts the integration graph modeling it at the source and global graphs.

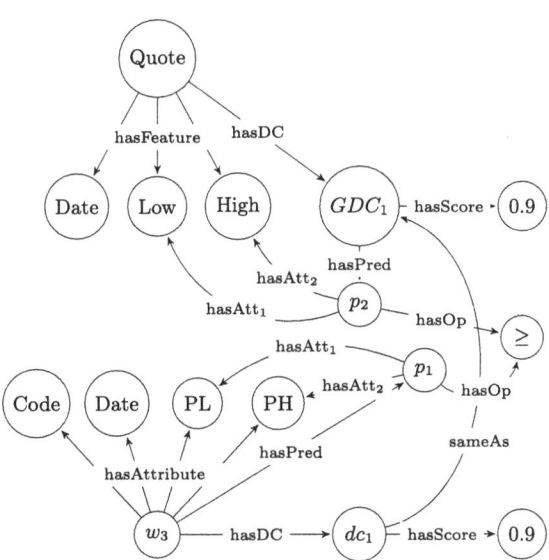

Fig. 3. Integration graph with DCs. Some edges, such as sameAs edges from attributes to features have been omitted for clarity.

4.2   Global DC Management

Once all local DCs have been propagated to the global graph, we perform two tasks to globally manage DCs: minimal DC maintenance and clash management.

Minimal DC Maintenance. It is essential to maintain minimal DCs in the global graph in order to reduce the number of valid constraint rules (as any minimal DC is still a valid DC if we add any other predicate). When a new DC φ is propagated from a source to the global graph and it happens to subsume a minimal global DC ϕ then, ϕ is considered redundant. Algorithm 1.1 updates the set of global minimal DCs when a new DC is propagated to the global schema. To that end, we initialize the set of DCs Σ_c that covers the same attributes c as the new global DC φ does (Line 1–2). For all the minimal DCs in Σ_c, we check whether the predicate sets composing the new global DC φ is a subset of any existing minimal DCs ϕ. If $\varphi.Pred \subset \phi.Pred$, we replace the ϕ with the new minimal DC φ. Otherwise, the new global DC is recognized as a valid DC but not labelled as minimal one (Line 3–9).

Algorithm 1.1: Minimal DC management.

Input   : Set of all GDCs Σ, new GDC φ
Output: Updated set of all GDCs Σ_i
```
/* If the new GDC is a valid DC and minimal, replace the previous
   GDC containing the same predicates.                          */
```
1 $c \leftarrow \varphi.atts$
2 $\Sigma_c \leftarrow \Sigma.contain(c)$
```
// check minimal GDCs from the impacted features.
```
3 **for** $\phi \in \Sigma_c.minimal$ **do**
4 $\quad$ **if** $\varphi.Pred \subset \phi.Pred$ **then**
5 $\quad\quad$ $\Sigma_c.minimal \leftarrow \Sigma_c.minimal - \phi$
6 $\quad\quad$ $\Sigma_c.minimal \leftarrow \Sigma_c.minimal + \varphi$
7 $\quad$ **else**
8 $\quad\quad$ $\Sigma \leftarrow \Sigma + \varphi$
9 $\quad$ **end**
10 **end**
11 **return** Σ

Example 3. Consider the following two DCs c_1, c_2 from the running example: $c_1 : \forall t_\alpha, t_\beta \in w_3, \neg(t_\alpha.Code = t_\beta.Code \wedge t_\alpha.Date = t_\beta.Date \wedge t_\alpha.PL = t_\beta.PL)$, and $c_2 : \forall t_\alpha, t_\beta \in w_3, \neg(t_\alpha.Code = t_\beta.Code \wedge t_\alpha.Date = t_\beta.Date)$. c_1 indicates that the combination of company *Symbol*, *Date* and *Low Price* can identify a stock quote. c_2 expresses the same constraint omitting the *Low Price*. Since $c_2.Preds \subset c_1.Preds$, the minimal DC will be updated to c_2.

DC Clash Management. Two global DCs clash if they refer to common features and contradict each other. Clashes happen when there is no instance that may satisfy both DCs. Given the definition of denial constraint, clashing DCs must be single-predicated over the same attribute and the logical conjunction of their predicates must be empty on the set of available instances. Algorithm 1.2 details how we detect clashes between two global DCs, which considers clashing and partially-clashing DCs.

Definition 4 (Clashing DCs). *Given two DCs c_1 c_2, they are clashing in T, if there does not exist any pair of tuples $\langle t_x, t_y \rangle \in T$, that can satisfy both DCs at the same time.*

Multi-predicated DCs can only partially clash. The partial clash would happen when there are contradictory tuple pairs in the sets of all satisfying tuple pairs from the two DCs due to conflictive predicates. Such predicates must hold on the same attribute and their logical conjunction must be empty.

Definition 5 (Partially clashing DCs). *Given two DCs c_i, c_j, they are partially clashing in T, if the set of all tuple pairs satisfying c_i contains the set of tuple pairs violating c_j.*

Algorithm 1.2: DC clash management

Input : Two global DCs φ_i, φ_j
Output: Updated set of all GDCs Σ_i

1 $c_i \leftarrow \varphi_i.att$, $c_j \leftarrow \varphi_j.att$
   `// check if the two GDCs are bounded to the same features`
2 **if** $c_i = c_j$ **then**
   | `// check if the predicate sets contains contradictory`
   | `pairs`
3 | **for** $p_i \in \varphi_i$, $p_j \in \varphi_j$ **do**
   | | `/* if the pair is contradictory, then a clash is found;`
   | | `   T(pᵢ) is the set of tuples satisfying predicate pᵢ */`
4 | | **if** $T(p_i) \in T(\overline{p_j})$ *or* $T(p_j) \in T(\overline{p_i})$ **then**
   | | | `// compare interestingness scores and choose the`
   | | | `highest`
5 | | | **if** $\varphi_i.score > \varphi_j.score$ **then**
6 | | | | $\Sigma_i \leftarrow \{\varphi_i\}$
7 | | | **else**
8 | | | | **if** $\varphi_j.score \geq \varphi_i.score$ **then**
9 | | | | | $\Sigma_i \leftarrow \{\varphi_j\}$
10 | | | | **end**
11 | | **end**
12 | **else**
13 | | $\Sigma_i \leftarrow \{\varphi_i, \varphi_j\}$
14 | **end**
15 | **end**
16 **else**
17 | $\Sigma_i \leftarrow \{\varphi_i, \varphi_j\}$
18 **end**
19 **return** Σ_i

---

5 Global Query Rewriting with Global DCs

Local DCs propagated to the target and accepted as global DCs may not hold in some sources. For example, consider Fig. 1. w_1, w_2 and w_3 contain attributes (e.g., *symb*, *s*, *code*) referring to the same global feature (*symbol*). Suppose a DC φ from w_1 on *symb*, which is then propagated to the global graph as DC ϕ on *symbol*. When querying the global graph, we must guarantee DC ϕ is guaranteed when performing query answering. This means we need to propagate ϕ to those sources where it originally did not hold. Algorithm 1.3 presents the method to construct DCs in the source graphs from a global DC. For a given global DC φ expressed in the graph, we denote $att(\varphi)$ and $op(\varphi)$ to distinguish the attributes and operators in φ. Following the rewriting algorithm in [16], we map the DC-linked features $feat(att(\vartheta))$ to the attributes in the wrappers. Given the relations from the global DCs E_ϑ, we form the DCs Σ_S in the source graph (the same attribute set can exist in various wrappers, where we form the DC for each wrapper). $dc(A_\theta, E_\vartheta, op(\vartheta))$ denotes the function to form valid DCs given the components.

Algorithm 1.3: Reconstruct a global DC

Input : Global DC ϑ
Output: Set of DCs in the source graph Σ_S
// Get nodes and edges from the global DC
1 $\langle V_\vartheta, E_\vartheta \rangle \leftarrow \vartheta$
// Map features to source attributes
2 $A_{dc} \leftarrow map(feat(att\vartheta))$
// Get the wrappers for the attributes
3 $W \leftarrow wrap(A_{dc})$
// Reconstruct DCs in source graph for each wrapper based on the edges and mapped attributes
4 **for** $w \in W$ **do**
5 | **for** $a_{dc} \in A_{dc}$ **do**
6 | | **if** $a_{dc} \in att(w)$ **then**
7 | | | $A_M \cup = a_{dc}$
8 | | **end**
9 | | $\Sigma_S \cup = dc(A_M, E_{dc}, op(\vartheta))$
10 | **end**
11 **end**
12 **return** Σ_S

5.1 Query with DCs

In [16], a query rewriting algorithm is presented for query answering over the global graph in terms of queries over the wrappers. Here, we extend the rewriting algorithm to enable the enforcement of global DCs within a global query. When rewriting a query Q, the method produces sets of rewritings for each concept in the query to further create a union of conjunctive queries. Then, here, we define $gdc(Q)$ to retrieve the global DCs covered by Q. Σ_G denotes the resulted set of $gdc(\varphi)$. V_Σ and E_Σ denote the composition of Σ, which are vertices and edges respectively. For each global DC in Σ_G, we form the set of DCs for each wrapper and generate the result set of DCs without any duplication. Algorithm 1.4 demonstrates the full method to derive the local DCs from a global query.

Algorithm 1.4: Reconstruct global DCs from a global query to the source graph.

   Input : Global query Q
   Output: Set of DCs in the source graph Σ_S
   `// Get the set of all global DCs covered by Q`
1 $\Sigma_G \leftarrow gdc(Q)$
2 **for** $\vartheta \in \Sigma_G$ **do**
3 | $\Sigma_S \cup = ReconstructGDC(\vartheta)$
4 **end**
5 **return** Σ_S

This way, when querying a global graph with DCs, we apply the constraint rules to all wrappers containing the restricted attributes. For example, consider an integration graph with feature Gender G and FirstName FN, below shows the SQL-like query to retrieve all the records with the global DCs $gdc(G)$: $\forall t_\alpha, t_\beta \in G, \neg(t_\alpha.FN = t_\beta.FN \wedge t_\alpha.G \neq t_\beta.G)$:

```
SELECT G.Gender, G.FirstName
FROM global_graph G WHERE gdc(G)
```

Consider the same integration graph with two wrappers w_1 and w_2 containing the mapped attributes G and FN. The LAV mapping will apply the $gdc(G)$ to both wrappers and join the results, as shown below.

```
SELECT w1.GD, w1.FN FROM wrapper_1 w1
UNION
SELECT w2.GD, w2.FN FROM wrapper_2 w2
WHERE gdc(G) /* Apply the gdc as the filter for all wrappers */
```

6   Validation

We implemented a case study based on financial data. We modeled the SEC Edgar database which releases the XBRL (eXtensible Business Reporting Language) taxonomies every year, given the annual update of U.S. GAAP (Generally Accepted Accounting Principles). Thus, we build wrappers for Edgar based on each release year (i.e., one wrapper per year). For each wrapper, DCFINDER produces sets of DCs. Given the high frequency of the Edgar schema version updating, the wrappers of Edgar share a large portion of attributes, which leads to the overlap of DCs for different wrappers. Following the global DC management algorithms, we are able to conciliate the different versions of local DCs and prune the total number of global DCs to a meaningful level. Following the minimal DC maintenance strategy, we are able to avoid redundant DCs at the global level and derive meaningful local DCs such as $dc_1 : \neg(t_0.Assets \leq t_1.Assets \wedge t_0.Equity \geq t_1.Equity)$ and $dc_2 : \neg(t_0.Liabilities \geq t_1.Liabilities \wedge t_0.Assets \leq t_1.Assets)$. Note dc_1 and dc_2 are valid minimal DCs since there does not exist valid DCs that can be derived from their subsets. Interestingly, the DC $dc_3 : \neg(t_0.Assets \leq t_1.Assets \wedge t_0.Equity \geq t_1.Equity \wedge t_0.Liabilities \geq t_1.Liabilities)$ implies the rule of financial reporting (assets equals to the sum of equities and liabilities).

We also apply the clash management algorithm to resolve contradicting DCs. For instance, the DC $dc_4 : t_0.PeriodEnding \neq t_1.PeriodEnding$ is generated due to the standardized release date in 2012. Then, $dc_5 : \neg(t_0.PeriodEnding = t_1.PeriodEnding)$ states the possible variance of release date for different companies in 2016. This is due to the U.S. GAAP update to allow flexibility for companies to define their own financial year. In this scenario, dc_4 and dc_5 are partially clashing because of the complementary predicates of $PeriodEnding$ feature. We first try to resolve this conflict by the interestingness scores, but both shown high statistical significance in each wrapper. Then, we applied each DC to the global graph and detect the #violations from all sources. dc_5 was valid in all wrappers, while dc_4 generated multiple violations. Thus, we rejected dc_4 and propagated dc_5 to the global schema. Overall, with the global DC management algorithms, we were able to prune the total number of global DCs to 21, which we manually validated as valuable business rules.

7   Conclusions and Future Work

We addressed the DQ problem for virtual data integration systems. The novelty of our approach lies on the consideration of a (potentially conflicting) set of data sources, as opposite to the traditional methods on data quality management that consider a single database instance. To that end, we first elicit DC rules from the data sources and express them in the integration graph to then define DC management methods that enable the global conciliation of DCs. We modified our query rewriting algorithm to guarantee global DCs, while query answering, in all data sources regardless of where they were generated. As future work, we aim to fully automate the process of generating, propagating and enforcing DCs. This requires the extension of traditional knowledge graph bootstrapping methods for quality rules. Another interesting line of work is that of automatically generating data flow operators such that they repair the data errors identified in some sources (instead of fixing them at query time).

Acknowledgements. This work was partly supported by the DOGO4ML project, funded by the Spanish Ministerio de Ciencia e Innovación under project PID2020-117191RB-I00. Sergi Nadal is partly supported by the Spanish Ministerio de Ciencia e Innovación, as well as the European Union - NextGenerationEU, under project FJC2020-045809-I.

References

1. Abedjan, Z., et al.: Detecting data errors: where are we and what needs to be done? Proc. VLDB Endow. **9**(12), 993–1004 (2016)
2. Batini, C., Rula, A.: From data quality to big data quality: a data integration scenario. In: SEBD, Volume 2994 of CEUR Workshop Proceedings, pp. 36–47. CEUR-WS.org (2021)
3. Batini, C., Rula, A., Scannapieco, M., Viscusi, G.: From data quality to big data quality. J. Database Manag. **26**(1), 60–82 (2015)
4. Bleifuß, T., Kruse, S., Naumann, F.: Efficient denial constraint discovery with hydra. Proc. VLDB Endow. **11**(3), 311–323 (2017)
5. Chomicki, J., Marcinkowski, J.: Minimal-change integrity maintenance using tuple deletions. Inf. Comput. **197**(1–2), 90–121 (2005)
6. Chu, X., Ilyas, I.F., Papotti, P.: Discovering denial constraints. Proc. VLDB Endow. **6**(13), 1498–1509 (2013)
7. Geerts, F., Mecca, G., Papotti, P., Santoro, D.: Cleaning data with Llunatic. VLDB J. **29**(4), 867–892 (2020). https://doi.org/10.1007/s00778-019-00586-5
8. Halevy, A.Y.: Answering queries using views: a survey. VLDB J. **10**(4), 270–294 (2001). https://doi.org/10.1007/s007780100054
9. Haug, A., Zachariassen, F., Van Liempd, D.: The costs of poor data quality. J. Ind. Eng. Manag. (JIEM) **4**(2), 168–193 (2011)
10. Heidari, A., McGrath, J., Ilyas, I.F., Rekatsinas, T.: HoloDetect: few-shot learning for error detection. In: SIGMOD Conference, pp. 829–846. ACM (2019)
11. Jarke, M., Jeusfeld, M.A., Quix, C., Vassiliadis, P.: Architecture and quality in data warehouses. In: Pernici, B., Thanos, C. (eds.) CAiSE 1998. LNCS, vol. 1413, pp. 93–113. Springer, Heidelberg (1998). https://doi.org/10.1007/BFb0054221
12. Kolaitis, P.G.: Schema mappings, data exchange, and metadata management. In: PODS, pp. 61–75. ACM (2005)
13. Laranjeiro, N., Soydemir, S.N., Bernardino, J.: A survey on data quality: classifying poor data. In: PRDC, pp. 179–188. IEEE Computer Society (2015)
14. Livshits, E., Heidari, A., Ilyas, I.F., Kimelfeld, B.: Approximate denial constraints. Proc. VLDB Endow. **13**(10), 1682–1695 (2020)
15. Loshin, D.: Evaluating the business impacts of poor data quality. Inf. Qual. J. (2011)
16. Nadal, S., Abello, A., Romero, O., Vansummeren, S., Vassiliadis, P.: Graph-driven federated data management. IEEE Trans. Knowl. Data Eng. (2021)
17. Pena, E.H.M., de Almeida, E.C., Naumann, F.: Discovery of approximate (and exact) denial constraints. Proc. VLDB Endow. **13**(3), 266–278 (2019)
18. Rekatsinas, T., Chu, X., Ilyas, I.F., Ré, C.: HoloClean: holistic data repairs with probabilistic inference. Proc. VLDB Endow. **10**(11), 1190–1201 (2017)
19. Sadiq, S.W., Papotti, P.: Big data quality - whose problem is it? In: ICDE, pp. 1446–1447. IEEE Computer Society (2016)
20. Schirmer, P., et al.: DynFD: functional dependency discovery in dynamic datasets. In: EDBT, pp. 253–264. OpenProceedings.org (2019)
21. Xiao, G., et al.: Ontology-based data access: a survey. In: IJCAI, pp. 5511–5519. ijcai.org (2018)

Time Series and Data Streams

Generating Comparative Explanations of Financial Time Series

Jacopo Fior$^{(\boxtimes)}$ ⓘ, Luca Cagliero ⓘ, and Tommaso Calò

Politecnico di Torino, Corso Duca degli Abruzzi, 24, 10129 Turin, Italy
jacopo.fior@polito.it

Abstract. Private and professional investors can easily access large amounts of financial data describing the temporal evolution of the stock prices. Making appropriate decisions about financial activities often entails performing comparative studies to get an increased comprehension of the underlying assets. The aim of this work is to automatically generate summarized explanations of financial stock series based on the most established fundamental indicators. Unlike any previous summary protoform, the newly proposed time series explanations (i) are suited to comparative analyses, i.e., they express a relative strength of the summary claim about a given stock compared to a reference stock cluster, and (ii) are based on a time series embedding representation indicating the level of similarity between different stocks/stock groups in various periods. The preliminary results demonstrate the usefulness and applicability of the proposed approach.

Keywords: Time series explanation · Time series embeddings · Data summarization

1 Introduction

One of the most labour-intensive activities of financial investors is to explore market-related data, such as financial reports, stock price series and macroeconomic indicators [3,11]. To get an increased comprehension of the underlying assets investors are very interested in getting readable explanations of time-variant events. The present work focuses on generating explanations of stock price series in textual form. Formulating the resulting summaries in natural language allows human users to better understand the temporal evolution of the analyzed series and to effectively support decision-making.

Standard protoforms are the most popular way to summarize time series in textual form. They exemplify relevant patterns in databases using explainable summary templates [8]. For instance, *"The samples of Time Series T acquired in the last week are very similar to those observed in most of the previous 10 weeks"* is an example of protoform, where *last week* is the time window under consideration, whereas *very similar* and *most* are respectively denoted as protofom *quantifier* and *summarizer*.

© Springer Nature Switzerland AG 2022
S. Chiusano et al. (Eds.): ADBIS 2022, LNCS 13389, pp. 121–132, 2022.
https://doi.org/10.1007/978-3-031-15740-0_10

Recently proposed protoform-based approaches automatically generate proto-forms from time series data by adopting clustering and fuzzy modelling [1,4]. The main drawbacks of state-of-the-art solutions are enumerated below:

- Standard protoforms are not suited to perform comparative analyses of time series data at multiple granularity levels (e.g., "*The samples of Time Series T acquired in the last week are very similar to those observed in the time series group G*").
- Since protoform quantifiers and summarizers are defined using static domain-specific rules, their values do not necessarily reflect the underlying data distribution.
- Protoforms are not tailored to the financial domain. Hence, they do not consider domain-specific aggregations (e.g., fundamental indicator levels, market sector and sentiment).

Paper Contribution. This work aims at generating *comparative explanations* of the financial stock price series by exploiting a self-supervised time series embedding representation. The key idea is to first encode the underlying stock series characteristics (e.g., price trend, seasonality, momentum, sentiment about the stock) into a unified vector space and then summarize the key differences between single stock vectors and the encoding of a stocks belonging to a reference group (e.g., the stocks of the same sector with highest operating profit). Quantifier and summarizer values are both dynamically defined based on a data-driven approach on top of the inferred latent space.

Running Example. The summary

> *Stock S is very similar to most of the most virtuous stocks of year 2020 for EBITDA indicator*

compares the historical price series of a specific stock S with those of a group of correlated stocks clustered by means of an established fundamental indicator (EBITDA). Notice that the comparative term *very similar*, i.e., the summarizer, synthesizes the observed level of similarity between S and the reference group, whereas the quantifier *most* indicates the required level of similarity. *year 2015* indicates the reference time period in which the statement holds.

The self-supervised procedure of time series encoding is applied to both historical stock series and stock-related news data. It synthetizes the key information about a stock on a daily basis. The purpose is to inherently capture not only the observed stock price trends, seasonality, and momentum but also the underlying market movers (e.g., the sentiment of the main market actors).

To empirically evaluate the effectiveness and applicability of the proposed approach we carry out both intrinsic and extrinsic evaluations. Specifically, in the intrinsic evaluation the generated summaries are first shortlisted using established protoform-based metrics and then evaluated with the help of a domain expert. In the extrinsic evaluation, we backtest the reliability of the generated stock recommendations. The preliminary results achieved on the U.S. stock market confirm the applicability of the proposed approach to support stock trading activities.

The rest of the paper is organized as follows. Section 2 overviews the prior works. Section 3 describes the analyzed financial data. Sect. 4 presents the proposed method. Section 5 summarizes the main empirical outcomes, whereas Sect. 6 draws the conclusions of the presented work.

2   Literature Review

The generation of explainable summaries can be based on static domain-specific rules, statistic/probabilistic approaches, or neural methods [10]. A joint effort of the Deep Learning community has been devoted to relieving experts of the definition of static rules by leveraging data and algorithms. Albeit state-of-the-art probabilistic/statistical approaches and neural methods generate the text automatically, the quality and readability of the output summaries is not always guaranteed. For these reasons, most existing time series summarization techniques still partly rely on rule-based methods, which generate standard summary templates called protoforms [1, 2, 4, 7]. For instance, [2, 7] generate narratives of data that summarize the key series trends (e.g., increasing, decreasing), whereas [1] uses Evolutionary Genetic Algorithms to explore the set of candidates summary templates and pick those meeting specific (user-specified) constraints. More recently, [4] adopt sequence pattern mining and clustering techniques to support the generation of protoforms. This work presents an hybrid approach to time series summarization that combines the reliability of rule-based strategies with the flexibility of neural NLP methods. Specifically, it leverages a high-dimensional vector representation of the time series, generated by an ad hoc embedding models [13], to dynamically construct summaries providing comparative explanations. To the best of our knowledge, the use of neural network-based approach to define comparative summaries of financial time series has never been proposed so far.

3   Data Overview

To generate explainable summaries of financial data we analyze stock-related data under multiple aspects, i.e., the raw time series of historical prices and exchange volumes, the most established price trend and volatility indicators, the news sentiment, and the values of main fundamental indicators.

Time Series Data. We focus on the time series T_s of the daily closing prices of each stock s belonging to the Standard&Poor (S&P500) index. In our study we consider historical stock data spanning from 2007 to 2018[1].

---

[1] We crawled data from AlphaVantage (https://www.alphavantage.co/). In the considered time span historical data are available for 468 out of the 500 firms.

Price-Related Indicators. We consider the following technical indicators describing the momentum, trend, and volatility of the stock prices [12].

- Exponential Moving Average (EMA) with 5, 20, 50, and 200 periods.
- Moving average convergence divergence (MACD) with the following EMA combinations: (5, 20), (20, 50), and (50, 200).
- Relative Strength index (momentum oscillator) with cutoff thresholds 30% (over-sold) and 70% (over-bought).
- Aroon oscillator (trend descriptor).
- Accumulation/distribution indicator (price and volume divergence).

News Sentiment. We analyze the sentiment ss of the news related to each stock and compute an average per-day and per-stock sentiment score between -1 and 1. A positive score ($ss(s) >> 0$) indicates a positive sentiment of the market about the stock s, whereas a negative score ($ss(s) << 0$) provides a negative feedback. In our experiments, we collect English-written news on the S&P500 stocks in the period 2007–2018 from Reuters[2] and apply VADER to perform rule-based sentiment analysis [6]. We consider around 5253 news articles per stock. Notice the number of daily news per stock is rather variable, as most popular stocks are more likely to be cited.

Fundamental Indicators. They describe the economical and financial factors that mainly influence the stock and the underlying assets. In this study we consider the following established indicators: Earnings Before Interests Taxes Depreciation and Amortization (EBITDA), Return On Equity (ROE), Return On Assets (ROA), Research & Development investments (R&D), Net Income [5].

4   Financial Data Summarizer

A sketch of the proposed method for financial time series summarization is depicted in Fig. 1. The summarization pipeline consists of the following steps:

1. *Financial data encoding*, whose goal is to transform the raw time series and news data into a unified vector representation of the stocks encoding both price-related trends, momentum and seasonality, and market sentiment.
2. *Quantifier/summarizer evaluation*, which entails estimating for each stock and reference time period the values of the corresponding summarizers and quantifiers on top of the encoded stock representation.
3. *Protoform generation*, whose goal is to compose the explainable summaries of the financial time series and compute the corresponding quality indices.

---

[2] https://www.reuters.com/.

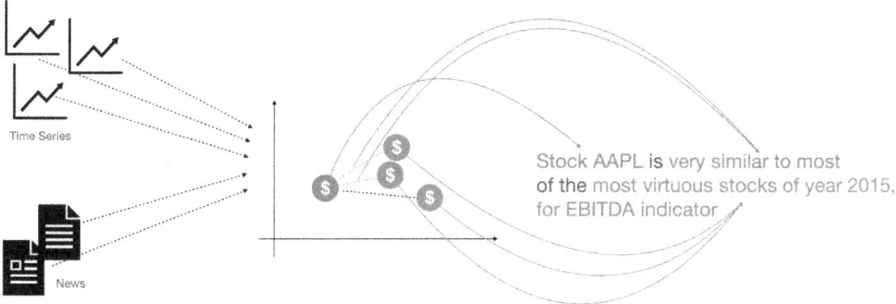

Fig. 1. Sketch of the time series summarization system.

Financial Data Encoding. This step entails encoding multimodal financial data into a unified, high-dimensional vector space. Each stock is represented by a vector in the latent space, which encodes both its most significant price-related features (i.e., trends, seasonalities, momentum) and the sentiment of the market extracted from the news articles.

Time series and news data are first transformed into a discrete sequence of symbols using a SAX representation [9] and then encoded the established Signal2Vec encoder [13][3]. In the SAX representation the daily samples of the series of stock prices, the technical indicators and news sentiment scores are mapped to a unique symbol to condense the daily multimodal information about each stock. Signal2Vec encodes discrete sequences of different time periods (e.g., yearly periods) annotated with the corresponding stock identifier. In such a way, sequences that refer to the same stock are used to describe the underlying behavior of the same stock.

Quantifier and Summarizer Estimation. Quantifiers and summarizers are the core elements of the comparative summaries. They express the level of adherence of the summary claim with the analyzed data.

We leverage the multimodal stock vector space trained at the first step to assign reliable quantifier/summarizer values. Specifically, let s_1, s_2 be the stocks under consideration for summary types *virtuos_stocks*, *year_to_stock*, and *virtuos_multivariate*. Let $v(s_1)$ and $v(s_2)$ be the corresponding vectors encoding the time series and news contents. The data-driven procedure instrumental for quantifier and summarizer estimation is described by the following procedure:

---

[3] In the experiments Signal2Vec is trained using the PV-DBOW architecture with vector size 100, 10 epochs, and a training window of 5 symbols.

Input: stock set S, fundamental indicator set I, reference time period T
for $\forall\, s \in S$ **do**
   for $\forall\, i \in I$ **do**
      $d(s_1, s_2) \leftarrow$ compute-distance$(v(s_1), v(s_2))$[4]
      $v_s^i \leftarrow$ value of indicator i for s within the reference time period T
      $q_s^i \leftarrow$ quantile of stock s according to i, depending on v_s^i
      if $q_s^i == 1st$ **then**
         $R^i = R^i \cup \{s\}$ (set of reference stocks according to i)
         for $\forall\, s \in S$ **do**
            $R_s^i \leftarrow$ Nearest neighbors of stock s according to i calculated
using set of distances d
         end for
      end if
   end for
end for
Output: R_s^i $\forall s \in S, i \in I$

For each fundamental indicator we first shortlist the top-ranked stocks (i.e., the stocks in the first quantile for the fundamental indicator. Then, for each stock in the vector space we compute the distance with each reference stock/group of stocks (e.g., the sector). Distances among vectors are used to quantify the similarity level with the reference group. Finally, quantifiers and summarizers in natural language are derived by uniformly discretizing the per-stock similarity scores.

Protoform Generation. We generate the five different types of comparative summaries reported in Table 1. Each summary is a sentence, called *protoform* [4], that provides a explanable comparison between time series data in natural language. Each protoform contains one or more fields denoting any of the following items:

- *Stock*: the name of the stock under consideration.
- *Sector*: the market sector under consideration.
- *Indicator*: the fundamental indicator under consideration.
- *Quantifier*: A word or phrase that specifies how often the summarizer is true.
- *Summarizer*: Word or phrase denoting a level of match between the compared items.
- *Time window*: A time window of interest for the given protoform.

Table 2 reports the possible values taken by each field and the summaries in which they appear. A more detailed description of the proposed protoforms is given below.

Given a fundamental indicator as reference metric of stock virtuosity, the summary type named *virtuous_stocks* compares a single stock with the most virtuous stocks, whereas the specular type *sectors* compares market sectors instead

---

[4] In the experiments we adopt the cosine distance in compliance with [13].

of single stocks. The type *virtuous multivariate* specifies a percentage of reference stocks with the given level of similarity and also allows the inclusion of multiple indicators for the definition of virtuous stocks.

Summaries *years_to_stock* and *year_to_years* perform time-based comparisons between time series. Specifically, the former type indicates for how many years one stock have been similar to all years of another one. The latter type compares a single year of the one stock with all the years of the other one defining a level of similarity and again how many years correspond to it.

Table 1. Proposed protoforms.

Summary type	Protoform template
virtuous_stocks	Stock [stock] is [summarizer] to most of the most virtuous stocks, for [indicator]
sectors	Sector [sector] is [summarizer] to most of the most virtuous stocks of [sector] sector, for [indicator]
years_to_stock	In [quantifier] years the stock [stock_1] has been [summarizer] to the stock [stock_2]
year_to_years	In year [period] the stock [stock_1] has been [summarizer] to [quantifier] years of the stock [stock_2]
virtuous_multivariate	Stock [stock] is [summarizer] to [quantifier] of the most virtuous stocks, for [indicator_1]..[indicator_n]

Table 2. Fields of the protoforms in Table 1.

Field	Values	Summaries
[stock]	Stock ticker choosen for the comparison	virtuous_stocks, years_to_stock, year_to_years, virtuous_multivariate
[summarizer]	very similar, fairly similar, not similar	all
[indicator]	EBITDA, ROE, ROA, R&D, Net Income	virtuous_stocks, sectors, virtuous_multivariate
[sectors]	Market sector (e.g. Energy, Healthcare, ...)	sectors
[quantifier]	none, few, many, all	years_to_stock, year_to_years
[period]	reference year	year_to_years
[quantifier]	percentage of reference stocks with given level of similarity	virtuous_multivariate

5 Experiments

Hardware and Code. We run the experiments on a hexa-core 2.67 GHz Intel Xeon with 32 GB of RAM, running Ubuntu Linux 18.04.4 LTS. The framework is written in Python and is available for research purposes upon request to the authors.

Execution Times. Table 3 summarizes the execution times taken by each phase of the time series summarization process. The computational time required to generate the time-dependent summaries (i.e., years_to_stock and years_to_years) is roughly one order of magnitude higher than all the other ones. The reason is that their generation entails partitioning stock series data into multiple time periods and recompute the indicator ranks separately for each reference period.

Table 3. Execution times per summary type.

Task	Execution time (avg ± std)
Time series and news data transformation	240 s ± 3 s
Sentiment Analysis	324 s ± 5 s
Multimodal data encoding	594 s ± 15 s
"virtuous_stocks" summary generation	245 s ± 5 s
"sectors" summary generation	220 s ± 3 s
"years_to_stock" summary generation	2793 s ± 23 s
"year_to_years" summary generation	5000 s ± 500 s
"virtuous_multivariate" summary generation	260 s ± 4 s

5.1 Intrinsic Evaluation

We characterize the generated summaries using a set of reference quality metrics first introduced in [4]. Metric values are normalized between zero and one. A brief description of the used metrics is given below.

- *Degree of truth (T1)*: it quantifies the truth of the quantifier-summarizer pair expressed by the summary. It is valid only for summary types year_to_years and years_to_stock.
- *Degree of Imprecision (T2)*: it measures the precision of the summary with respect to the whole data collection.
- *Degree of covering (T3)*: it indicates the percentage of data instances that are covered by the summary statement.
- *Degree of Appropriateness (T4)*: it quantifies the gap between the observed summarizers' values and the expected ones. This metric is valid only for the virtuous_multivariate type.

Table 4 reports some representative summary examples belonging to different type and the corresponding quality metric values. The generated summaries can be ranked by decreasing coverage and precision to shortlist the most reliable stock explanations. For example, the summaries of type *sectors* allow end-users to compare the market sector *Energy* with *Utilities* and *Industrial*, respectively. The Degree of covering (T3) indicates that the reported *sectors* summaries are supported by roughly half of the covered data instances. According to the Degree of Imprecision (T2), their precision is almost maximal (99%) in both cases.

Table 4. Examples of generated summaries.

Summary type	Summary	T1	T2	T3	T4
virtuous_stocks	Stock DU is not similar to most of the most virtuous stocks, for the ROA indicator		1	0.99	
virtuous_stocks	Stock AAPL is very similar to most of the most virtuous stocks, for the ROA indicator		0.99	0.41	
sectors	Most of the stocks of Energy sector, has been not similar to most of to the most virtuous stocks of Utilities sector for the ROA indicator		0.99	0.47	
sectors	Most of the stocks of Energy sector, has been very similar to most of to the most virtuous stocks of Industrial sector for the R&D indicator		0.99	0.65	
years_to_stock	In few years the stock AAPL has been fairly similar to the stock SYF	0.83	0.90	0.17	
years_to_stock	In most years the stock AAPL has been very similar to the stock ANSS	0.77	0.74	0.42	
year_to_years	In year 2015 the stock HPE has been very similar to few years of the stock JEF	0.71	0.71	0.14	
year_to_years	In year 2015 the stock HPE has been fairly similar to most years of the stock FDX	0.93	0.75	0.33	
virtuous_multivariate	Stock FITB is fairly similar to 22% of the most virtuous stocks, for the ROE and the Net Income indicator		0.93	0.22	0.24
virtuous_multivariate	Stock FITB is fairly similar to 32% of the most virtuous stocks, for the ROE and the ROA indicator		0.9	0.32	0.17

5.2 Extrinsic Summary Validation

We validate the usability of the information provided by per-stock summaries via extrinsic evaluation. Specifically, Table 5 reports two summary examples of type *Sectors* that compare the performance of the *Industrials* sector with that of the *Communication Services* and to the *Materials* sectors, respectively. In Fig. 2 we show the corresponding temporal price variations. The summaries are coherent with the observed price series trends: the *Industrials* sector is highly similar to the most virtuous *Communication Services* stocks, whereas is weakly similar to those of the *Materials* sector.

Table 5. Sectors-type summary examples.

Summary	T1	T2	T3	T4
Most of the stocks of Industrials sector, has been very similar to most of to the most virtuous stocks of Communication Services sector for the ROE indicator		0.99	0.44	
Most of the stocks of Industrials sector, has been not similar to most of to the most virtuous stocks of Materials sector for the ROE indicator.		0,99	0,41	

Table 6 reports two summaries of type *years_to_stock* that compare the performance of the Apple stock with that of the Vertex Pharmaceuticals and CME Group stocks. According to the generated summaries, the price movements of the stocks AAPL are expected to be more similar to those of stock VRTX than those of CME. The expected result is confirmed by historical time series depicted in Fig. 3 (see, for example, years 2014–2016).

Table 6. years_to_stock summary examples.

Summary	T1	T2	T3	T4
In most years the stock AAPL has been very similar to the stock VRTX	0,77	0,74	0,42	
In most years the stock AAPL has been not similar to the stock CME	0.77	0.74	0.42	

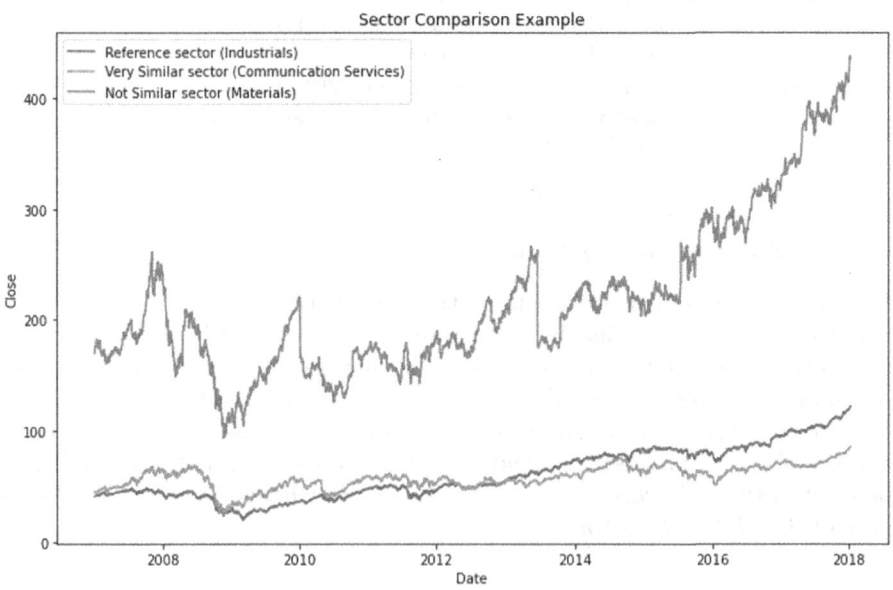

Fig. 2. Comparison between the Energy Sector and the most virtuous stocks for Industrial and Utilities Sectors.

Fig. 3. Comparison between APPL, VRTX, and CME stocks.

6   Conclusions and Future Works

The paper proposed a new approach to generate explainable summaries of financial time series in textual form. The key idea is to represent the key information about the stocks into a unified latent space, among which price-related time series data and news sentiment scores. By leveraging the vector representation to get reliable stock and stock group similarities we are able to automatically estimate the quantifiers and summarizers needed to generate the protoforms.

The preliminary results show that the provided summary examples (1) achieve satisfactory quality levels according to the metrics defined in [4], (2) are coherent with the expectation, and (3) can be exploited by domain experts to support decision-making.

We plan to extend the empirical validation by designing and test a dedicated mobile application through which private and professional investors can access and evaluate the generated summaries and the corresponding evaluation metrics. We will collect subjective user feedbacks with the twofold aim at improving the robustness of the empirical validation and exploiting the relevance feedback in order to selectively filter the generated summaries.

Acknowledgements. The research leading to these results has been partly supported by the SmartData@PoliTO center for Big Data and Machine Learning technologies.

References

1. Catillo-Ortega, R.M., Marín, N., Sánchez, D.: A fuzzy approach to the linguistic summarization of time series. J. Multiple-Valued Logic Soft Comput. **17** (2011)
2. Gatt, A., et al.: From data to text in the neonatal intensive care unit: using NLG technology for decision support and information management. AI Commun. **22**(3), 153–186 (2009)
3. Goyal, K., Kumar, S.: Financial literacy: a systematic review and bibliometric analysis. Int. J. Consum. Stud. **45**(1), 80–105 (2021). https://doi.org/10.1111/ijcs.12605. https://onlinelibrary.wiley.com/doi/abs/10.1111/ijcs.12605
4. Harris, J.J., Chen, C.H., Zaki, M.J.: A framework for generating summaries from temporal personal health data. ACM Trans. Comput. Healthcare **2**(3), 1–43 (2021). https://doi.org/10.1145/3448672
5. Huang, Y., Capretz, L.F., Ho, D.: Machine learning for stock prediction based on fundamental analysis. In: 2021 IEEE Symposium Series on Computational Intelligence (SSCI), pp. 1–10 (2021). https://doi.org/10.1109/SSCI50451.2021.9660134
6. Hutto, C., Gilbert, E.: VADER: a parsimonious rule-based model for sentiment analysis of social media text. In: Proceedings of the International AAAI Conference on Web and Social Media, vol. 8, no. 1, pp. 216–225, May 2014. https://ojs.aaai.org/index.php/ICWSM/article/view/14550
7. Kacprzyk, J., Wilbik, A., Zadrozny, S.: An approach to the linguistic summarization of time series using a fuzzy quantifier driven aggregation. Int. J. Intell. Syst. **25**(5), 411–439 (2010). https://doi.org/10.1002/int.20405
8. Kacprzyk, J., Zadrony, S.: Linguistic database summaries and their protoforms: towards natural language based knowledge discovery tools. Inf. Sci. **173**(4), 281–304 (2005)
9. Keogh, E., Lin, J., Fu, A.: HOT SAX: efficiently finding the most unusual time series subsequence. In: Proceedings of the Fifth IEEE International Conference on Data Mining, ICDM 2005, pp. 226–233. IEEE Computer Society, USA (2005). https://doi.org/10.1109/ICDM.2005.79
10. van der Lee, C., Krahmer, E., Wubben, S.: Automated learning of templates for data-to-text generation: comparing rule-based, statistical and neural methods. In: Proceedings of the 11th International Conference on Natural Language Generation, pp. 35–45. Association for Computational Linguistics, Tilburg University, The Netherlands, November 2018. https://doi.org/10.18653/v1/W18-6504
11. Liapis, C.M., Karanikola, A., Kotsiantis, S.: A multi-method survey on the use of sentiment analysis in multivariate financial time series forecasting. Entropy **23**(12), 1603 (2021). https://doi.org/10.3390/e23121603. www.mdpi.com/1099-4300/23/12/1603
12. Murphy, J.J.: Technical Analysis of the Financial Markets: A Comprehensive Guide to Trading Methods and Applications. Penguin, New York (1999)
13. Nalmpantis, C., Vrakas, D.: Signal2Vec: time series embedding representation. In: Macintyre, J., Iliadis, L., Maglogiannis, I., Jayne, C. (eds.) EANN 2019. CCIS, vol. 1000, pp. 80–90. Springer, Cham (2019). https://doi.org/10.1007/978-3-030-20257-6_7

Summarizing Edge-Device Data via Core Items

Damjan Gjurovski$^{(\boxtimes)}$ [ID], Jan Heidemann, and Sebastian Michel [ID]

TU Kaiserslautern (TUK), Kaiserslautern, Germany
{gjurovski,j_heideman18,michel}@cs.uni-kl.de

Abstract. In this work, we consider the problem of summarizing a data stream through an item-based summary using core items. We consider an IoT setting, where computing such summaries at the edge devices instead of emitting the whole data stream can drastically reduce the network traffic and speed up further processing. Core items of a data stream are the items with the highest values for a given monotone submodular utility function. To create stream summaries, we propose the SoftSieving approach for parallel processing with low memory consumption and fast execution time while attaining acceptable utility gain. Through extensive experiments with real-world datasets, we show the suitability of our approach and its superiority over state-of-the-art competitors.

1 Introduction

The amount of data that is continuously generated by different applications and devices, like social networks and deployed sensing devices, is increasing at an unprecedented speed. Processing such a vast amount of data is still commonly done at centralized compute clusters where high computational power and network bandwidth are available for deep analytical tasks. The required data transfer from originating sources to such a centralized instance creates significant network traffic. This is especially visible when considering data sources on edge devices that collect and transmit data in real-time. We aim to minimize network traffic to gather data in centralized locations and, simultaneously, to speed up subsequent data processing. We propose to do so not by pushing the entire analytical processing to the edge devices —that often have limited compute power— but by compacting the stream using item-based summaries that represent the original data in an optimal way given an objective (utility) function.

As a motivating example, consider the task of determining outliers in camera surveillance footage to detect wildfire outbreaks in a widespread national park. When the camera footage has high resolution and high frame rate, and there are multiple cameras at multiple locations, it is easy to imagine how the amount of data to be transferred and further processing become difficult to manage, specifically in areas with only rudimentary wireless network coverage. A summary with only the most useful frames can drastically reduce the amount of data while keeping the important information to be further evaluated.

© Springer Nature Switzerland AG 2022
S. Chiusano et al. (Eds.): ADBIS 2022, LNCS 13389, pp. 133–147, 2022.
https://doi.org/10.1007/978-3-031-15740-0_11

The most obvious and straightforward solution to summarize data are random sampling techniques [2,27]. However, while they are understandable and easy to implement, they can result in important information being lost [8]. To overcome this problem, one solution is to use a **utility function**, which can measure the informativeness [4], representativeness [3,30], coverage [13], etc., of a selected data subset. With a utility function, the information gain for any new data point can be calculated and data points can accordingly be added to the summary. Typically, such functions belong to the class of *monotone submodular functions*, which means they are *non-decreasing*, and possess the *diminishing returns property*. Following Zhao et al. [29], we refer to these functions as *core item functions*, and the items identified by them are called *core items*.

As the problem of finding an optimal subset according to one of the functions is NP-hard [10], the main focus has been on finding good approximation algorithms [3,6,7,15]. In data stream settings, the basis for most state-of-the-art approaches is the Sieve-Streaming algorithm, introduced by Badanidiyuru et al. [3], where a data stream summarization is performed by maximizing a submodular set function subject to a cardinality constraint. However, when considering stream processing at the edge, Sieve-Streaming and the related variants Sieve-Streaming++ [15] and ThreeSieves [6] exhibit limitations. We develop our approach by considering the existing algorithms and addressing their limitations.

The related research investigating submodular function maximization typically focuses on text data [9,18,19] which enables the usage of natural coverage functions. However, such data is rarely encountered in edge applications. Additionally, the datasets are often not realistic representations of data generated on the edge but rather come from machine learning and bring diverse data points for classification purposes [3,6,7,15]. Furthermore, most evaluations are performed on setups with high processing power, not representative of edge devices.

Problem Statement, Contributions, and Outline. In our work, data arrives in the form of data stream D on k edge devices. To avoid transferring every item e_i from D through the network, we aim at computing item-based summaries using core items directly at the edge devices. The core items will be computed as the items that have the highest value for a given monotone submodular utility function f. The task of computing the core items will be done in parallel on the $k-1$ edge devices such that 1 edge device will be responsible for consolidating and generating the global core item set.

In this paper, we make the following **contributions**.

- We introduce an approach for fast core items computation in a data stream at the edge, which we call *SoftSieving*.
- We have tailored Sieve-Streaming [3], Sieve-Streaming++ [15], and Three-Sieves [6] to be applicable inside an Apache Storm topology [1].
- We performed extensive experiments by using two real-world datasets measuring the processing times, latency, and utility values of the approaches.

The remainder of the paper is organized as follows. Section 2 discussed related work, followed by background information on data stream summarization and submodular functions in Sect. 3. Section 4 reviews the shortcomings of existing approaches and presents our approach. Section 5 reports on the result of the experimental evaluation before Sect. 6 concludes the paper.

2   Related Work

Although different notions of *core sets* exist in related work on data streams [13, 21], the terminology in this work is based on the one used by Zhao et al. [29]. The core items are representative items chosen from a data stream. The problem of *core items tracking* is to continuously maintain core items in a streaming setting. Their approach is based on probabilistic decay. Since this work deals with a setting where all items within a window are equally important, these approaches are not considered in the comparisons.

Sieve-Streaming [3] is a popular algorithm for submodular function maximization of data streams, and the basis for our approach. The authors formalize the problem of summarizing a data stream to the maximization of a submodular set function subject to a cardinality constraint. The algorithm, unlike some previous approaches [17,23], uses a single pass over the data. This approach achieves the best possible approximation for this setting [11].

Sieve-Streaming++ [15] is motivated by the observation that Sieve-Streaming unnecessarily maintains sieves that cannot achieve the best utility value over all sieves anymore. These sieves can be safely discarded and replaced by sieves based on the new lower bound.

Dynamic Sieve Streaming [7], unlike the previous approaches, is explicitly aimed at a distributed IoT setting. It proves improved upper and lower bounds to be used for the active set utility function. However, the experiments were not done in an IoT-representative setting but on a machine with higher processing power and memory size. Moreover, the improvements are specific to active set, and we aim to compare approaches on multiple utility functions. Thus, this approach will not be included in our experiments.

The two previous algorithms do not weaken the theoretical guarantees of Sieve-Streaming. In contrast, ThreeSieves [6] ignores the theoretical worst case and aims for better practical performance. ThreeSieves maintains only one threshold, drastically reducing memory cost, and dynamically changes that threshold.

The related research on approaches that focus on submodular function optimization is vast [9,18,19,23,29]. However, these approaches will not be explored in more detail, as they either do not offer significant improvement over the mentioned approaches or do not apply to our edge streaming setting.

3   Preliminaries

3.1   Apache Storm

Apache Storm [1] is a popular real-time, distributed stream processing framework. Data processing in Storm is done through *topologies*. A topology consists of two types of components, *spouts* and *bolts*. Data flows through them in the form of tuple streams. Tuples are first emitted by the spouts. Bolts receive and process the tuples by executing an arbitrary function and can send as output a new tuple to another bolt for more complex operations. Apache Storm can require too many resources that might not be available in typical edge devices. Thus, in our work, we use EdgeWise [12]. In Apache Storm, each operation is assigned to a thread and scheduling is handled by the operating system, which is not aware of congestion inside the topology. On the edge, this can lead to high latency and backpressure. In EdgeWise, there is a congestion-aware scheduler assigning the operations to threads from a fixed-size worker pool. Queue lengths are balanced and backpressure is minimized.

3.2 Stream Summarization

To form the summaries, we consider utility functions which can measure the informativeness, representativeness, coverage, etc., of the selected subset.

Utility Functions: To mathematically determine the quality of a summary, subsets of the data set need to be assigned a function value and then compared. This is the purpose of a utility function. For any new data item, the new value of the utility function can be used to determine whether to add it to the summary or not. A utility function f is any function $2^D \Rightarrow \mathbb{R}_{\geq 0}$ that assigns all subsets a nonnegative value [29]. The objective is to find the subset of size at most K with the best utility value. Formally put: $OPT = S^* = \underset{S \subseteq D, |S| \leq K}{\arg\max} \; f(S)$.

Submodular Functions: Suitable utility functions often belong to the class of *monotone submodular functions*. They have a diminishing returns property, i.e., $f(\{e\} \cup A) - f(A) \geq f(\{e\} \cup B) - f(B)$ for $A \subseteq B$. Additionally, they are monotone, so $f(\{e\} \cup A) - f(A) \geq 0$ for all e and A. The best approximation ratio for OPT in existing algorithms is $\mathcal{O}(\frac{1}{2} - \epsilon)$, which is also the theoretical maximum approximation ratio for the streaming setting [11].

Using the submodular functions as utility functions, we create data summaries, i.e., sets of core items. For a utility function f, the *marginal gain* for adding item e to the core item set S can be calculated as $\Delta_f(e \mid S) = f(S \cup \{e\}) - f(S)$. We assume $f(\emptyset) = 0$. In this work, we consider two different monotone submodular utility functions based on *active sets* and *exemplar-based clustering*. **Active set** stems from *Gaussian Processes* (GP), which are used in nonparametric regression [28]. In GP the active set is used for efficiency and one way to choose an active set is the Informative Vector Machine (IVM) [16]. IVM is monotonic and submodular, as shown by Seeger [24]. The **K-medoids** problem aims to build clusters around exemplars from the set of data points [14]. To compute distances, it requires a nonnegative distance function. However, when working with data streams a problem arises since the distance to all data points needs to be known. Fortunately, the function is *additively decomposable* [22]. Thus, in our setting, we can continuously generate samples needed for the computation.

4   Algorithms for Core Item Sets Computation

In the following, we present our proposed approach by first analyzing the existing algorithms and identifying their limitations. The Sieve-Streaming algorithm [3] and its variants, Sieve-Streaming++ [15] and ThreeSieves [6], provide solutions of high quality for data stream summarization. However, when considering data stream summarization on edge devices, we have identified that all of these approaches have certain limitations and can be further improved.

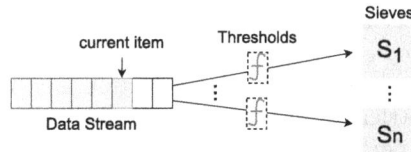

Fig. 1. Computing sieves in a data stream

The *Sieve-Streaming* algorithm manages multiple sieves with different thresholds in parallel, as shown in Fig. 1. For a new data item, if it satisfies the marginal gain, it will be added to any sieve for which it exceeds the specific threshold. In other words, each sieve filters items based on its threshold. At least one of the sieves is expected to have a fitting threshold and produce a good core item set. The elements from the sieve with the best utility value are returned as a result. The thresholds are approximated using the maximum singleton value $m = \max_{e \in D} f(\{e\})$, i.e., the maximum value of the submodular utility function f for any single item e. The lower bound is m since that value has already been reached. If the summary has size K, the best case would be K items increasing the function value by m, resulting in upper bound of $K \cdot m$. The number of sieves depends on the parameter $\epsilon > 0$, which also influences the quality of the result.

If Sieve-Streaming is implemented in an Apache Storm topology without modifications, no parallelism is possible since the best utility value over all sieves is needed for the output. This will lead to increased processing load and increased latencies. To overcome this, we split the algorithm into a part where sieves are processing tuples and finding core items and a part where the core item sets of the different sieves are collected and compared.

When analyzing the Sieve-Streaming algorithm and topology, several limitations are immediately observable. First, the maximum singleton value m can be updated for every new entry, which also means that the number of sieves can change when this update occurs. Hence, in the proposed topology, there will be additional communication overhead between the bolts responsible for computing sieves and the collector bolt, especially when the maximum singleton value m changes frequently. Furthermore, the large number of created sieves can directly lead to problems with both storage and computation times.

Although *Sieve-Streaming++* [15] improves this approach by modifying the lower bound m for the optimal solution OPT as the algorithm is running, still the same limitations are present. The number of sieves remains a problem for low ϵ. Additionally, since the Storm topology remains the same, the problem of additional communication overhead when m changes frequently persists.

ThreeSieves [6] maintains only one single sieve and decreases its threshold over time. The rules for adding an item to a sieve are the same as in Sieve-Streaming, however, when an item e is not added, a counter t is incremented by one. If the counter reaches the value of parameter T, the threshold value is decreased to the next-biggest estimate. This results in a threshold that is lowered more and more over time. The implementation of ThreeSieves in a Storm topology needs only one bolt. Although this eliminates the unnecessary communication overhead, it leads to an obvious limitation, i.e., lack of parallelization. This single bolt is a bottleneck, leading to increased processing latencies. Additionally, the algorithm performance depends heavily on the choice of parameter T. If it is too small, the threshold will get too low and the summary will be filled quickly, resulting in lower utility. If it is too large, useful items can be missed.

4.1   SoftSieving

Considering the limitations of the existing approaches, we build our new approach called *SoftSieving* by keeping the following design goals in mind.

- Use fewer sieves while keeping a high solution quality.
- Facilitate parallelization as much as possible.
- Make the processing as fast as possible by apt assignment of computations.

The first design goal steers in the direction of ThreeSieves [6] since its dynamic sieve can achieve high utility for many settings. The question then becomes how to increase the number of sieves. There should be bolt instances running in parallel that are responsible for computing their own sieves. Each instance should be independent of the others to minimize the communication overhead. Thus, there are two general methods that we can take.

1. Split the stream over n bolts and have all of them use the same threshold(s).
2. The n bolt instances process all tuples, each with different threshold(s).

Since one of our design goals is to minimize the number of sieves, it is natural to consider the first method. However, we cannot expect much gain from having additional sieves in every bolt. This comes directly from the ThreeSieves algorithm since the solution quality is already high while using only one sieve. The sieve is built starting from the highest possible threshold and lowering the threshold over time. As a result, high-utility items at the beginning of the data stream can be wrongly discarded. To better include these items, we can employ the reverse strategy by starting at the lowest threshold and increasing it. In this way, when the core item set of the sieve is filled with low utility items, it will be difficult to replace them later with higher utility items. Thus, we will include

Algorithm 1 SoftSieving

1: **for** $i = 1, ..., n$ **do**
2:     $S_i \leftarrow \emptyset$, $t_i \leftarrow 0$, $b_i \leftarrow 0$
3: **while** $(e \leftarrow D.next())$!= *null* **do**
4:     $m \leftarrow \max(m, f(\{e\}))$, $O \leftarrow \{(1 + \epsilon)^i \mid m \leq (1 + \epsilon)^i \leq K \cdot m\}$
5:     $V_n \leftarrow n$ equidistant samples from O, and $\max(O)$
6:     Delete all S_i such that $v_i \notin O$ **and** Create new S_i all new $v_i \in O$
7:     **for** $(S_i, v_i) \mid i = 1, \ldots, n$ and $v_i = V_n[i]$ **do**
8:         **if** $\Delta_f(e \mid S_i) \geq \frac{\frac{v_i}{2} - f(S_i)}{K - |S_i|}$ **then**
9:             **if** $|S_i| < K$ **then**
10:                 $S_i \leftarrow S_i \cup \{e\}$
11:             **else if** $b_i < K$ **then**
12:                 $e_s \leftarrow \arg\min_{e \in S_i} \Delta_f(e \mid S_i\{e\})$
13:                 **if** $\Delta_f(e \mid S_i \backslash \{e_s\}) > \Delta_f(S_i)$ **then**
14:                     $S_i \leftarrow S_i \cup \{e\} \backslash \{e_s\}$
15:                     $b_i \leftarrow b_i + 1$
16:             $t_i \leftarrow 0$
17:         **else**
18:             $t_i \leftarrow t_i + 1$
19:             **if** $t_i \geq T$ **then**
20:                 $v_i \leftarrow$ next lowest threshold from O
21:                 $O \leftarrow O \backslash v_{next}$, $t_i \leftarrow 0$
22: **return** $\arg\max_{S_i \mid i=1,\ldots,n} f(S)$

---

sieves that start from lower thresholds but still decrease them. The number of sieves depends on how much the thresholding calculations throttle the topology. Considering the first design goal, we will prefer a small, fixed number of sieves. Fixing the number of sieves prevents a rapidly increasing number of sieves and will enable fast inner-bolt processing. Since the data stream is split, we would require a collector bolt responsible for building the core item sets. However, when adding a new item, every sieve bolt would need the current set of core items for their sieves at all times. Thus, this will result in an extensive amount of updates.

To avoid numerous updates, we propose the usage of preemption [5]. Preemption means that once an item is added to the summary, it does not necessarily stay there but instead can be replaced by newer items at any point in time. The first K items are always added to the summary. Further items are swapped if they improve the solution by more than a defined threshold. If the summary is not fixed, we do not need to update the sieve bolts. They can continuously send the core items to the collector, which then checks if they improve the global solution when using swapping. The collector does not need to process as many items as the sieve bolts but still checking for every item in the summary quickly becomes infeasible with increasing K. Thus, the marginal gain of every item in the summary is stored. If the summary is full, new items are compared only to the item with the lowest gain. Consequently, we avoid the updates and fulfill our third design goal. To avoid excessive load in the collector and longer computations in the sieve bolts, the summaries will be restricted to size K.

Following these design decisions, as next, we present the **SoftSieving algorithm** as shown in Algorithm 1. In Lines (1–2), we initialize the algorithm. O is the set of thresholds, satisfying the defined lower and upper bounds. Compared to existing algorithms, not all thresholds are chosen from the set O, nor the maximum is only used. Instead, V_n gets n equidistant samples from O (Line 5). For the items from the data stream, we repeat steps 3 to 21. Similarly to the related approaches, we update m and the sieves (Lines 4 – 6). Next, we add items to the

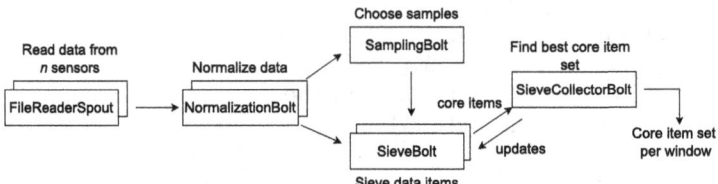

Fig. 2. SoftSieving storm topology

sieves (Lines 8–16) such that an item will be added if Lines 8 and 9 are satisfied. The amount by which adding e to S_i increases the utility function must be big enough such that sieve S_i can still reach the approximated optimal value v_i, i.e., increase by $v_i - f(S_i)$ after adding $K - |S_i|$ items to the summary. To account for some items in the summary influencing the utility value more that others, $\frac{v_i}{2}$ is used instead of v_i. If a sieve is full, we compare the next K items above the threshold against e_s, the item with the lowest utility gain from the core item set S_i (Lines 12–13). If item e offers higher utility, e and e_s are swapped. If an item is not added and Line 19 is satisfied, the sieve-thresholds are decreased (Lines 17–21). Finally, the sieve with the highest utility is returned (Line 22).

The actual Apache Storm **topology** responsible for realizing the SoftSieving algorithm is depicted in Fig. 2. The topology consists of five main components. First, the multiple instances of the *FileReaderSpout* are responsible for emitting the data from the data stream. Since for some data streams the data needs to be normalized, we introduce the *NormalizationBolt*. As explained in Sect. 3.2, we might need samples for computing the utility functions. For that reason, we introduce the *SamplingBolt*. The *SieveBolt* is responsible for maintaining and updating the sieves. The *SieveCollectorBolt* receives the core items from the *SieveBolts* and updates the core item set with the best utility. Although there can be several *SieveBolts*, there can be only one instance of the *SieveCollectorBolt*.

Theoretical Analysis. We will now analyze the time and space cost as well as the utility for SoftSieving, with a focus on the Storm implementation. For m bolt instances and n sieves per bolt, SoftSieving stores $\mathcal{O}(n \cdot (m + 1) \cdot K)$ items. Each bolt instance has its own n summaries with up to K items.

For the time analysis, consider a single item e. For the bolt it is assigned to, it takes n evaluations of the utility function, which take time T_K for a summary of size K. If instead the core item set and an additional K items have been sent, no further evaluations are performed. In the collector bolt, there are two possibilities. If the global summary S for item e is not full yet, e is added to S in $\mathcal{O}(1)$ and the new utility value is calculated, taking T_K time. Otherwise, a swap is considered, which takes T_K time for the utility evaluation of $S \backslash \{e_{min}\} \cup \{e\}$. Overall, the collector receives a maximum of $2K$ core items from each summary, i.e., a total of $m \cdot n \cdot 2K$. For a data stream of size N split between m bolt tasks, the time complexity is $\mathcal{O}(\frac{N}{m} \cdot (n \cdot T_K) + 2nmK \cdot T_K)$. Considering tumbling windows of size w, we get $\mathcal{O}(\frac{N}{m} \cdot (n \cdot T_K) + 2nwmK \cdot T_K)$. When considering the updates of the maximum singleton value, since there is no upper bound on the number of updates, this increases to $\mathcal{O}(\frac{N}{m} \cdot (n \cdot T_K) + N \cdot T_K)$, or since $m, n << N$, $\mathcal{O}(N \cdot T_K)$. In practice, there will be significantly faster runtimes since the updates will not happen for every data item. For other approaches, the number of queries per element is often used [3,6,15], which is $\mathcal{O}(n)$ here since we always use n sieves.

Regarding the quality, we can consider the result in ThreeSieves [6] as a lower bound since we always include the sieve with threshold $K \cdot m$. The authors prove that the solution S achieves an approximation to the optimal value OPT of $(1-\epsilon)(1-\frac{1}{\exp(1)})OPT$ with probability $(1-\alpha)^K$. When using n sieves, we expect to reach the optimal threshold v^* for one of them after $\frac{|O|}{2n} \cdot T$ items, instead of $\frac{|O|}{2} \cdot T$ for one sieve. This follows from choosing the thresholds equidistant from O, thus partition O into partitions of size $\frac{|O|}{n}$. In their proof, the probability of $(1 - \alpha)^K$ comes from $P(v_1 = v_1^*, ..., v_K = v_K^*) > (1 - \frac{-\ln(\alpha)}{T})^K$, where, v_i^* are the thresholds of the sieve and $v_i = \Delta(S \mid e_i)$ are the marginal gains achieved by the greedy algorithm (for details see [6], appendix). For SoftSieving, any of the n sieves can achieve these values, increasing the probability. We can approximate it with $1 - ((1-\alpha)^K)^n$ for n sieves, but recall that the n thresholds are not chosen at random, but to be equidistant. The swapping yields no such improved theoretical guarantees since we try to swap with the item with the lowest marginal gain. This is done for performance reasons but can result in swapping new item e only with its most similar item from the core item set, bringing minimal improvement.

5   Experimental Evaluation

The proposed topology and the considered competitors are implemented in Edge-Wise [12]. The experiments were performed on a Raspberry Pi 4 with a 1.5 Ghz Quad-Core-processor and 8 GB of main memory. The Raspberry Pi runs an Apache Storm Cluster and ZooKeeper, where the topologies were submitted. We carried out experiments with the goal of answering the following questions.

1. Can we find summaries in a reasonable timeframe which is preferable to sending all data items to the core without summarization?
2. How well can the approaches handle increasing load from the IoT-device(s)?
3. Does the SoftSieving approach offer significantly better processing times or solution quality compared to existing algorithms?

Thus, we measure the total latency over the topologies (questions 1 and 3). We measure the processing times in the SieveBolts (questions 2 and 3) and the utility for all approaches (question 3). We perform experiments with varying summary sizes $K \in \{5, 20, 100\}$, window sizes $w \in \{1000, 10000, 100000\}$, and the parameter impacting the number of sieves $\varepsilon \in \{0.1, 0.01, 0.001\}$. For processing times and total latency, the average over all tuples is taken, the utility is the average value over all windows.

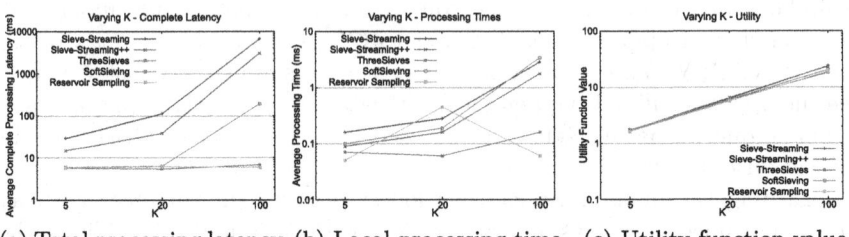

(a) Total processing latency (b) Local processing time (c) Utility function value

Fig. 3. Varying summary size ($K = 5$, $K = 20$, $K = 100$) - log scaled

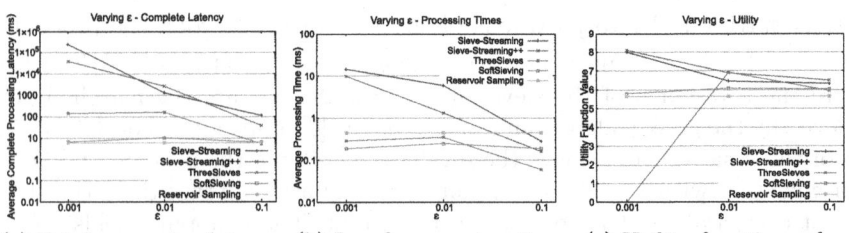

(a) Total processing latency (b) Local processing time (c) Utility function value

Fig. 4. Varying number of sieves ($\varepsilon = 0.001$, $\varepsilon = 0.01$, $\varepsilon = 0.1$) - a, b log-scaled

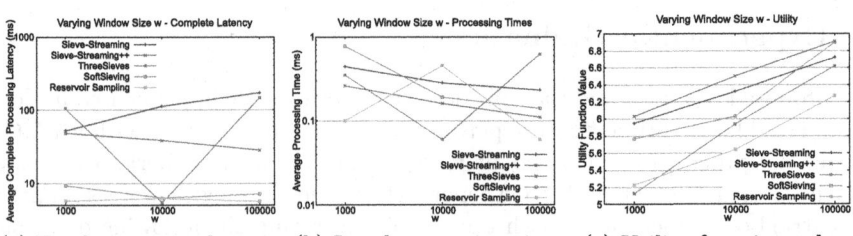

(a) Total processing latency (b) Local processing time (c) Utility function value

Fig. 5. Varying window size ($w = 1000$, $w = 10000$, $w = 100000$)

Competitors: We considered the Sieve-Streaming [3], Sieve-Streaming++ [15], ThreeSieves [6], and Reservoir Sampling [27]. All the competitors were implemented in a Storm topology. All sieve-based algorithms are executed with parameters $K = 20, w = 10000, \varepsilon = 0.1$, unless otherwise specified. ThreeSieves and SoftSieving additionally use $T = 100$ as a standard parameter and SoftSieving uses $n = 4$. The utility functions used are Active Set and k-medoids (Sect. 3.2). The topologies are configured to have 10 SieveBolts and one SieveCollectorBolt. The NormalizationBolts have 5 instances. The SamplingBolt uses one instance.

Datasets: The approaches were evaluated on two real-world datasets, specifically chosen to simulate a real IoT scenario. The first dataset is a *telemetry dataset* from Stafford [26]. The data is collected from three IoT-devices, reading environmental sensor data in regular intervals. The measurements include temperature, humidity, CO, liquid petroleum gas, smoke, light, and motion, having 405184 entries. The distance between two items was implemented as sum of the euclidean distances of all measurements. The data is normalized. The second *images dataset* is an RGB-D dataset [20, 25]. It contains image frames from three Kinect sensors in an university hall. The images in the dataset are stored as 8 bits, 3 channels PPM images with 640×480 pixels. We preprocess the data and transform it into feature vectors of size 804. Distances are calculated as sum of the euclidean distances of all feature vector entries. Since the feature vector is normalized the NormalizationBolt is not necessary.

First, we performed experiments on the telemetry dataset. When one parameter is varied, the other remain on their standard values.

Varying Summary Size (K): Considering the total latency (Fig. 3a), all approaches are affected by the summary size, but Sieve-Streaming and Sieve-Streaming++ perform the worst, with SoftSieving being up to an order of magnitude below them. Reservoir Sampling stays consistent since the sampling calculations are not affected by K. ThreeSieves is also barely affected, which means that the increase in computation cost is not notable when using only one sieve. The processing time of the SieveBolt (Fig. 3b) increases as well. Reservoir Sampling, as it is not affected by the summary size, does not show a correlation to K. Figure 3c depicts the average utilities where a higher value corresponds to a better approach. Clearly the results differ more for larger K. SoftSieving lies between Sieve-Streaming and ThreeSieves, but it constantly outperforms ThreeSieves. As expected, increasing the summary size leads to increased latency and processing time, while increasing the utility. Sieve-Streaming and Sieve-Streaming++ are the most dependent on K having latencies orders of magnitude larger compared to SoftSieving. ThreeSieves and Reservoir Sampling are barely affected by K and produce the best latencies. However, they produce the lowest utility.

Varying Number of Sieves (ε): By increasing ε, the number of sieves decreases and with that the latency for all sieve-based approaches (Fig. 4a). This is most notable for Sieve-Streaming and Sieve-Streaming++. SoftSieving achieves lower total latency than ThreeSieves, which is a direct consequence of the improved parallelization. The local processing times follow the same trend

(Fig. 4b). Sieve-Streaming and Sieve-Streaming++ have high processing times for low ε. The utility values in Fig. 4c show the dependence of Sieve-Streaming and Sieve-Streaming++ on the number of sieves. Their utility decreases when ε increases and fewer sieves are used. SoftSieving shows a slight increase with increasing ε, while ThreeSieves has lower utility for both $\varepsilon = 0.001$ and $\varepsilon = 0.1$. The number of sieves does not change for ThreeSieves, but a lower ε means that, with constant T, it takes longer to decrease the threshold to a suitable value. The extreme case is shown for $\varepsilon = 0.001$, where no items were added to the core item set since the threshold remained too high.

Varying Window Size (w): Different window sizes show how quickly the approaches build a high-quality summary and how much they improve on their

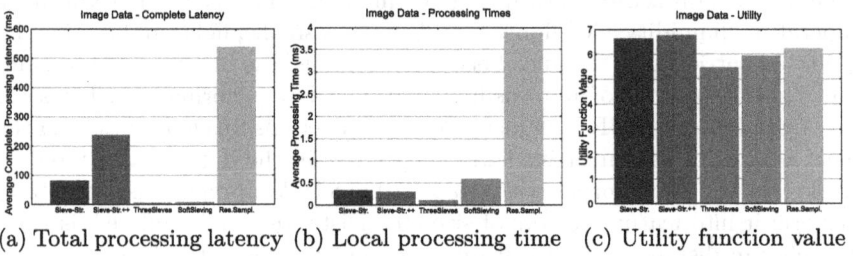

(a) Total processing latency (b) Local processing time (c) Utility function value

Fig. 6. Image dataset (standard parameters)

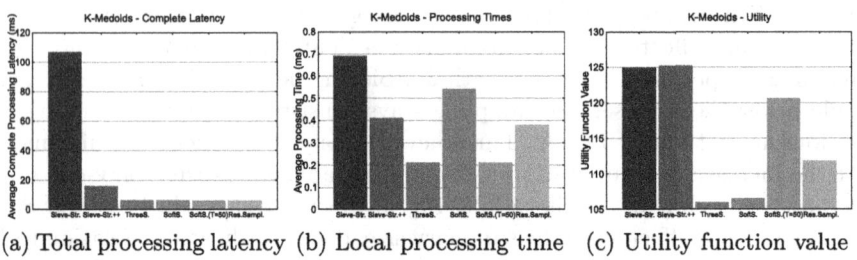

(a) Total processing latency (b) Local processing time (c) Utility function value

Fig. 7. K-Medoids utility function (standard parameters)

summaries over time. The total latency stays consistent for most approaches when varying w, with the exception of ThreeSieves for $w = 10000$ (Fig. 5a). The local processing times (Fig. 5b) of Sieve-Streaming/++ and SoftSieving show a decrease when increasing w. The cause for this can be the increase in the number of sieves that have completed their summary and no longer require processing over the window. For the utility (Fig. 5c), there is an increase for larger w for all approaches. Interestingly, this includes reservoir sampling. This may indicate that for our dataset, the data changes over time, and evenly distributed sampling naturally includes these changes. SoftSieving achieves higher utility as w increases such that it reaches the utility of Sieve-Streaming and Sieve-Streaming++ for $w = 100000$ while constantly outperforming ThreeSieves.

Image Dataset: We evaluate on the images dataset using the standard parameters and the active set utility function. For the complete processing latency (Fig. 6a), the approaches show a similar distribution to the sensor data, with the exception of Reservoir Sampling. When looking at the local processing times (Fig. 6b) it is apparent that the cause for this is the slow sampling in the respective bolt. The utility values (Fig. 6c) show SoftSieving being lower than Reservoir Sampling, with ThreeSieves being the lowest by a large margin. This shows the impact of tuning T, or parameters like ε. Consequently, they are surpassed by random sampling in utility since they are too restrictive with their thresholds.

K-Medoids Utility Function: We evaluated the k-medoids utility function on the telemetry dataset with the standard parameters. All topologies include a SamplingBolt, collecting sample sets of size 50 to use for the per window utility calculations. The complete processing latency (Fig. 7a) is low for all approaches except Sieve-Streaming. This is because the effects of having more sieves, and therefore more utility function evaluations, are stronger with k-medoids. The local processing times (Fig. 7b) of SoftSieving are higher than ThreeSieves. Since the total latency remains low, this indicates effective partitioning of the data stream. SoftSieving with $T = 50$ was included to highlight the importance of T. The utility of SoftSieving (Fig. 7c), although higher than ThreeSieves, is low compared to the other approaches. However, the high utility for $T = 50$ shows that this is a result of the choice of T. In conclusion, the results are comparable to active sets, but the choice of T can greatly influence our approach.

6   Conclusion

We investigated the problem of continuously extracting core-item–based summaries from a data stream. Specifically, we looked at an edge setting, where finding such summaries can save network load and speed up centralized applications that depend on the edge data. We proposed a new algorithm for data stream summarization using core items, called SoftSieving. It uses a fixed, low number of dynamic sieves and enables parallelized processing. The summaries are soft, meaning that core items can be swapped for ones with greater utility gain. We compared the performance to the state-of-the-art sieve-based algorithms in an extensive experimental evaluation and showed that our approach achieves acceptable balance between fast processing and high utility.

Acknowledgments. This work has been partially funded by the German Federal Ministry of Education and Research under grant number 28DE113C18 (DigiVine). The responsibility for the content of this publication lies with the authors.

References

1. Apache storm (2011). https://storm.apache.org/. Accessed 09 Apr 2022
2. Aggarwal, C.C.: On biased reservoir sampling in the presence of stream evolution. In: VLDB, pp. 607–618 (2006)
3. Badanidiyuru, A., Mirzasoleiman, B., Karbasi, A., Krause, A.: Streaming submodular maximization: massive data summarization on the fly. In: KDD (2014)
4. Brown, G., Pocock, A.C., Zhao, M., Luján, M.: Conditional likelihood maximisation: a unifying framework for information theoretic feature selection. J. Mach. Learn. Res. **13**, 27–66 (2012)
5. Buchbinder, N., Feldman, M., Schwartz, R.: Online submodular maximization with preemption. In: SIAM, pp. 1202–1216 (2014)
6. Buschjäger, S., Honysz, P.-J., Pfahler, L., Morik, K.: Very fast streaming submodular function maximization. In: Oliver, N., Pérez-Cruz, F., Kramer, S., Read, J., Lozano, J.A. (eds.) ECML PKDD 2021. LNCS (LNAI), vol. 12977, pp. 151–166. Springer, Cham (2021). https://doi.org/10.1007/978-3-030-86523-8_10
7. Buschjäger, S., Morik, K., Schmidt, M.: Summary extraction on data streams in embedded systems. In: IOTSTREAMING@ PKDD/ECML (2017)
8. Chen, J., Zhang, Q.: Distinct sampling on streaming data with near-duplicates. In: PODS, pp. 369–382 (2018)
9. Dasgupta, A., Kumar, R., Ravi, S.: Summarization through submodularity and dispersion. In: ACL, vol. 1, pp. 1014–1022 (2013)
10. Feige, U.: A threshold of ln n for approximating set cover. J. ACM (JACM) **45**(4), 634–652 (1998)
11. Feldman, M., Norouzi-Fard, A., Svensson, O., Zenklusen, R.: The one-way communication complexity of submodular maximization with applications to streaming and robustness. In: SIGACT, pp. 1363–1374 (2020)
12. Fu, X., Ghaffar, T., Davis, J.C., Lee, D.: {EdgeWise}: a better stream processing engine for the edge. In: USENIX ATC, pp. 929–946 (2019)
13. Indyk, P., Mahabadi, S., Mahdian, M., Mirrokni, V.S.: Composable core-sets for diversity and coverage maximization. In: PODS, pp. 100–108 (2014)
14. Kaufman, L., Rousseeuw, P.J.: Partitioning around medoids (program pam). In: Finding Groups in Data: An Introduction to Cluster Analysis. Wiley Series in Probability and Statistics, vol. 344, pp. 68–125 (1990)
15. Kazemi, E., Mitrovic, M., Zadimoghaddam, M., Lattanzi, S., Karbasi, A.: Submodular streaming in all its glory: tight approximation, minimum memory and low adaptive complexity. In: ICML, pp. 3311–3320 (2019)
16. Lawrence, N., Seeger, M., Herbrich, R.: Fast sparse gaussian process methods: the informative vector machine. In: NIPS, vol. 15 (2002)
17. Leskovec, J., Krause, A., Guestrin, C., Faloutsos, C., VanBriesen, J., Glance, N.: Cost-effective outbreak detection in networks. In: SIGKDD, pp. 420–429 (2007)
18. Lin, H., Bilmes, J.: Multi-document summarization via budgeted maximization of submodular functions. In: HLT-NAACL, pp. 912–920 (2010)
19. Lin, H., Bilmes, J.: A class of submodular functions for document summarization. In: HLT, pp. 510–520 (2011)
20. Luber, M., Spinello, L., Arras, K.: People tracking in RGB-D data with on-line boosted target models. In: IROS, pp. 3844–3849 (2011)
21. Mirrokni, V.S., Zadimoghaddam, M.: Randomized composable core-sets for distributed submodular maximization. In: STOC, pp. 153–162 (2015)

22. Mirzasoleiman, B., Karbasi, A., Sarkar, R., Krause, A.: Distributed submodular maximization: Identifying representative elements in massive data. In: NIPS, vol. 26 (2013)
23. Nemhauser, G.L., Wolsey, L.A., Fisher, M.L.: An analysis of approximations for maximizing submodular set functions-i. Math. Program. **14**(1), 265–294 (1978)
24. Seeger, M.: Greedy forward selection in the informative vector machine. Technical report, UC Berkeley (2004)
25. Spinello, L., Arras, K.O.: People detection in RGB-D data. In: RSJ, pp. 3838–3843 (2011)
26. Stafford, G.A.: Environmental sensor telemetry data (2020). https://www.kaggle.com/datasets/garystafford/environmental-sensor-data-132k. Accessed 28 Apr 2022
27. Vitter, J.S.: Random sampling with a reservoir. ACM TOMS **11**(1), 37–57 (1985)
28. Williams, C.K., Rasmussen, C.E.: Gaussian Processes for Machine Learning, vol. 2. MIT press, Cambridge (2006)
29. Zhao, J., Wang, P., Tao, J., Zhang, S., Lui, J.C.: Continuously tracking core items in data streams with probabilistic decays. In: ICDE, pp. 769–780 (2020)
30. Zhuang, H., Rahman, R., Hu, X., Guo, T., Hui, P., Aberer, K.: Data summarization with social contexts. In: CIKM, pp. 397–406. ACM (2016)

Parallel Techniques for Variable Size Segmentation of Time Series Datasets

Lamia Djebour$^{(\boxtimes)}$, Reza Akbarinia, and Florent Masseglia

Inria, University of Montpellier, CNRS, LIRMM, Montpellier, France
{lamia.djebour,reza.akbarinia,florent.masseglia}@inria.fr

Abstract. Given the high data volumes in time series applications, or simply the need for fast response times, it is usually necessary to rely on alternative, shorter representations of these series, usually with loss. This incurs approximate comparisons of time series where precision is a major issue. In this paper, we propose a new parallel approach for segmenting time series before their transformation into symbolic representations. It can reduce significantly the error incurred by possible splittings at different steps of the representation calculation, by taking into account the sum of squared errors (SSE). This is particularly useful for time series similarity search, which is the core of many data analytics tasks. We provide theoretical guarantees on the lower bound of similarity measures, and our experiments illustrate that our technique can improve significantly the time series representation quality.

Keywords: Time series · Representations · Information retrieval

1 Introduction

Time series have attracted an increasing interest due to their wide applications in many domains. The continuous flow of emitted data may concern personal activities (*e.g.*, through smart-meters or smart-plugs for electricity or water consumption) or professional activities (*e.g.*, for monitoring heart activity or through the sensors installed on plants by farmers). This results in the production of large and complex data, usually in the form of time series [1,2,4–7,11] that challenges knowledge discovery.

As a consequence of the high data volumes in such applications, similarity search can be slow on raw data. One of the issues that hinder the analysis of such data is the high dimensionality. This is why time series approximation is often used as a means to allow fast similarity search. SAX [8] is one of the most popular time series representations, allowing dimensionality reduction on the classic data mining tasks. SAX constructs symbolic representations by splitting the time domain into segments of equal size where the mean values of segments represent the time series intervals (PAA approach). This approximation technique is effective for time series having a uniform and balanced distribution over the time domain. However, we observe that, in the case of time series having high variation over given time intervals, this division into segments of fixed length is not efficient. Our main contribution is to provide an adaptive interval distribution, rather than an equal distribution in time. However, the number of possible segmentations of k segments with n can be very high. Furthermore, when searching for the best variable-size segmentation, a large number of computation is involved in case of large sets of time series. Therefore, we propose efficient parallel techniques using GPUs for improving the execution time of our segmentation algorithm. In this paper, we make the following contributions:

- We propose a new representation technique, called ASAX_SSE, that allows obtaining a variable-size segmentation of time series with better precision in retrieval tasks thanks to its lower information loss. Our representation is based on SSE measurement for detecting what time intervals should be split.
- We propose a lower bounding method that allows approximating the distance between the original time series based on their representations in ASAX_SSE.
- We propose efficient parallel algorithms for improving the execution time of our segmentation approach using GPUs.
- We implemented our approach and conducted empirical experiments using more than 120 real world datasets. The results suggest that ASAX_SSE can obtain significant performance gains in terms of precision for similarity search compared to SAX. They illustrate that the more the data distribution in the time domain is unbalanced (non-uniform), the greater is the precision gain of ASAX_SSE. For example, for the *ECGFiveDays* dataset that has a non-uniform distribution in the time domain, the precision of ASAX_SSE is 93% compared to 55% for SAX.

The rest of the paper is organized as follows. In Sect. 2, we define the problem we address. In Sect. 3, we describe the details of ASAX_SSE representation, and in Sect. 4 we present parallel versions of ASAX_SSE. In Sect. 5, we present the experimental evaluation of our approach. Finally, we discuss the related work in Sect. 6 and give our conclusion in Sect. 7.

2   Problem Definition and Background

We first present the background about SAX representation, and then define the problem we address. A time series X is a sequence of values $X = \{x_1, ..., x_n\}$. We assume that every time series has a value at every timestamp $t = 1, 2, ..., n$. The length of X is denoted by $|X|$.

SAX allows a time series T of length n to be reduced to a string of arbitrary length w.

2.1   SAX Representation

Given two time series $X = \{x_1, ..., x_n\}$ and $Y = \{y_1, ..., y_n\}$, the Euclidean distance between X and Y is defined as [6]: $ED(X, Y) = \sqrt{\sum_{i=1}^{n}(x_i - y_i)^2}$. The

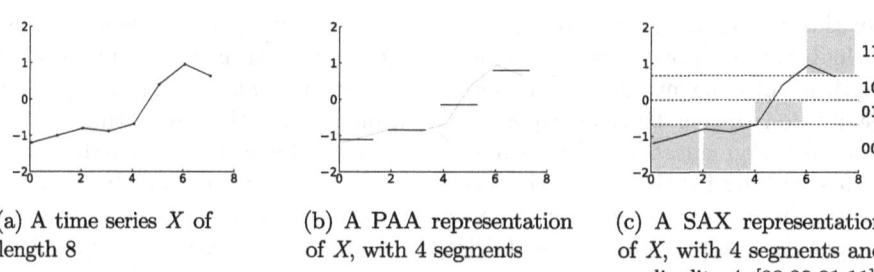

(a) A time series X of length 8

(b) A PAA representation of X, with 4 segments

(c) A SAX representation of X, with 4 segments and cardinality 4, [00,00,01,11]

Fig. 1. A time series X is discretized by obtaining a PAA representation and then using predetermined break-points to map the PAA coefficients into SAX symbols. Here, the symbols are given in binary notation, where 00 is the first symbol, 01 is the second symbol, etc. The time series of Fig. 1a in the representation of Fig. 1c is [first, first, second, fourth] (which becomes [00, 00, 01, 11] in binary).

Euclidean distance is one of the most popular similarity measurement methods used in time series analysis.

The SAX representation is based on the PAA representation [8] which allows for dimensionality reduction while providing the important lower bounding property as we will show later. The idea of PAA is to have a fixed segment size, and minimize dimensionality by using the mean values on each segment. Example 1 gives an illustration of PAA.

Example 1. Figure 1b shows the PAA representation of X, the time series of Fig. 1a. The representation is composed of $w = |X|/l$ values, where l is the segment size. For each segment, the set of values is replaced with their mean. The length of the final representation w is the number of segments (and, usually, $w << |X|$).

By transforming the original time series X and Y into PAA representations $\overline{X} = \{\overline{x}_1, ..., \overline{x}_w\}$ and $\overline{Y} = \{\overline{y}_1, ..., \overline{y}_w\}$, the lower bounding approximation of the Euclidean distance for these two representations can be obtained by:
$$DR_f(\overline{X}, \overline{Y}) = \sqrt{\frac{n}{w}} \sqrt{\sum_{i=1}^{w} (\overline{x}_i - \overline{y}_i)^2}$$
The SAX representation takes as input the reduced time series obtained using PAA. It discretizes this representation into a predefined set of symbols, with a given cardinality, where a symbol is a binary number. Example 2 gives an illustration of the SAX representation.

Example 2. In Fig. 1c, we have converted the time series X to SAX representation with size 4, and cardinality 4 using the PAA representation shown in Fig. 1b. We denote SAX(X) = [00, 00, 01, 11].

The lower bounding approximation of the Euclidean distance for SAX representation $\hat{X} = \{\hat{x}_1, ..., \hat{x}_w\}$ and $\hat{Y} = \{\hat{y}_1, ..., \hat{y}_w\}$ of two time series X and Y is defined as: $MINDIST_f(\hat{X}, \hat{Y}) = \sqrt{\frac{n}{w}} \sqrt{\sum_{i=1}^{w} (dist(\hat{x}_i, \hat{y}_i))^2}$ where the function $dist(\hat{x}_i, \hat{y}_i)$ is the distance between two SAX symbols $\hat{x}_i$ and $\hat{x}_i$. The lower bounding condition is formulated as: $MINDIST_f(\hat{X}, \hat{Y}) \leq ED(X, Y)$

2.2   Similarity Queries

The problem of similarity queries is one of the main problems in time series analysis and mining. In information retrieval, finding the k nearest neighbors (k-NN) of a query is a fundamental problem. Let us define *exact* and *approximate* k nearest neighbors.

Definition 1 (Exact k nearest neighbors). *Given a query time series Q and a set of time series D, let $R = ExactkNN(Q, D)$ be the set of k nearest neighbors of Q from D. Let $ED(X, Y)$ be the Euclidean distance between two time series X and Y, then the set R is defined as follows:*

$$(R \subseteq D) \wedge (|R| = k) \wedge (\forall a \in R, \forall b \in (D - R), ED(a, Q) \leq ED(b, Q))$$

Definition 2 (Approximate k nearest neighbors). *Given a set of time series D, a query time series Q, and $\epsilon > 0$. We say that $R = AppkNN(Q, D)$ is the approximate k nearest neighbors of Q from D, if $ED(a, Q) \leq (1+\epsilon)ED(b, Q)$, where a is the kth nearest neighbor from R and b is the true k^{th} nearest neighbor.*

2.3   Problem Statement

The SAX representation proceeds to an approximation by minimizing the dimensionality: the original time series are divided into segments of equal size. This representation does not depend on the time series values, but on their length. It allows SAX to perform the segmentation in $O(n)$ where n is the time series length. However, for a given reduction in dimensionality, the modeling error may not be minimal since the model does not adapt to the information carried by the series.

Our goal is to propose a variable-size segmentation of the time domain that minimizes the loss of information in the time series representation. Formally, the problem we address is stated as follows. Given a database of time series D and a number w, divide the time domain into w segments of variable size such that the representation of the times series based on that segmentation lowers the error of similarity queries.

3   Adaptive SAX Based on the Representation's Sum of Squared Errors (ASAX_SSE)

We propose ASAX_SSE, a variable-size segmentation technique for time series representation. To create a segmentation with minimum information loss on time series approximation, ASAX_SSE divides the time domain based on the Sum of squared errors (SSE) value of the representation.

In the rest of this section, we first describe the notion of Sum of Squared Errors (SSE) for the time series representation. Then, we describe our algorithm for creating the variable-size segments based on this measurement. Finally, we present our method for measuring the lower bound distance between time series in the proposed representation. This lower bounding is useful for efficient evaluation of kNN queries.

3.1 Sum of Squared Errors (SSE)

In Statistics, Sum of Squared Errors (SSE) is defined as the sum of the squares of the errors. In other words, SSE is the sum of the squared differences between the actual and the estimated values. Formally, SSE is defined as follows.

Definition 3. *Given a vector X of n elements and a vector $\tilde{X}$ being the estimated values generated from X, SSE of the estimation is given by: $SSE(X, \tilde{X}) = \sum_{i=1}^{n} (x_i - \tilde{x}_i)^2$*

In our context, we calculate the SSE on the PAA representation obtained from the transformation of the original time series of a dataset according to a given segmentation. The SSE computed on this representation allows to measure the approximation error on the time series by the PAA representation compared to the original time series. The lower the SSE, the closer is the PAA representation to original data.

By transforming a time series $X = \{x_1, ..., x_n\}$ into a PAA representation $\overline{X} = \{\overline{x}_1, ..., \overline{x}_w\}$, X is reduced to the PAA representation composed of w segments. For each segment, the set of values is replaced with their mean. We can compute the SSE for each segment, that is in this case, the sum of the squared differences between each value (actual value) and its segment's mean (estimated value). In the next subsections, we show how to compute the SSE of a PAA representation considering only one segment (called LSSE) or all segments (called GSSE). As shown by experiments, using these two different SSE measurements may lead to different results in terms of precision and execution time.

3.2 SSE of PAA Representation Considering One Segment (LSSE)

Let $\overline{X}$ be the PAA representation of X with w segments. The LSSE (local SSE) of $\overline{X}$ for a particular segment is the sum of the squared errors for the time series values in this segment. Formally, LSSE of $\overline{X}$ for a segment s_i is computed as: $LSSE(s_i, \overline{x}_i) = \sum_{j=LB(s_i)}^{UB(s_i)} (x_j - \overline{x}_i)^2$ where s_i is the selected segment, $LB(s_i)$ and $UB(s_i)$ are the start and end time points of s_i respectively.

3.3 SSE of PAA Representation Considering All Segments (GSSE)

The global SSE (GSSE), is computed by taking into account all segments of the PAA representation $\overline{X}$: $GSSE(X, \overline{X}) = \sum_{i=1}^{w} \sum_{j=LB(s_i)}^{UB(s_i)} (x_j - \overline{x}_i)^2$ where $LB(s_i)$ and $UB(s_i)$ are the start and end time points of the segment s_i respectively.

Algorithm 1: Variable-size segmentation

Input: D: time series database; n: the length of time series; $size$: the starting size of segments; w: the required number of segments

Output: w variable-size segments

1   $k = \lceil \frac{n}{size} \rceil$

2   $segments = \{\bigcup_{i=0}^{k-1}[size \times i, size \times (i+1) - 1]\}$ // split time domain into k segments of size $size$

3   **while** $k \neq w$ **do**

4      $segmentsToMerge = null$

5      $msse = \infty$

6   **for** $i=1$ to $k-1$ **do**

7         $s = $ merge (s_i, s_{i+1})

8         $tempSegments = segments - \{s_i, s_{i+1}\}$

9         $tempSegments = tempSegments \bigcup s$

10   //merge segment i and segment $i+1$ in $tempSegments$

11        $sse = 0$

12   **foreach** ts in D **do**

13           $sse = sse + SSE(ts)$

14   **if** $sse < msse$ **then**

15           $segmentsToMerge = i$

16           $msse = sse$

17      $s = $ merge $(s_{segmentsToMerge}, s_{segmentsToMerge+1})$

18      $segments = segments - \{s_{segmentsToMerge}, s_{segmentsToMerge+1}\}$

19      $segments = segments \bigcup s$

20      $k = k\text{-}1$

21   **return** $segments$

3.4   Variable-Size Segmentation Based on SSE Measurement

Given a database of time series D, and a number w, our goal is to find the k variable size segments that minimize the loss of information in time series representations by minimizing the approximation error of these representations.

Intuitively, our algorithm works as follows. Based on a starting segment size value $size$, it firstly splits the time domain into k segments of length $size$. The default value of $size$ is 2. The algorithm performs $k - w$ iterations, and in each iteration it finds the two adjacent segments s_i and s_{i+1} whose merging gives the minimum SSE (MSSE) on the representations, and merges them. By doing this, in each iteration the two selected segments are merged to form a single segment which replaces them in the set of segments, reducing the number of segments by one. This continues until having w segments.

Let us now describe our algorithm in more details. The pseudocode is shown in Algorithm 1. It first sets the current number of segments, denoted as k, to $\frac{n}{size}$. Then, it splits the time domain into k segments of length $size$ that are included to the set $segments$ (Line 2).

Afterwards, in a loop, until the number of segments is more than w the algorithm proceeds as follows. For each segment s_i (i from 1 to $k - 1$), s_i is merged with segment s_{i+1} to form a single segment denoted as s (Line 7). Then, a temporary set of segments $tempSegments$ is created including the new segment and all previously created segments except s_i and s_{i+1} i.e., except the two that have been merged (Lines 8, 9). Then, for each time series ts in the database D, the algorithm generates its PAA representation and calculates the corresponding

SSE (Line 13) calling either GSSE function in the case that the entire PAA representation is considered for the error calculation, or LSSE function if the error is computed on segment s. Then, it adds the result of the computed SSE to sse (Line 13). After having calculated the sum of the SSE for the PAA representation of all the time series contained in D, if the SSE is less than the MSSE (minimum SSE) obtained so far, the algorithm sets i as the segment to be merged with the next one, and keeps the SSE of the representation (Lines 15, 16). This procedure continues by trying the merging of every two adjacent segments of $segments$ at each time, and computing the SSE. The algorithm selects the merging whose SSE is the lowest, and updates the set of the segments by removing the selected segments, and inserting its merging to $segments$ (Lines 17–19). Then, k, which stands for the number of current segments, is decremented by one (Line 20). The algorithm ends when k gets equal to the required number, $i.e.$, w.

Let us illustrate the principle of our algorithm using an example. For simplicity, we consider a dataset containing only a single time series and we calculate the approximation error on the entire time series representation using GSSE approach.

Example 3. Let us apply our algorithm on the time series X in Fig. 2 by taking the initial size of 2 for the segments. The algorithm starts by dividing the time domain into 4 segments of size 2. The next step is to reduce the number of segments from 4 to 3. To this purpose, the algorithm tests the merging of every two adjacent segments of the 4 existing segments, in order to find the one that has the minimum SSE. Three different scenarios are possible:

Scenario 1: The first scenario is shown in Fig. 3a where s_1 and s_2 of the initial segmentation (shown in Fig. 2) are merged into one segment. We generate the PAA representation of X using the 3 segments, and then compute the SSE of this representation that is $SSE_1(X, \overline{X}) \approx 1.167$.

Scenario 2: This scenario is shown in Fig. 3b in which s_2 and s_3 of the initial segmentation are merged. As for Scenario 1, we generate the PAA representation of X using the current segmentation. Here, $SSE_2(X, \overline{X}) \approx 1.915$.

Scenario 3: The last scenario is shown in Fig. 3b, where we merge s_3 and s_4. For this segmentation, $SSE_3(X, \overline{X}) \approx 1.745$.

We have calculated the SSE for the three scenarios. Since we aim to minimize the SSE, we have to choose the minimum SSE value (MSSE), that is $MSSE = 1.167$ corresponding to the segmentation generated in Scenario 1. The latter is chosen for this iteration of our algorithm and we continue the next iterations, until the number of segment reaches w.

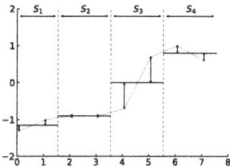

Fig. 2. PAA representation of a time series X of length 8 with 4 segments.

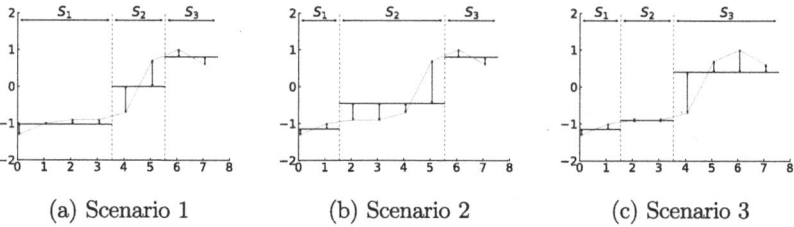

(a) Scenario 1   (b) Scenario 2   (c) Scenario 3

Fig. 3. The three different scenarios of ASAX_SSE segmentation with 3 segments. Scenario 1 is the one chosen because it provides the MSSE.

3.5 Lower Bounding of the Similarity Measure

SAX [9] defines a distance measure on the PAA representation of time series as described in Sect. 2.1. Given the representation of two time series, the DR_f function allows obtaining a lower bounding approximation of the Euclidean distance between the original time series. By the following theorem, we propose a lower bounding approximation formula for the case of variable size segmentation in ASAX_SSE.

Theorem 1. *Let X and Y be two time series. Suppose that by using ASAX_SSE we create a variable size segmentation with w segments, such that the size of the ith segment is l_i. Let $\overline{X}$ and $\overline{Y}$ be the representations of X and Y in ASAX_SSE. Then, $DR_v(\overline{X}, \overline{Y})$ gives a lower bounding approximation of the Euclidean distance between X and Y: $DR_v(\overline{X}, \overline{Y}) = \sqrt{\sum_{j=1}^{w} ((\overline{x}_j - \overline{y}_j)^2 \times l_j)}$*

Proof: The proof has been removed due to lack of space.

4   Parallel Versions of ASAX_SSE

We propose efficient parallel techniques using GPUs for improving the execution time of our segmentation algorithm. In our approach, the CPU controls the main loop of the segmentation computation process and does light operations, while the time-consuming tasks are parallelized on GPU, particularly the SSE computation on a dataset for a given segmentation. We propose two parallel versions of the algorithm using CUDA framework to provide a fast computation of the variable-size segmentation over long time series and/or large number of time series: 1) ASAX_DP that performs the parallelization on data; 2) ASAX_SP that makes the parallelization on segments.

4.1   Parallelization on Data

The main idea of our first parallel algorithm, called ASAX_DP (ASAX Data Parallel), is to divide the dataset into blocks (partitions), and to assign the SSE computation for the time series of each block to a core of the GPU.

Let us describe the proposed algorithm. Initially, the host (CPU) sends the whole dataset D to the GPU (this data transfer between the CPU and the GPU is done only once). Then, the host creates the initial segmentation *segments* by splitting the time domain into the k starting segments. Afterwards, in a loop, until the number of segments is more than w, it generates a candidate segmentation by merging 2 segments of the last validated segmentation. For each candidate segmentation, the GPU is used for computing SSE on D. For this, the host calls the GPU kernel that computes SSE in parallel operating on different time series of the different dataset blocks. In the kernel, each thread calculates the SSE on the time series of its block and stores the result in a shared array, called *sseArray*, that is sent back to the CPU. The host calculates the sum of the received results to get the SSE on D, and updates the MSSE (minimum SSE) if the SSE obtained in this iteration is less than the MSSE obtained until now. After testing all possible segmentations, it chooses the one that has the minimum SSE, updates the set of segments *segments* and decrements the current number of segments k by one. This process continues until k reaches the required number of segments w.

4.2   Parallelization on Segments

Here, we propose ASAX_SP (ASAX Segment in Parallel), a parallel algorithm in which the computations related to each possible merging of segments is done by a different GPU core. As shown by our experiments, this algorithm can be more efficient than the one presented previously in the cases where the time series are long (e.g., more than 1000 values per time series).

The initialization of this algorithm is the same as the algorithm presented in the previous subsection. The host starts by sending the dataset D to the GPU, and dividing the time domain into k starting segments to form the set *segments*.

Then, until the number of segments has not reached w, the host calls the GPU kernel to compute SSE on D of each possible segmentation in parallel. The number of launched threads is equal to the number of possible segmentations obtained when reducing the number of segments from k to $k - 1$. In the kernel, each thread calculates its segmentation by merging two segments s_i and s_{i+1} where i is the thread position. The thread computes the SSE of the segmentation on the dataset D and stores the result in a shared array, called *sseArray*, according to its position. The result array is sent back to the CPU. Each element of the array represents the SSE for a candidate segmentation. The host selects the one having the lowest SSE value, and then updates *segments* and k. This process continues until k reaches w.

5   Experiments

In this section, we present the experimental evaluation of ASAX_SSE. We first describe the experimental setup. Then, in Subsect. 5.2, we compare the precision of ASAX_SSE representation with that of the existing SAX representation. Finally, in Subsect. 5.3, we evaluate the performance of the parallel versions of ASAX_SSE by measuring the execution time of the variable-size segmentation using GPUs.

5.1   Setup

All approaches are implemented with Python programming language. ASAX_SSE and SAX implementations use Numba JIT compiler to optimize machine code at runtime. The GPU-based part of the parallel algorithms is written in Numba[1].

The ASAX_SSE and SAX experiments were conducted on a machine using Ubuntu 18.04.5 LTS operating system with 20 Gigabytes of main memory, and an Intel Xeon(R) 3,10 GHz processor with 4 cores. The parallel experimental evaluation was conducted on an NVIDIA GeForce RTX 2080 Ti GPU card, equipped with 4 352 CUDA cores and 11 GB of memory installed in the same machine. We compare the proposed ASAX_SSE and SAX in terms of precision on all the real-world datasets available in the UCR Time Series Classification Archive[2]. We evaluate the performance of the parallel algorithms on two datasets taken from the same archive, the size of the datasets is increased to reach 1M by repeating the contained time series multiple times. For each approach, the length w of the approximate representations is reduced to 10% of the original time series length and the variable-size segmentation algorithms are initialized by splitting the time domain into segments of length 2.

5.2   Precision of k-Nearest Neighbor Search

We compare the quality of ASAX_SSE and SAX representation on all 128 datasets of the UCR Time Series Classification Archive. For each dataset, we measure the precision of the approximate k-NN search as the average precision for a set of arbitrary random queries taken from this dataset. The search precision for each query Q from a dataset D is calculated as: $p = \frac{|AppkNN(Q,D) \cap ExactkNN(Q,D)|}{k}$ where $AppkNN(Q,D)$ and $ExactkNN(Q,D)$ are the sets of approximate k nearest neighbors and exact k nearest neighbors of Q from D, respectively. $AppkNN(Q,D)$ is obtained using the DR_f distance measure for SAX and DR_v for the ASAX_SSE representation, while $ExactkNN(Q,D)$ contains the exact k-NN results of Q using the euclidean distance ED. $AppkNN(Q,D)$ and $ExactkNN(Q,D)$ use a linear search that consists in computing the distance from the query point Q to every other point in D, keeping track of the "best so far" result.

---

[1] Our code is available at: https://github.com/lamiad/ASAX_SSE.
[2] https://www.cs.ucr.edu/eamonn/time_series_data_2018/.

The precision results are reported in Fig. 4 where the precision gain/loss (as percentage) for ASAX_SSE compared to SAX precision is measured for each dataset. The integer part of the obtained precision is taken into consideration to compare the two methods. Figure 4a shows the precision results for ASAX_GSSE (*i.e.*, ASAX_SSE using GSSE) and Fig. 4b those for ASAX_LSSE (*i.e.*, ASAX_SSE using LSSE). The results are illustrated using a scatter chart where the horizontal axis represents the dataset number and the vertical axis shows the precision gain/loss obtained. We observe a gain in precision for the large majority of datasets. We obtained a gain in precision for 80% of the datasets with ASAX_GSSE and 84% with ASAX_LSSE.

The distribution of time series over the time domain varies from one dataset to another. There are some for which the distribution is quite balanced, those which undergo some variations and others whose variation increases a lot. Figure 4 does not allow explaining the precision gain or loss since we need to have the visualisation of the time series for each datasets, for this, an analysis is done regarding the precision results obtained and the shape of data. We have noticed that the more the distribution of the data is unbalanced the more the gain is important. The maximum gain achieved is a significant 38% for both ASAX_GSSE and ASAX_LSSE methods, obtained for the *ECGFiveDays* dataset. This high gain is due to the unbalanced data distribution over the time domain on this dataset. We were able to achieve a precision of 93% for ASAX_SSE while it is 55% for SAX, because ASAX_SSE performed a better distribution of the segments according to information gain by creating several segments in the parts that undergo a significant variation that produces more accurate times series representations leading to a better result for the approximate k-NN search. We can see that for some datasets the computed gain is zero meaning equivalent precision for ASAX_SSE and SAX due to the balanced shape of the time series over the time domain. Regarding the few datasets where we obtain lower precision, the loss is relatively low (mostly near zero).

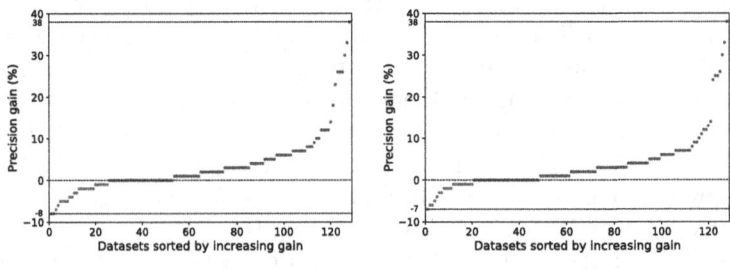

(a) Precision gain for ASAX_GSSE   (b) Precision gain for ASAX_LSSE

Fig. 4. The precision gain for ASAX_GSSE and ASAX_LSSE compared to SAX. The obtained gain is up to 38% for both methods

Globally, our results suggest the effectiveness of our approach and its advantage over the state of the art when applied to time series especially those with unbalanced distribution over the time domain.

5.3   Scalability

This subsection presents the time cost of the variable-size segmentation for our proposed algorithms. We measure the variable-size segmentation time costs of the parallel algorithms ASAX_DP and ASAX_SP, and compare them to that of the variable-size segmentation for the sequential algorithm ASAX_SSE. The percentage of precision gain computed in the experiments described in the previous subsection shows that the gain obtained with the ASAX_GSSE approach is less than the one obtained with ASAX_LSSE. Furthermore, the evaluation of the time cost for ASAX_GSSE approach (sequential and parallel methods) showed that this approach is more time consuming than ASAX_LSSE. For these reasons, we present the results of our parallel algorithms, ASAX_DP and ASAX_SP, only using the LSSE measurement.

Figure 5 and Fig. 6 report the performance gains of our parallel approaches compared to the sequential version of ASAX_LSSE. Figure 5 reports the variable-size segmentation time for the ASAX_DP and ASAX_LSSE with varying dataset size. The computation time increases with the number of time series for both algorithms. But, it is much lower in the case of ASAX_DP than that of the sequential ASAX_LSSE. The performance gains vary significantly depending on the number of time series. As seen, the gain reaches ×45 for 1M of time series.

Figure 6 reports the computation time of variable-size segmentation for the ASAX_SP and ASAX_LSSE. Here we vary the time series length. The running time increases with the length of time series and, as one could expect, the sequential ASAX_LSSE takes much more time than ASAX_SP. Depending on time series length, ASAX_SP shows performance gains that can reach ×24 for 1000 time series of length 2700.

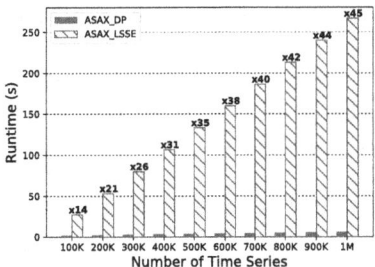

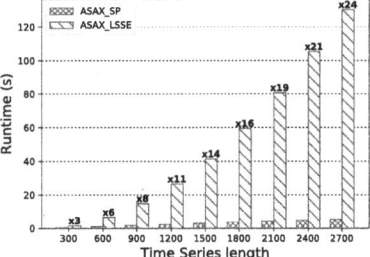

Fig. 5. Variable-size segmentation time for ASAX_DP and ASAX_LSSE as a function of dataset size. The original time series are of length 130.

Fig. 6. Variable-size segmentation time for ASAX_SP and ASAX_LSSE as a function of time series length. The dataset size is fixed to 1000.

Figure 7 and Fig. 8 compare the parallel segmentation computation time
of our approaches. In Fig. 7, we evaluate the two approaches with varying
dataset size (number of time series) and fixed time series length. For this case,
we observe that ASAX_DP is always faster than ASAX_SP. The results show
that using ASAX_DP is advantageous in the case of databases of many small
time series. In Fig. 8, we vary the time series length and we fix the dataset size
for the evaluation. We notice that when time series length $n = 100$, ASAX_DP
is a little faster than ASAX_SP, but when the length of time series increases,
ASAX_SP becomes faster than ASAX_DP. The performance gain reaches $\times 7.5$
for time series of length 1000. ASAX_SP allows better performance gains when
the database consists of few and long time series.

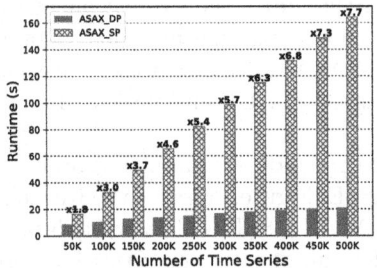 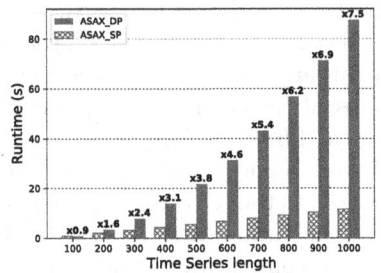

Fig. 7. Comparison of parallel segmentation time using ASAX_DP and ASAX_SP, as a function of dataset size. The original time series are of length 300.

Fig. 8. Comparison of parallel segmentation time using ASAX_DP and ASAX_SP, as a function of time series length. The dataset size is fixed to 10 000.

6   Related Work

Several techniques have been yet proposed to reduce the dimensionality of time
series. Examples of such techniques that can significantly decrease the time and
space required for similarity search are: singular value decomposition (SVD) [6],
the discrete Fourier transformation (DFT) [1], discrete wavelets transformation
(DWT) [4], piecewise aggregate approximation (PAA) [7], random sketches [5],
Adaptive Piecewise Constant Approximation (APCA) [3], and symbolic aggregate approXimation (SAX) [9].

SAX [9] is one of the most popular techniques for time series representation.
It uses a symbolic representation that segments all time series into equi-length
segments and symbolizes the mean value of each segment.

Some extensions of SAX have been proposed for improving the similarity
search performance via indexing [2,11]. For example, iSAX [11] is an indexable
version of SAX designed for indexing large collections of time series. iSAX 2.0 [2]
proposes a new mechanism and also algorithms for efficient bulk loading and
node splitting policy, which is not supported by iSAX index.

There have been SAX extensions designed to improve the representation of each segment, while using the SAX fixed-size segmentation, e.g., [10,12,15]. For example, SAX_TD improves the representation of each segment by taking into account the trend of the time series. It uses the values at the starting and ending points of the segments to measure the trend. TFSA [14] and SAX_CP [13] are other trend-based SAX representation methods. TFSA proposes a representation method for long time series based on the trend, and SAX_CP considers abrupt change points while generating the symbols in order to capture time series' trends.

To increase the quality of time series approximation, we propose an adaptive approach ASAX_SSE based on variable-length segmentation of time series by taking into account the sum of absolute error. Our approach is complementary to the existing SAX extensions, *e.g.*, in indexing based techniques or those that use the trend for representing the segments. For example, our variable-size segmentation can be used in iSAX, SAX_TD and SAX_CP for segmenting the time series.

7   Conclusion

We addressed the problem of approximating time series, and proposed ASAX_SSE, a new technique for segmenting time series before their transformation into symbolic representations. ASAX_SSE can reduce significantly the error incurred by possible splittings at different steps of the representation calculation, by taking into account the sum of squared errors (SSE). We also proposed two parallel algorithms for improving the execution time of ASAX_SSE using GPUs. We evaluated the performance of our segmentation approach through experimentation using more than 120 real world datasets. The results suggest that the more the data distribution in the time domain is unbalanced (non-uniform), the greater is the precision gain of ASAX_SSE. For example, for the *ECGFiveDays* dataset that has a non-uniform distribution in the time domain, the precision of ASAX_SSE is 93% compared to 55% for SAX. Furthermore, the results illustrate the effectiveness of our parallel algorithms, *e.g.*, up to $\times 45$ faster than the sequential algorithm for 1M time series.

References

1. Agrawal, R., Faloutsos, C., Swami, A.: Efficient similarity search in sequence databases. In: Lomet, D.B. (ed.) FODO 1993. LNCS, vol. 730, pp. 69–84. Springer, Heidelberg (1993). https://doi.org/10.1007/3-540-57301-1_5
2. Camerra, A., Shieh, J., Palpanas, T., Rakthanmanon, T., Keogh, E.J.: Beyond one billion time series: indexing and mining very large time series collections with i SAX2+, pp. 123–151. Knowl. Inf, Syst (2014)
3. Chakrabarti, K., Keogh, E., Mehrotra, S., Pazzani, M.: Locally adaptive dimensionality reduction for indexing large time series databases. ACM Trans. Database Syst. **27**(2), 188–228 (2002)

4. Chan, K., Fu, A.W.: Efficient time series matching by wavelets. In: Proceedings of the ICDE (1999)
5. Cole, R., Shasha, D., Zhao, X.: Fast window correlations over uncooperative time series. In: KDD Conference, pp. 743–749 (2005)
6. Faloutsos, C., Ranganathan, M., Manolopoulos, Y.: Fast subsequence matching in time-series databases. In: Proceedins of the SIGMOD (1994)
7. Keogh, E.J., Chakrabarti, K., Pazzani, M.J., Mehrotra, S.: Dimensionality reduction for fast similarity search in large time series databases. Knowl. Inf. Syst. **3**(3), 263–286 (2001)
8. Lin, J., Keogh, E., Lonardi, S., Chiu, B.: A symbolic representation of time series, with implications for streaming algorithms. In: SIGMOD (2003)
9. Lin, J., Keogh, E., Wei, L., Lonardi, S.: Experiencing SAX: a novel symbolic representation of time series, Data Min. Know. Discov. **15**, 107–144. (2007)
10. Lkhagva, B., Suzuki, Y., Kawagoe, K.: New time series data representation ESAX for financial applications. In: ICDE Workshops (2006)
11. Shieh, J., Keogh, E.: iSAX: indexing and mining terabyte sized time series. In: KDD Conference, pp. 623–631 (2008)
12. Sun, Y., Li, J., Liu, J., Sun, B., Chow, C.: An improvement of symbolic aggregate approximation distance measure for time series. Neurocomputing **138**, 189–198 (2014)
13. Yahyaoui, H., Al-Daihani, R.: A novel trend based SAX reduction technique for time series. Exp. Syst. Appl. **130** (2019)
14. Yin, H., Yang, S.q., Zhu, X.q., Ma, S.d., Zhang, L.m.: Symbolic representation based on trend features for knowledge discovery in long time series. Front. Inf. Technol. Electr. Eng. **16**, 744–758 (2015)
15. Zhang, H., Dong, Y., Xu, D.: Entropy-based symbolic aggregate approximation representation method for time series. In: IEEE Joint Int. Information Technology and Artificial Intelligence Conference (ITAIC), pp. 905–909 (2020)

On Line Analytical Processing

ORTree: Tuning Diversified Similarity Queries by Means of Data Partitioning

João Victor de Oliveira Novaes[1(✉)] , Lúcio Fernandes Dutra Santos[2] ,
Agma Juci Machado Traina[1] , and Caetano Traina Jr.[1]

[1] Institute of Mathematics and Computer Sciences, University of São Paulo,
São Carlos, SP 13566-590, Brazil
`novaes.jvo@usp.br`, {`agma,caetano`}`@icmc.usp.br`
[2] Federal Institute of Technology of North of Minas Gerais, Montes Claros,
MG 39404-058, Brazil
`lucio.santos@ifnmg.edu.br`

Abstract. As modern applications gather more and more data, the data types also become more complex. Traditional retrieval operations based on identity and order comparisons are not suitable for those types. Instead, similarity operators are much more interesting for querying complex data and are gaining increasing attention. Similarity queries retrieve the elements most similar to a query center but, they tend to return elements that are very similar to others in the result set, reducing users' interest in the answer. To overcome this problem, researchers have considered incorporating a diversity degree in the similarity operators. Unfortunately, diversified similarity queries are computationally expensive, as they need to assess the relationship between each pair of elements in the result. Several works in the literature present techniques to speed up diversity in similarity queries, but they are either not scalable or only consider the diversity property. In this paper, we propose an index data structure, called the Omni-Range Tree (ORTree), that partitions the query space into a small subset of similar elements to a query element and prospect representative candidates aiming at dispatch diversified similarity queries. Our experimental evaluation shows that our index structure can reduce the query execution by time up to 95% without harming the quality of the results concerning other literature methods.

Keywords: Diversified similarity queries · Metric spaces indexing · Pivot-based space partitioning

1 Introduction

With the evolution of data acquisition and of the applications domains employing Database Management Systems (DBMS), it has become needed to store and retrieve more complex data, such as images, audio, videos, and long texts. Classic

---

FAPESP (grants No. 2016/17078-0, 2020/07200-9), CAPES (grant 001) and CNPq.

S. Chiusano et al. (Eds.): ADBIS 2022, LNCS 13389, pp. 165–178, 2022.
https://doi.org/10.1007/978-3-031-15740-0_13

comparisons performed by current DBMSs are mainly based on identity and order relationships, which are not adequate for complex data. On the other hand, similarity queries are the information process that evaluates a element given by the user, called the query center (s_q), and retrieves of a set of elements that are alike, but not equal, to the reference element (s_q) [9,10].

The metric space model [8] is an efficient approach for handling similarity queries, in which data elements are mapped into a known domain where they are compared by a distance function (δ) to assess how similar are the elements, assuming that smaller distances correspond greater similarity among elements. The two most useful similarity retrieval operators are the similarity range (R_q) and the k-nearest neighbor (kNN_q) queries. A range query retrieves the elements from the dataset that are farther apart than a radius (r) from the query center s_q. A k-nearest neighbors query retrieves the k elements most similar to s_q. However, when similarity criteria are applied to large datasets with high cardinality and density, they tend to lose expressiveness and, consequently, quality. A major semantic-driven problem related to increasing data volume is that the similarity query operators are unable to filter the result set elements similar to each other [4]. The problem with too similar objects in result sets is that they can mislead users to believe that the database does not store the required information [4,9,11,14].

Several researchers have considered including diversity in query results, aiming at returning elements that are both similar to the query center and diverse from each other. Several diversification strategies can be found in the literature. They are classified as based on coverage or on novelty (also called as distance-similarity) [9]. The former return elements enforcing a dissimilarity threshold, returning only elements that respect a given distance between them. In this approach, the goal is to find elements that cover different information. The later search for elements that maximize a double criteria objective function, where similarity and diversity are balanced according to a user's defined preference parameter. Their goal is to find elements that are not redundant with the elements already found [4,14,16]. Depending on the data domain, one approach may be more suitable than the others although are important for result diversification. This work focuses on novelty-based diversification approaches.

Searches for diversified similarity are intrinsically costlier than searches seeking only similarity. This is due to two facts: more elements need to be loaded and compared, and each candidate needs to be compared not only to the query center but also to the other elements already in the answer. Novelty-based algorithms usually consider diversification as an NP-Hard optimization problem [4,14], but an exact solution is not usually obtained in a feasible time. An alternative to reduce computational costs relies either on heuristics or on metaheuristics [10]. However, both algorithms have scalability problems. The most common approach [11,14,15] is to select a subset of candidates to be processed by the diversification algorithms. However, not only the execution time but also the answer quality is directly impacted on how such selection is performed [11,15]. One of the most impacting problem of these approaches occurs when elements that do not maximize the objective function are selected. This often happens when applied approaches focus only on diversity [5,15].

In this work we present a strategy and corresponding algorithms to speed up similarity with diversity based on novelty. We seamlessly integrated an index structure with candidate selection methods to allow selecting elements considering both similarity and diversity. Our main contributions are: (i) An index structure that partitions the search space considering the distance among elements. (ii) Two algorithms to speed up similarity with diversity query algorithms to be employed in conjunction with our structure.

The remainder of this paper is organized as follows. Section 2 presents related works and basic concepts. Section 3 presents our proposal for space partitioning and selecting elements. Section 4 presents the evaluation environment and the results obtained. Finally, Sect. 5 presents our conclusions.

2   Background

Here we present related index structures and existing works, which allows partitioning elements of the dataset to perform similarity queries efficiently and foster obtaining diversified results.

2.1   Range-Tree

The Range-Tree (RT) [3,15] aims at quickly find the elements contained within the range of the query. It partitions the dataset considering the full range of values in every dimension. Figure 1a illustrates a RT storing a two-dimensional data, where the range of the first dimension spans from 0 to 2. In this structure, the root stores all elements within the range of the first dimension. The other non-leaf (intermediate) nodes store the partitions generated by the subranges of the root. To handle spaces with dimensionality greater than one, each node has a child pointing to a subrange tree that partitions the elements in the next dimension. During a search, whenever a node within the range of the current dimension is found, a search in the next dimension is performed, until there are no more dimensions to search. A d-dimensional range query Rd_r is expressed as a sub-range for each dimension of the dataset, such as $Rd_r = \{x_{init}, x_{finish}, y_{init}, y_{finish}\}$[1]. The time complexity to query a RT is $O(\log^d(n))$ and its space complexity is $O(n\log^d(n))$, where d is the dataset dimensionality and n is its cardinality.

2.2   MAM - Omni-Technique

A similarity query can be executed by a sequential scan, where every element is compared to the query center s_q. However, calculating every distance slow down the process, due to the high computational cost of similarity calculations. To

---

[1] Notice that a Rd_r correspond to elements within a sequence of values in each d-Dimension, thus it is distinct from a similarity range query R_q.

speed up the queries, many Metric Access Methods (MAM) were proposed to store and efficiently retrieve data based on the Metric Spaces properties, which allow reducing the number of similarity comparisons [8,13]. Among the several well-known structures we highlight those based on the Omni-Technique [13].

Given a dataset with a fractal dimension $\mathbb{D}$, the Omni-Technique is a pivot-based indexing approach that assumes that $\lceil\mathbb{D}\rceil$ is the ideal cardinality for a set of pivots $\mathcal{P}$ employed to accelerate query execution. The omni-technique aims at pre-computing the distances of every element to every pivot. The distances are called the omni-coordinates of each data element and they are employed to reduce the number of distance comparisons during a query execution. The process has two steps: filter and refine. During the filtering step, the omni-coordinates of the query center s_q is calculated and the triangular inequality property is used to find the regions that contain the query results, which can include false positives. The refinement step removes the false positives and generates the final answer.

The omni-coordinates generate a new search space, more compact than the original one. Thus, they can be used both as an indexing and a dimensionality reduction strategy. We take into account both benefits for developing our proposed method, as described in Sect. 3.

2.3   The Diversity Problem

A diversified similarity query can be defined as an optimization problem that looks for elements R that are both similar to the query center but also diverse from each other. This goal can be expressed as a double-criteria objective function that targets to maximize similarity and diversity, as follows. Given a dataset S, a query element s_q, an integer k, a function δ_{Sim} that measures how similar each element s_i is from s_q, and a function δ_{div} that measures how diverse two elements are, the diversification problem can be expressed as [9,11,14]:

$$R = argmax(\mathcal{F}(s_q, R)), \forall R \subseteq S : |R| = k, \qquad (1)$$

$$\mathcal{F}(s_q, R) = (1-\lambda)\cdot\sum_{i=0}^{k}\delta_{sim}(s_q, r_i) + \frac{2\lambda}{(k-1)}\sum_{i=1}^{k-1}\sum_{j=i+1}^{k}\delta_{div}(r_i, r_j) : r_i, r_j \in R \quad (2)$$

Parameter $\lambda[0,1]$ defines how much diversity the user expects. When $\lambda = 0$, the problem is reduced to a kNN_q. When $\lambda > 0$, the problem becomes NP-Hard with time complexity $O(n^k)$ (where $n = |S|$), as it must evaluate every subset $R(|R| = k)$ to find the one with the largest $\mathcal{F}$.

Several approximate algorithms have been proposed to generate a good query answer in a feasible time, some executing in $O(n^2)$ time. However, even considering this complexity reduction, they still can take a long time to get an answer. Thus, approximate algorithms typically have two phases: Candidate selection and Diversification. The candidate selection phase extracts a subset S' with cardinality $m = |S'| << |S|$. In this way, the search space is reduced to $m = |S'|$. In the diversification phase, a similarity with diversity algorithm is applied to the set of candidates S' [9,11,14].

2.4   Diversity Algorithms

One of the first and best-known diveristy algorithm in the literature is the *Maximal Marginal Relevance* (MMR) [2]. An element is marginally more important if it is as similar to the query element as it is diverse from the elements already inserted in the answer set. Thereafter, other elements that maximize the MMR function are incrementally inserted. The time complexity of MMR is $O(kn)$, however, other algorithms can generate better results.

The *Greedy Marginal Contribution* (GMC) [14] is an incremental algorithm that basically follows the same steps of MMR but uses another objective function, the *maximum marginal contribution* (MMC). The MMC function evaluates the contribution of the element $s_i \in S$ considering the similarity between s_i and s_q, the diversity between s_i and the elements already in R and the diversity between s_i and the elements of the candidate set S' that are not yet in R.

The *Greedy Randomized with Neighbor Expansion* (GNE) [14] is based on the GRASP meta-heuristic (*Greedy Randomized Adaptive Search Procedure*). The GNE can be divided in two phases: construction and local search. In the construction phase, the algorithm iteratively generates an initial solution to maximize MMC. In the local search phase, the algorithm improves the initial solution, looking for a higher quality solution in the neighborhood of the current solution. If no better solution is found, the current one is returned.

The *Max-Sum Dispersion* (MSD) [7] algorithm incrementally builds the answer R, selecting the pair of elements that maximizes the objective function. Basically, at each iteration it chooses two elements s_i and $s_j \in S$ that are both similar to s_q and different from each other. For cases where k is odd, MSD randomly chooses the last element to be inserted.

GMC, GNE, and MSD are capable of generating better results than MMR. However, their time complexity are $O(n^2)$, which makes the process of analyzing many elements even longer [11,14].

2.5   Candidate Selection

Several approaches to select/filter elements were developed to improve efficiency whereas also finding good answers. The most common use a similarity search (kNN_q or R_q) to return the subset S' with the m elements closest to s_q, but other candidate selection strategies have been considered too [5,11,14,15].

For example, in [11] were conducted an evaluation of distinct filter approaches combined to novelty algorithm, in which RDI standed out. RDI, returns m elements using the concept of Result Diversification based on Influence [12]. Although very fast, it is based on a method that does not guarantee that the selected elements actually maximize the objective function ($\mathcal{F}$). Another point is that RDI does not restrict the search space, so it is not uncommon that the entire dataset is analyzed, which can sometimes make the selection process slower than other approaches.

A modification of the algorithm for the Cover-Tree construction [1] was presented in [5], which here we call CT, aiming at efficiently find diversified sets

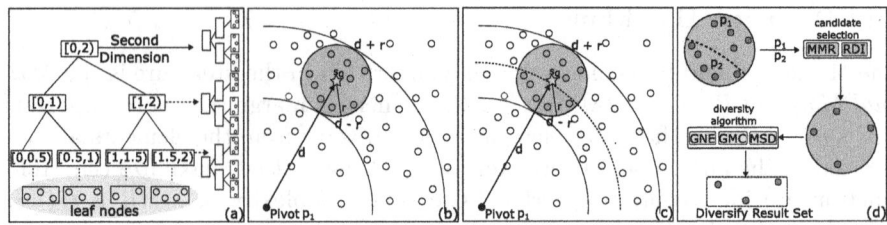

Fig. 1. Two-dimensional Range Tree and the process of candidate selection to partition the search space using one pivot. (a) Each node in the first dimension leads to another Range Tree in the second dimension. (b) Range query defined by $R_q(s_q, r)$. (c) The query range is partitioned (dotted line), generating two subranges. (d) Considering the generated partition, a candidate selection approach is applied on each partition. The elements selected from each partition are joined for the diversity algorithm.

in data streams. The proposed algorithm transforms each level of the Cover Tree into a possible solution for a diversification heuristic. Following up, several search algorithms were proposed, one of them returns the k elements contained in one of the upper nodes of the Cover Tree, which can be quickly obtained. However, the proposed algorithm builds the tree considering only the diversity between the elements. Furthermore, the construction algorithm has complexity $O(n^2)$, which makes the whole process as expensive as the algorithms previously presented.

The RC-Index [15] selects candidates using two data structures: a Range Tree and a Cover Tree. The Range Tree partitions the dataset and, for each partition, creates a corresponding cover tree. Thereafter, given a search range, the cover trees within the query range are used to extract a subset of candidates. The candidates are extracted in a way similar to the CT approach, the main difference being that, given a level (L_k) that contains k elements, the candidates at some lower level are returned: by default, three levels below L_k. However, this approach is ineffective for high-dimensional datasets, since its building complexity is $O(\gamma^6 log^{d+1}(n))$, in addition, its space complexity is the same as the Range Tree: $O(n \log^d(n))$.

The next section presents a novel structure which, as the RC-Index, partition the search space also using a Range Tree, but coupled to the Omni-technique to reduce the space complexity, which allows a much faster similarity query execution. Distinctly from RC-Index, we also present an algorithm to convert a similarity range query into a distance range query, which can be executed in an RT.

3   Methodology

Here we propose an efficient method to answer diversified similarity queries, based on two concepts: Spatial partition and Candidate selection. The first aims at partitioning the data into small subsets, so that the elements in each subset are similar to each other and the cost of selecting diversified candidate elements is as small as possible. The second concept aims at quickly selecting the elements from each subset that can maximize both similarity and diversity. To partition the dataset, our approach extends the Range Tree with the Omni techniques, creating the **Omni-Range Tree** (the ORTree). It partitions the dataset using omni-coordinates, which improves the similarity range query (R_q) efficiency. Moreover, as the amount of evaluated omni-coordinates is defined by the fractal dimension $(\mathbb{D})$ of the dataset, it almost always reduces the data dimensionality too and, consequently, redux the structure time and space complexity. The Range Tree is employed due to its low query complexity of $O(\log^{\mathbb{D}}(n))$.

Building a ORTree is very similar to build a RT, but instead of the original attributes, the omni-coordinates are used. Algorithm 1 show the process for building a ORTree. After choosing the pivots, the omni-coordinates of every element are calculated (lines 1–8) in time complexity of $O(n)$, generating a $\mathbb{D}$-dimensional space. Thereafter, a RT is built as follows. The elements are sorted following the first dimension and inserted into the root node (lines 9–10). At this point, the elements are partitioned into two subsets considering the median of the first dimension of the omni-coordinates array. Next, the 'left' and 'right' children are built recursively for the current dimension (lines 16–17) and the 'next' child for the next dimension (lines 18–20), repeating recursively until there are no more dimensions.

Algorithm 1. Building the ORTree
___

Input: Set of pivot elements $\mathcal{P}$, δ a metric distance function and the dataset S.
Output: ORTree.

1: $\mathcal{O}_S \leftarrow \emptyset$ ▷ set of omni-coordinates
2: **for** $\forall \, s_i \in S$ **do**
3:    $\mathcal{O}_{si} \leftarrow \emptyset$
4:    **for** $\forall \, p \in \mathcal{P}$ **do**
5:       $coord \leftarrow \delta(s_i, p)$ ▷ coordinate corresponding to p.
6:       $\mathcal{O}_{si} \leftarrow \mathcal{O}_{si} \cup coord$
7:    **end for**
8:    $\mathcal{O}_S \leftarrow \mathcal{O}_S \cup \mathcal{O}_{si}$
9: **end for**
10: $\text{Sort}(\mathcal{O}_S, 0)$
11: $ORTree.root = Construct(\mathcal{O}_S, 0)$
12: **return** ORTree
13: **function** CONSTRUCT(set, dim)
14:    **if** set.size $== 0$ **then**
15:       **return** NULL
16:    **end if**
17:    $node.left = Construct(left, dim)$
18:    $node.right = Construct(right, dim)$
19:    $dim = dim + 1$
20:    $\text{Sort}(set, dim)$
21:    $node.next = Construct(set, dim)$
22:    **return** node
23: **end function**

Given a $R_q(s_q, r)$, the ORTree retrieves the nodes that store elements within the query range. To obtain the RD_r range, the following steps are executed. The omni-coordinates of s_q and then the ranges for each dimension are generated, defining the search range r as the omni-coordinate of s_q (its distance to each pivot): $range : \{O_{sq_i} - r, O_{sq_i} + r\}$. The same procedure is applied to all other dimensions. Given the ranges, the query finds and returns the nodes contained within the ranges. The distances from s_q to each element in nodes is calculated and only those within r are returned: $\delta(s_q, s_i) \leq r$. Our approach is able to support both the $k - NN_q$ and R_q similarity queries, but here we are going to focus on R_q.

Aiming at achieving a selection of better candidates than those provided by other methods in the literature, we took into account the data structure provided by ORTree to create a novel process that allows a more efficient selection of candidates, maintaining a quality equivalent to the other approaches. Our method uses the nodes retrieved by an ORTree to extract a set of m elements from each node within the range defined by $R_q(s_q, r)$. Thus, we use the partitions generated by a ORTree to quickly select the candidate elements.

Figure 1 illustrates our strategy. Initially, a range query is performed using the ORTree (Fig. 1b), then, considering that the returned elements will be in different nodes (partitions) (Fig. 1c), we apply, at each node, an algorithm to select m different elements. The selected elements are joined to form the final candidate set, which is passed to the diversifying algorithm (Fig. 1d). One of the main advantages of this approach is that the number of comparisons tends to be much smaller, speeding up the selection algorithm. Based on this principle, we developed two methods(RT_MMR and RT_RDI), based on different algorithms, which can select the candidates, that are both similar to s_q and diverse from others.

The Range Tree MMR (**RT_MMR**) aims at using MMR to select the candidates. It is faster than any of the GMC, GNE, and MSD algorithms, although, it follows the same diversification strategy, allowing for select elements considering the diversity preference (λ). When using MMR, it is expected that the elements selected from each node, be in smaller quantity than it would originally be, but with equivalent diversity.

The Range Tree RDI (**RT_RDI**) seeks the candidate considering the influence-based diversification approach. As it is coverage-based, this approach tends to be very faster than MMR. However, this approach tends to analyze more elements than the previous approaches and therefore may be slower. To get around this problem, we use this approach on each of the nodes returned by ORTree, allowing reducing the number of comparisons between the elements, and thus making the selection faster. Unlike the original approach that may analyze the full dataset, our approach reduces the search space to at most $R_q(s_q, r)$.

Every approach selects m elements from each node (or all elements when the node has less than m elements), so the number of elements returned from each node is expected to be much less than the original amount. Consequently, the number of elements selected by the approaches tends to be smaller than that defined by the $R_q(s_q, r)$ approach.

4   Experiments

We performed several experiments to validate our proposal and evaluate whether the execution of the diversity algorithms were indeed faster and whether the quality of the results remains equivalent when compared to the base algorithms from the literature. Three datasets were selected for this purpose: US_cities, NASA [6] and Corel[2]. The datasets were selected to evaluate the behavior of the proposed approaches, exploiting data with different cardinalities and dimensionalities, mainly in relation to the fractal dimension, which impacts the construction and querying times. From each dataset, 50 elements were randomly selected to be used as queries centers following a hold-out strategy. Table 1 summarizes the information about each dataset and the query parameters. We define the query range (r) so that the number of elements analyzed in each dataset is approximately the same.

Table 1. Datasets statistics.

Dataset	$|S|$	d	$\mathbb{D}$	Pivot number	$\delta_{sim}/\delta_{div}$	Range	Elements retrieved	Dataset description
US_Cities	25,374	2	1.62	2	L_2	$\{2.0, \mathbf{4.0}\}$	$\{560, 1860\}$	Geographic coordinates of American cities
Nasa	40,150	20	2.63	3	L_2	$\{0.6, \mathbf{0.7}\}$	$\{716, 1514\}$	Feature vectors generated from NASA images
Corel	20,000	9	4.8	5	L_2	$\{1.9, \mathbf{2.3}\}$	$\{680, 1572\}$	Feature vectors generated from common images

We compared the results of ORTree with the following approaches from the literature: Rq, RC-Index and CT. Rq uses all elements returned by $R_q(s_q, r)$. To ensure that the elements returned are within the same range, the RC-Index was implemented using the ORTree instead of the original Range Tree. CT is built using the results of a $R_q(s_q, r)$ performed over the ORTree.

For each experiment, the following values were used as query parameters for the similarity with diversity queries (in bold are the default values): $\lambda = \{0.3, \mathbf{0.5}, 0.7\}$ and $k = \{10, \mathbf{20}, 30\}$. For the RT_MMR and RT_RDI approaches, we define by default that the number of elements to be selected from each ORTree node is the number of elements to be returned by the query, thus $m = k$.

4.1   Index Creation Time

Figure 2 shows the ORTree creation time compared to the RC-Index. Figure 2(a), shows the time for US_Cities and Nasa datasets. While Fig. 2(b) shows the creation time for the Corel dataset when we vary the number of pivots (1–5). Both figures show that the RC-Index has a much higher construction cost

---

[2] Sample extract from https://archive.ics.uci.edu/ml/datasets/corel+image+features, accessed at: 06/05/2022.

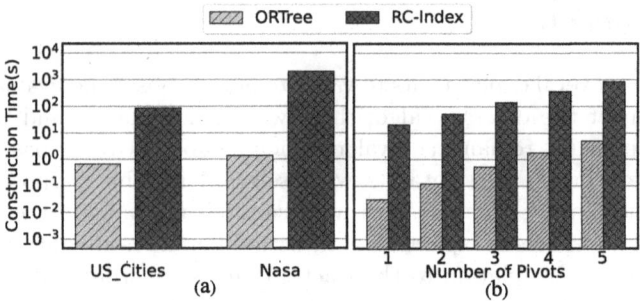

Fig. 2. ORTree and RC-Index build time in log scale for the US_Cities, Nasa and Corel dataset. (a) Time for US_Cities and Nasa. (b) Time for Corel dataset, with varying number of pivots.

than the ORTree. This is due to the difference in the complexity and the larger number of distance calculations performed by the RC-Index. It is also much slower than the ORTree, following the time complexity of $O(\gamma^6 nlog^{d+1}(n)) > O(nlog^d(n))$. Furthermore, the ORTree construction does not depend on distance calculations, whereas the RC-Index builds a cover tree, which requires several distance calculations, to build each node of the range tree, which greatly increases the execution time.

4.2   Quality Experiments

For each approach, the queries were performed using the GMC, GNE and MSD algorithms. Figure 3 shows the values returned by the objective function (Eq. 2) of each approach and algorithm. The Fig. 3 (a, b and c) show the search results for each of the algorithms using the US_Cities dataset for range = 4, Fig. 3 (d, e and f) show the results for Corel dataset with range = 2.4 and, Fig. 3 (g, h and i) show the results for Nasa dataset with range = 0.7.

To the US_Cities dataset, several approaches ties when $\lambda = 0.3$, but CT proved to be inferior to the other approaches. For $\lambda = 0.5$, all approaches are practically tied, except CT that achieves better results, for the GMC and GNE algorithms, including those generated by the traditional approach Rq. For $\lambda = 0.7$, we have a tie between the RT_MMR, Rq, and RC-Index approaches, the other approaches achieve lower results, with RT_RDI being better than CT. For the Corel dataset, again, many approaches tie, with CT achieving the worst results. For $\lambda = 0.5$, all approaches tie, and CT achieves better results for the GMC and GNE algorithms, but worse results for the MSD which is the best result in this case. At $\lambda = 0.7$ all approaches tie, except CT which achieves the worst results in both algorithms. In the Nasa dataset, for $\lambda = 0.3$, all approaches have very similar results, with CT showing lower results. For $\lambda = 0.5$, there is a tie between the approaches, making it difficult to point out an approach that is better in all cases. At $\lambda = 0.7$, the RT_MMR, Rq, and RC-Index approaches tie

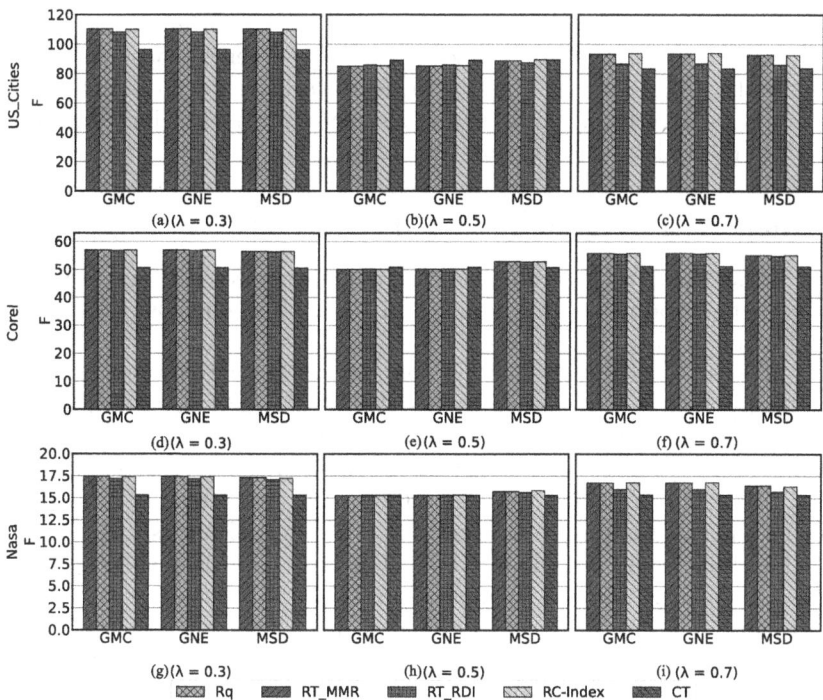

Fig. 3. Quality results according to the diversity objective function ($\mathcal{F}$), for dataset US_Cities (a, b and c) with $r = 4$, Corel (d, e and f) with $r = 2.3$ and Nasa (g, h and i) with $r = 0.7$

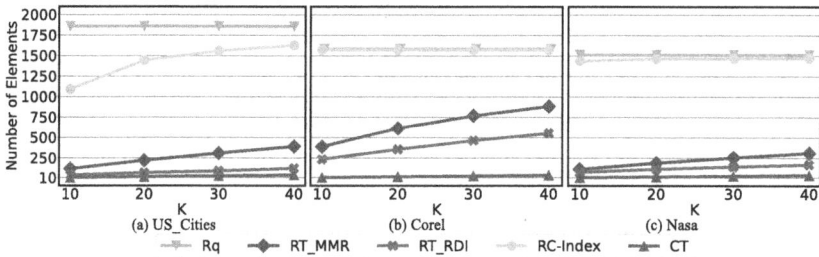

Fig. 4. Number of elements retrieved. (a) Number of elements for dataset US_Cities with $r = 4$. (b) Number of elements for dataset Corel with $r = 2.3$. (c) Number of elements for Nasa dataset with $r = 0.7$

in the results achieved, RT_RDI and CT achieve inferior results, with CT being the worst approach between the two.

In some datasets, the CT approach was able to generate better results than the traditional approach. In this case, the candidate selection process is likely to remove some solutions that are local optimal. Therefore, the candidate selection process can not only reduce the execution time of the algorithms but also provide higher quality. However, candidate selection can also remove optimal solutions. Therefore, the RT_RDI and CT approaches to achieve better results in some cases and worse in others. This situation did not happen with the RT_MMR and RC-Index approaches, they remain equivalent to Rq.

4.3 Number of Elements Retrieved

Figure 4 shows the average number of candidates selected by each approach. Figure 4a show the results for dataset US_Cities, Fig. 4b show the results for Corel dataset and Fig. 4c show the results for dataset Nasa. Rq always select the fixed number of elements, defined by a parameter. For all datasets, the number of candidates returned by our approaches (RT_MMR and RT_RDI), like the RC-Index, grows as k grows. However, in every case, our approaches retrieve fewer elements than Rq and RC-Index. For the US_Cities dataset (Fig. 4a), RT_MMR and RT_RDI always retrieve less than 30% of the elements from Rq. The RC-Index, on the other hand, retrieves around 50% fewer elements but as k grows, this value drops by approximately 80%. For the Corel dataset (Fig. 4b), the results are similar, with our approaches retrieving 57% fewer elements. However, RC-Index retrieves almost the same amount of elements as Rq, (discussed later). The Nasa dataset exhibits the same behavior of the other datasets, with RT_MMR and RT_RDI recovering less than 25% of the elements in relation to Rq, while RC-Index recovers closely the same amount. In all datasets, CT retrieves fewer elements because it always returns k elements.

Regarding the number of candidates returned by the RC-Index, because of the strategy of selecting the three-level elements below the cover trees (Sect. 2), number of elements is always greater than k. Also, depending on the distribution of the data, going down three levels may be enough to select all elements of the node. Another point that contributes to this situation is that the more pivots, the more partitions in the search space, which implies fewer elements per node in the ORTree (and RT). However, this situation does not happen with RT_MMR, and RT_RDI as they always return k or fewer elements from each node.

4.4 Query Time Evaluation

Figure 5 shows the execution time of the query algorithms using the proposed approaches. The figures show respectively the run-time for US_Cities (a and b), Corel (c and d) and NASA (e and f) datasets. The execution time using Rq and RC-Index tends to be longer than the execution time of every other approach, which is expected, due to the number of elements retrieved. In some cases, the RC-Index has a slightly higher cost than Rq, this happens because the RC-Index performs more operations and retrieves more or less the same number of elements as Rq. In all graphs, it is possible to see that RT_RDI is by far the fastest approach, followed by RT_MMR and CT approaches.

Figure 5 (a and b), shows that both RT_MMR and RT_RDI are faster than Rq, RC-Index, and CT. RC-Index is faster than Rq and CT faster than both. Figure 5

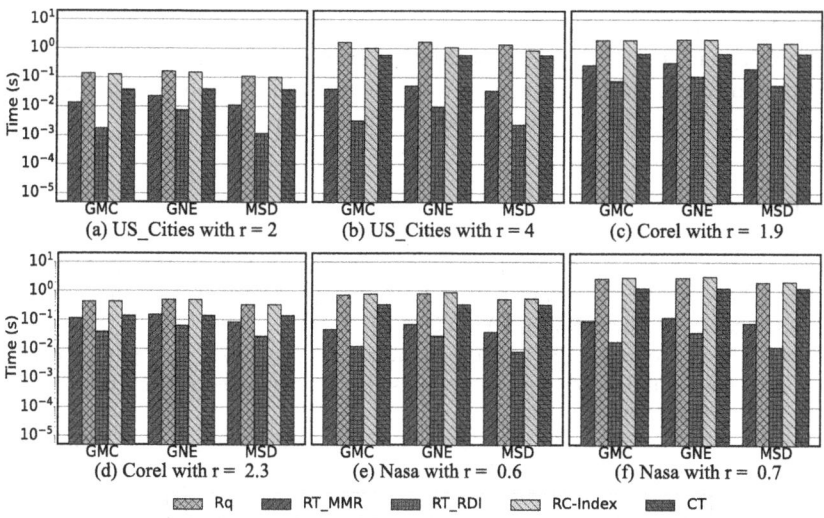

Fig. 5. Query time in log scale. (a - b) US_Cities, $r \in \{2, 4\}$. (c - d) Corel, $r \in \{1.9, 2.3\}$. (e - f) Nasa , $r \in \{0.6, 0.7\}$.

(c and d), shows that the results are similar to the previous figures, except that RC-Index has the same execution time as Rq, and in some cases (Fig. 5c), CT turns out to be faster than RT_MMR. In this case, the smallest amount of candidates compensate for the quadratic CT construction time. Finally, Fig. 5 (e and f), show that the execution time of RC-Index can be longer than that of Rq. For the other approaches, the results are as before, RT_RDI being the fastest, followed by RT_MMR and CT, with RT_MMR being many times faster.

The results show that RT_MMR and RT_RDI are by far the fastest approaches. regarding the Nasa dataset, the approaches are, respectively, 95% and 97% faster than the traditional approach (Rq).

5   Conclusions and Future Work

In this work, we present the ORTree, a new indexing structure based on the Range Tree and on the Omni-Technique, which allows performing diversified similarity queries much faster without reducing the quality of the answers. Along with this novel framework, we presented two approaches that use the ORTree to efficiently select candidates for the diversification process. Our experiments show that the proposed approaches can significantly reduce the number of elements that are analyzed in the process of diversification. Consequently, the query time was significantly reduced, in some cases being 95% faster. In addition, the quality results show that even reducing the number of candidate elements, the quality of the results remains equivalent to the traditional approach. Because of this, the experiments shows that ORTree is the best candidate selection approach for the analyzed aspects.

As a future work, we plan to develop new candidate selection approaches that use the ORTree and to develop alternative approaches for large search spaces. We also plan to extend the presented approaches, which are based on a similarity range query, to also handle the k-nearest neighbor queries.

References

1. Beygelzimer, A., Kakade, S., Langford, J.: Cover trees for nearest neighbor. In: Proceedings of the 23rd International Conference on Machine Learning. pp. 97–104. ACM, New York, NY, USA (2006)
2. Carbonell, J., Goldstein, J.: The use of mmr, diversity-based reranking for reordering documents and producing summaries. In: Proceedings of the 21st SIGIR. pp. 335–336. ACM (1998)
3. De Berg, M., Van Kreveld, M., Overmars, M., Schwarzkopf, O.: Computational Geometry, pp. 1–17. Springer, Heidelberg (1997). https://doi.org/10.1007/978-3-540-77974-2
4. Drosou, M., Jagadish, H., Pitoura, E., Stoyanovich, J.: Diversity in big data: a review. Big Data **5**(2), 73–84 (2017)
5. Drosou, M., Pitoura, E.: Diverse set selection over dynamic data. IEEE Trans. Knowl. Data Eng. **26**(5), 1102–1116 (2014)
6. Figueroa, K., Navarro, G., Chávez, E.: Metric spaces library (2007). http://www.sisap.org/Metric_Space_Library.html
7. Gollapudi, S., Sharma, A.: An axiomatic approach for result diversification. In: WWW2009, pp. 381–390 (2009)
8. Hetland, M.L.: The basic principles of metric indexing. In: Coello, C.A.C., Dehuri, S., Ghosh, S. (eds.) Swarm Intelligence for Multi-objective Problems in Data Mining. Studies in Computational Intelligence, vol. 242, pp. 199–232. Springer, Heidelberg (2009). https://doi.org/10.1007/978-3-642-03625-5_9
9. Novaes, J.V.O., et al.: J-EDA: a workbench for tuning similarity and diversity search parameters in content-based image retrieval. J. Inf. Data Manag. **12** (2021)
10. Lopes, C.R., Santos, L.F.D., Jasbick, D.L., de Oliveira, D., Bedo, M.: An empirical assessment of quality metrics for diversified similarity searching. J. Inf. Data Manag. **12**(3) (2021)
11. Santos, L.F.D., Oliveira, W.D., Carvalho, L.O., Ferreira, M.R.P., Traina, A.J.M., Traina, C.: Combine-and-conquer: improving the diversity in similarity search through influence sampling, Proceedings of the 30th SAC, pp. 994–999 (2015)
12. Santos, L.F.D., Oliveira, W.D., Ferreira, M.R.P., Traina, A.J.M., Traina, C.: Parameter-free and domain-independent similarity search with diversity. In: Proceedings of the 25th SSDBM. ACM, New York, NY, USA (2013)
13. Traina, C., Filho, R.F., Traina, A.J., Vieira, M.R., Faloutsos, C.: The Omni-family of all-purpose access methods: a simple and effective way to make similarity search more efficient. VLDB J.l **16**(4), 483–505 (2007)
14. Vieira, M.R., et al.: On query result diversification. In: Proceedings of the 27th ICDE, 11–16 April 2011, Hannover, Germany, pp. 1163–1174. IEEE (2011)
15. Wang, Y., Meliou, A., Miklau, G.: RCIndex: diversifying answers to range queries. Proc. VLDB Endow. **11**(7), 773–786 (2018)
16. Zheng, K., Wang, H., Qi, Z., Li, J., Gao, H.: A survey of query result diversification. Knowl. Inf. Syst. **51**(1), 1–36 (2016). https://doi.org/10.1007/s10115-016-0990-4

A Knowledge-Based Approach to Support Analytic Query Answering in Semantic Data Lakes

Claudia Diamantini, Domenico Potena, and Emanuele Storti[(✉)]

DII, Polytechnic University of Marche, Ancona, Italy
{c.diamantini,d.potena,e.storti}@univpm.it

Abstract. The increased flexibility brought by Data Lake technologies, along with size and heterogeneity of quickly changing data sources, bring novel challenges to their management. Making sense of disparate data and supporting users to identify the most relevant sources for a given analytic request are indeed critical requirements to make data actionable. This is particularly relevant in data science applications, where users want to analyse statistical measures from a variety of data sources. To this aim, in the paper we introduce a knowledge-based approach for a Semantic Data Lake, capable of supporting efficient integration of data sources and their alignment to a Knowledge Graph representing indicators of interest, their mathematical formulas and dimensions of analysis. By leveraging manipulation of indicator formulas, a query-driven discovery approach is exploited to dynamically identify the sources, along with the needed transformations, to respond a given .

Keywords: Data Lake · Query-driven discovery · Knowledge Graph · Multidimensional model

1 Introduction

Data Lakes (DL) have recently emerged as schema-agnostic repositories for storing data in their native format, providing centralized access and the capability to apply data transformation when needed according to an ELT (Extraction, Load, Transformation) approach. This increased flexibility, along with size and heterogeneity of growing data sources bring novel challenges related to data management. In particular, the lack of a global schema and the need to make sense of disparate raw data require proper modeling of their metadata, to make data actionable and avoid data swamps (see also [15]). As recognized by recent literature (e.g., [13]), how to integrate heterogeneous data sources and help users to find the most relevant data are still open issues in this setting and are often seen as intertwined operations. This is particularly relevant in data science applications, where users want to analyse statistical measures from a variety of data sources. Examples include Open Data Lakes managed by public bodies, e.g., to monitor the effectiveness of governmental initiatives like a vaccination campaign, or analysing outcomes from Open Science collaborative projects.

To address such challenges, a novel paradigm called *query-driven discovery* was proposed to combine the two aspects [12], following the idea to find datasets

© Springer Nature Switzerland AG 2022
S. Chiusano et al. (Eds.): ADBIS 2022, LNCS 13389, pp. 179–192, 2022.
https://doi.org/10.1007/978-3-031-15740-0_14

that are similar to a query dataset and that can be integrated in some way (either by joins, unions or aggregates).

In the literature, a variety of solutions have been proposed for DL integration, ranging from approaches based on raw data (and related metadata) management to semantic-enriched frameworks, the latter being intrinsically more suitable to address issues related to data variety/heterogeneity and data quality. Among them, some work focused on holistic models capable of addressing a variety of structures, e.g., in [8] a DL system is proposed, that discovers, extracts, and summarizes the structural metadata from the (semi)structured data sources, and annotates (meta)data with semantic information to avoid ambiguities, while in [2] a network-based model to represent technical metadata of structured, semi-structured and unstructured data sources is proposed. Knowledge graphs are exploited in [5] to drive integration, relying on information extraction tools, e.g., Open Calais, that may assist in linking metadata to uniform vocabularies, while in [6] a graph is built by a semantic matcher, leveraging word embeddings to find links among semantically related data sources.

In this work, we propose a query-driven knowledge-based approach for integration and discovery in a Data Lake. The approach builds on a Knowledge Graph including a formal model of measures (also named indicators) and their computation formulas [4], in which concepts are used to enrich source metadata. On top of the model, the contributions of this work are multifold:

- We define mechanisms for *integration* of data sources into the Semantic Data Lake and *mapping discovery*, based on efficient evaluation of set containment [16] between a source domain and a concept in the Knowledge Graph.
- We define an ontology-based and math-aware *query answering* function, capable of identifying the set of sources collectively capable of responding the user request, and the proper transformation rules to make the needed calculation. For instance, let us suppose a user is interested in analysing measure $CO_2PerPerson$, but it is not available in any source. Given that such a measure can be calculated as $\frac{TotalCO_2}{Population}$, a response can be obtained by combining sources providing the two components $TotalCO_2$ and $Population$ measures.
- To quantitatively estimate the quality of such results, we define a *degree of joinability* index, that evaluates to what extent the sources are joinable, i.e., how much they share the same values over the same attributes.

With respect to the content-driven notion of query-driven discovery that was proposed in [12], our approach also considers metadata (i.e., mappings to indicators concept in the Knowledge Graph and their formulas) as a support to reformulate the query and determine which sources can be used to respond. This helps in reducing the search space by identifying the most semantically relevant data sources according to the discovery need. The rest of the paper is structured as follows: in Sect. 2 a case study is introduced that will be used throughout the paper. Section 3 is devoted to introduce the Semantic Data Lake model. The approach for source integration is discussed in Sect. 4, while query answering mechanisms are detailed in Sect. 5. Section 6 discusses an evaluation of the approach. Finally, Sect. 7 concludes the work and draws future research lines.

2 Case Study: Azure COVID-19 Data Lake

In this work we take as example a set of COVID-19 related datasets from the Microsoft Azure Covid-19 Data Lake [11] and Our World in Data repository:

S1) *Bing COVID-19 Data*[1], which includes confirmed, fatal, and recovered cases per country/region, updated daily for years 2020–2021.

S2) *COVID Tracking Project*[2], with numbers on tests, confirmed cases, hospitalizations daily from every US state for years 2020–2021.

S3) *European Centre for Disease Prevention and Control (ECDC) Covid-19 Cases*[3], which includes the latest available public data on COVID-19 cases worldwide from the European Center for Disease Prevention and Control (ECDC), reported per day and per country for year 2020.

S4) *Oxford COVID-19 Government Response Tracker (OxCGRT)* [9], which contains systematic information on measures against COVID-19 taken by governments, for years 2020–2021.

S5) *Open World in Data*[4], which contains data on the number of people in hospitals and ICU per day and country, for years 2020–2021.

In Table 1 we summarize relevant detail about the sources, that are derived from the source metadata provided by the publishers.

Table 1. Details of sources S1–S5.

Source	# Rows	# Cols	Measures	Dimensions
S1	3051712	17	confirmed, confirmed_change, deaths, deaths_change, recovered, recovered_change	updated, country_region, admin_region, iso2, iso3, iso_subdivision
S2	22261	31	positive, negative, death, recovered, hospitalized_currently, in_icu_currently, in_icu_cumulative, on_ventilator_currently, on_ventilator_cumulative, pending	date, iso_country, state, iso_subdivision
S3	61900	14	cases, deaths	date_rep, continent_exp, countries_and_territories, iso_country, geo_id, country_and_territory_code
S4	231192	38	confirmedcases, confirmeddeaths	countryname, countrycode, date, ISO_country
S5	28661	8	daily ICU occupancy, daily ICU occupancy per million, daily hospital occupancy, daily hospital occupancy per million	entity, ISO_code, date

---

[1] https://www.bing.com/covid.

[2] https://github.com/COVID19Tracking/covid-tracking-data.

[3] https://www.ecdc.europa.eu/en/covid-19/data-collection.

[4] https://github.com/owid/covid-19-data.

3   Semantic Data Lake: Data Model

In this Section, we briefly review the model for a Semantic Data Lake that was discussed in [4], on top of which the source integration, mapping discovery and query answering mechanisms will be defined, as discussed in next sections.

We define a Semantic Data Lake as a tuple $SDL = \langle S, \mathcal{G}, \mathcal{K}, m \rangle$, where $S = \{S_1, \ldots, S_n\}$ is a set of data sources, $\mathcal{G} = \{G_1, \ldots, G_n\}$ is the corresponding set of metadata, $\mathcal{K}$ is a Knowledge Graph and $m \subseteq \mathcal{G} \times \mathcal{K}$ is a mapping function relating metadata to knowledge concepts. Our approach is agnostic w.r.t. both the degree of structuredness of the sources, ranging from structured datasets to semi-structured (e.g., XML, JSON) documents, and the specific DL architecture at hand, e.g., based on ponds vs. zones (see also [7,15]). If the architecture is pond-based, in fact, the approach is applied to datasets in a single stage, while in zone-based DLs the approach can be applied on any stage of the platform, although it is best suited to the staged area for data exploration/analysis. As a minimum requirement, we assume a data ingestion process to wrap separate data sources and load them into a data storage. The model for a Semantic Data Lake is detailed in the following.

3.1   Metadata Layer

Different typologies of metadata can be related to a resource, depending on how they are gathered [14]. Hereby, we refer to *technical* metadata, i.e., related to data format and, whenever applicable, to their schema. Since the representation of metadata is highly source-dependent (e.g., the schema definition for a relational table), a uniform representation of data sources in a *metadata layer* is required for the management of a data lake. The procedure to represent technical metadata of a given source depends on the typology of data source, e.g., a relational database has tables with attributes, while XML/JSON documents include complex/simple elements and their attributes. For each source S_k, metadata are represented as a directed graph $G_k = \langle N_k, A_k, \Omega_k \rangle$, where N_k are nodes, A_k are edges and $\Omega_k : A_k \rightarrow \Lambda_k$ is a mapping function s.t. $\Omega_k(a) = l \in \Lambda_k$ is the label of the edge $a \in A_k$. The graph is built incrementally by a *metadata management* system [2], starting from the definition of a node $n \in N_k$ for each metadata element. An edge $(n_x, n_y) \in A_k$ is defined to represent the *structural* relation existing between the elements o_x, o_y, e.g., this corresponds to the relations between a table and a column of a relational database, or between a JSON complex object and a simple object. Further details on this modeling approach are available in [2].

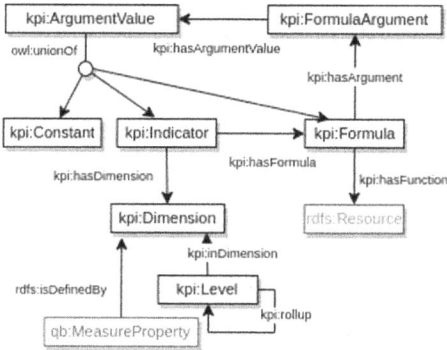

Fig. 1. Main classes and properties in KPIOnto ontology.

3.2 Knowledge Layer

The knowledge layer of the Semantic Data Lake comprises:

– *KPIOnto*[5] an OWL2-RL ontology aimed to provide the terminology to model an indicator in terms of description, unit of measurement and mathematical formula for its computation. The ontology also provides classes and properties to fully represent multidimensional hierarchies for dimensions (e.g., level *Province* rolls up to *Country* in the *Geo* dimension) and members. The main classes and properties, including those aimed at representating a formula in terms of operands and operator, are shown in Fig. 1.

– a *Knowledge Graph* $\mathcal{K} = \langle K_N, K_A, K_\Omega \rangle$, where K_N and K_A respectively represent nodes and edges, while K_Ω is a mapping function assigning labels to edges. It provides a representation of the domain knowledge in terms of definitions of indicators, dimension hierarchies and dimension members. Concepts are represented in RDF as Linked Data according to the KPIOnto ontology, thus enabling standard graph access and query mechanism.

– *Logic Programming rules*, which are enacted by a logical reasoner (namely, XSB[6]) to automatically provide *algebraic services*, capable of performing mathematical manipulation of formulas (e.g., equation solving), which are exploited to infer all formulas for a given indicator. This functionality is used to support query answering (see Sect. 5).

Figure 2 shows (a) a fragment of the Knowledge Graph for the case study representing dimensions *Time* and *Geo* with the corresponding levels, and (b) highlights the mathematical relations among a set of indicators. The full

---

[5] KPIOnto specifications are available at http://w3id.org/kpionto.
[6] http://xsb.sourceforge.net/.

list of measures defined in the Knowledge Graph is as follows: *Positive,
Cumulative_Positive, Negative, Deaths, Cumulative_Deaths, Recovered,
Cases, ICU, Cumulative_ICU, ICU_on_Positive_Rate.*

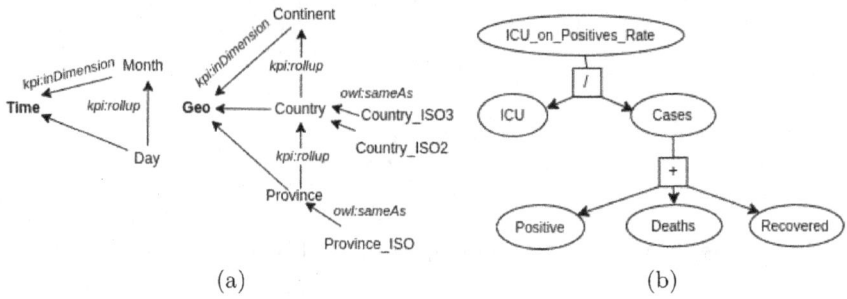

(a)   (b)

Fig. 2. Case study: (a) dimensions, levels and (b) indicators with their formulas.

4   Integration and Mapping Discovery

This section is aimed to discuss (a) how to identify, given a new data source,
dimensions and measures, and (b) how to properly map them to the Knowledge
Graph. In the following, we refer to data *domain* as a set of values from a
data source. If the data source is a relation table, a domain can be seen as the
projection of one attribute. Conversely, if the data source is a JSON collection, a
domain is the set of values extracted from all the included documents according
to a given path (e.g., using JSONPath expression). As a result, a data source
corresponds to a set of domains.

Identification of Dimensions. In order to identify whether a given domain
from a data source (e.g., the attribute *countryname* in S4) and a dimensional
level (e.g., *Geo. Country*) represent the same concept, a matching step is required.
One of the most widely adopted index for comparing sets is the Jaccard similarity
coefficient, aimed at measuring the similarity between finite sets as the ratio
between their intersection and their union. When sets are skewed, i.e., have very
different cardinality, this index is however biased against the largest one. In such
contexts an asymmetric variant can be used, namely the *set containment*, that
is independent on the dimension of the second set.

Definition 1 *(Set Containment). Given two sets X,Y, the set containment is
given by* $c(X,Y) = \frac{|X \cap Y|}{|X|}$.

Given that the cardinality of a domain (without duplicates) is typically much lower than that of a dimensional level, this index is better suited than Jaccard to evaluate whether a domain has intersection with a given level[7]. For this reason, we rely on set containment in our work. As an example, let us consider a domain $A = \{Rome, Berlin, Paris\}$ and a dimensional level $Geo.City$ including 100 cities in Europe. In this case, $c(A, Geo.City) = \frac{3}{3}$, meaning that the domain perfectly matches the dimensional level, while $Jaccard(A, Geo.City) = \frac{3}{100}$. We formalize the problem of mapping a domain of a data source to a dimensional level as a reformulation of the *domain search* problem [16], which belongs to the class of R-nearest neighbor search problems. We first give the definition of relevant dimensional level for a given domain as follows.

Definition 2 *(Relevant dimensional levels for a domain). Given a set of dimensional levels $\mathcal{L}$, a domain D, and a threshold $t \in [0, 1]$, the set of relevant dimensional levels from $\mathcal{L}$ is $\{X : c(D, X) \geqslant t, X \subseteq \mathcal{L}\}$.*

The number of relevant dimensional level for a domain may be greater than one, although in practice we are interested in the level with the greatest threshold t, i.e., the most relevant dimensional level. As an example, the most relevant dimensional level for domain *country_ region* in data source S1 is *Geo.Country*, while for *iso_ subdivision* is *Geo.Province_ISO*.

Comparing a given domain to a dimensional level involves a linear time complexity in the size of the sets. Given the target scenario, which may include data sources with hundred of thousands or even millions of tuples, the computation of the index may often be not scalable in many practical cases. An improvement discussed in the literature as for the Jaccard index consists in its estimation using MinHash computation [1], which involves performing the comparison on their MinHash signatures instead of on the original sets. Given a hash function h, a domain can be mapped to a corresponding set of integer hash values of the same length. For a domain X, let $h_{min}(X)$ be the minimum hash value. Given two sets X,Y, the probability of their minimum hash values being equal is the Jaccard index, i.e., $P(h_{min}(X) == h_{min}(Y)) = J(X, Y)$. Since the comparison can only be true or false, this estimator has a too high variance for a useful estimation of the Jaccard similarity. However, an unbiased estimate can be obtained by considering a number of hashing functions and averaging results: this is done by counting the number of equivalences in the corresponding minimum hash values and dividing by the total number of hash values for a set.

If data sources have high dimensionality, however, pair-wise comparison is still highly time consuming. In our scenario, for a source with N domains and M dimensional levels the time complexity is in $O(N * M)$. For such a reason, in practice MinHash is used with a data structure capable of significantly reducing the running time, named Locality Sensitivity Hashing (or LSH) [10], a sub-linear approximate algorithm.

---

[7] Under this assumption, the set containment is equivalent to the *overlap* (or Szymkiewicz–Simpson) coefficient, i.e., $\frac{|X \cap Y|}{|min(X,Y)|}$.

While the previous approach is targeted to the Jaccard index, an estimation of the set containment can be obtained through LSH Ensemble [16], which is proved to be suitable for skewed sets and more performing than alternative solutions in terms of accuracy and execution time. In our approach, we rely on LSH Ensemble to obtain the dimensional levels with which a given domain of a source is estimated to have a containment score above a certain threshold.

Definition 3 *(LSH Ensemble). Given a domain D from a data source S, given a set of dimensional levels $\mathcal{L}$, and a threshold $t \in [0,1]$, $LSH\_Ensemble$ is a function returning the set of relevant dimensional levels for D.*

Identification of Measures. In terms of dataset attributes, measures are particular domains which are purely quantitative. As such, unlike dimensional levels, a measure belongs to a certain data type but is not constrained to a finite number of possible values. For this reason, solutions for evaluating domain similarity through containment such as LSH Ensemble cannot be applied.

In this work we rely on a string-similarity approach, namely LCS (Longest Common Subsequence) in the comparison of the attribute names of a data source with the list of measure names in the Knowledge Graph. For each domain, the measure names that have the highest value of LCS, i.e., that are most similar, are returned. This is useful to propose only a subset of the measures defined in the Knowledge Graph to the DL Manager. To make an example, for S2 the measure *in_icu_cumulative* is mapped to the Knowledge Graph measure *Cumulative_ICU*, which is the closest syntactically.

We'd like to note that however a manual revision is ultimately required, as the recognition can be affected by homonyms and unclear or ambiguous wording of the domain names. For instance, for S3 the measure *cases* gets mapped to the Knowledge Graph measure *Cases*, but its meaning is different: indeed, by reviewing the publisher metadata, it is clear that instead it actually accounts for the number of positive cases. As such, it needs to be mapped to the measure *Positive*. More advanced approaches could be considered for this step, including some based on dictionary, semantic similarity (e.g., [2]) or frequency distribution and will be discussed in future work.

Representation of Mappings. Given a domain of a data source and the most relevant dimensional level with respect to a given threshold, the domain is mapped to the corresponding level in the Knowledge Graph.

Definition 4 *(Set of mappings). Let $\mathcal{K}$ be the Knowledge Graph, G_S be a metadata graph for a source S, $D \subseteq S$ be a domain, $L \in \mathcal{L}$ be the most relevant dimensional level for D, the mapping between D and L is defined as a tuple $m = (n_D, n_L, c(D, L))$. The set of mappings M_{G_S} includes all mappings for dimensions in S.*

Similarly, given a domain and a related identified measure, a mapping between the corresponding nodes is created. In the following, we represent by

Dim_S the set of dimensional levels available in a source S and by $Ind_S \subseteq \mathcal{I}$ the set of measures available in S, where $\mathcal{I}$ is the set of indicators defined in $\mathcal{K}$.

5   Query Answering

The mappings defined between the metadata graphs and the Knowledge Graph are exploited to support query-driven discovery and query answering in the Data Lake context. This requires to determine what data sources are needed and how to combine them for a given request. A user query Q is expressed as a tuple $Q = \langle ind, \{L_1, \ldots, L_n\}\rangle$, where ind is an indicator and $\{L_1, \ldots, L_n\}$ is a set of levels, each belonging to a different dimension. A data source S has a compatible dimensional schema with respect to a query if S contains a subset of the levels in the query.

Definition 5 *(Compatible dimensional schemas). Given a data source S, given a query $Q = \langle ind, \{L_1, \ldots, L_n\}\rangle$, the dimensional schema of S is compatible with Q iif $Dim_S \subseteq \{L_1, \ldots, L_n\}$.*

For all dimensions of the query that are not included in S, the source is assumed to supply such dimensions at the most aggregate level. For instance, if a query requires indicator *Positive* for *Country*, *Day* and *Age group*, data source $S2$ has a compatible dimensional schema: the missing level *Age group* is assumed to be aggregated at the highest level and therefore not reported. A data source can respond a query if its dimensional schema is compatible and if it provides the requested indicator. On the other hand, if the indicator is not provided by any source but it can be calculated from other indicators, a set of data sources may collectively answer the query if they have a compatible dimensional schema and provide all the component indicators. In the latter case, the actual calculation of the indicator requires to join the needed data sources.

Definition 6 *(Existence of a solution). Given a query $Q = \langle ind, \{L_1 \ldots, L_n\}\rangle$ and a set S of data sources, Q has a solution iif: either (1) $\exists S_x \in S$ such that $ind \in Ind_{S_x} \wedge Dim_{S_x} \subseteq \{L_1, \ldots, L_n\}$, or (2) $\exists$ a formula $f_\alpha = ind = f(ind_1, \ldots, ind_m)$ such that $\forall ind_i$ $(\exists S_i \in S$ such that $ind_i \in Ind_{S_i} \wedge Dim_{S_i} \subseteq \{L_1, \ldots, L_n\})$.*

In the current framework, the derivation of a formula for an indicator relies on the reasoning services introduced in Sect. 3. A detailed discussion of the working mechanism for the services is available in [3]. Query answering is aimed to retrieve a (sorted set of) solution(s) from a user query and involves the following steps, that are summarized in Algorithm 1:

- (Line 1) the algorithm takes as input the query $Q = \langle ind, \{L_1, \ldots, L_n\}\rangle$ and an integer k representing the number of solutions to retrieve.
- (Line 2) the *find_rewriting(ind,$\{L_1, \ldots, L_n\}$)* function is executed, which returns a formula for ind such that all its component measures $\{ind_1, \ldots, ind_m\}$ are provided by some data source according to Definition 6,

and such sources are compatible with the dimensional schema of the query. For this task, reasoning services are exploited that are capable of manipulating the mathematical relations among indicators to retrieve alternative rewritings of a formula. The function returns the retrieved formula f_α and, for each component of the formula ind_i, a set $\Phi_i \subseteq S$ of sources that provide ind_i. In other terms, Φ_i includes the (alternative) data sources from which ind_i can be retrieved.

- (Line 3) the cartesian product of all the sets Φ_i is computed in order to list all combinations of data sources that can be used to calculate the formula, where a combination is a tuple $\langle S_1, \ldots, S_m \rangle$. The set $\Phi_{\bowtie}$ includes all alternative sets of sources capable of providing a solution.
- (Lines 4–7) given that more than a single solution may be available, due to the fact that multiple sources can provide the same measure, sorting them according to a quality index is needed. Although a set of sources may provide the needed measures, their join does not necessarily produce a non-empty result. Here, we refer to the *degree of joinability*, discussed below, which measures the likelihood to produce a result out of a join between two (set of) domains. For each tuple $\langle S_1, \ldots, S_m \rangle$ in $\Phi_{\bowtie}$, a new tuple $\langle x, \{S_1, \ldots, S_n\} \rangle$ is produced, where $x \in [0, 1]$ is the degree of joinability among sets $S_1, \ldots, S_m$.
- (Line 8) the set Ψ is sorted in descending order by the degree of joinability.
- Finally, the formula f_α and the k-top solutions in Ψ_{sort} are returned.

Algorithm 1. Query answering

1: **function** FINDSOLUTION($\langle ind, \{L_1, \ldots, L_n\}\rangle$,k)
2:   $(f_\alpha(ind_1, \ldots, ind_m), \{\Phi_1, \ldots, \Phi_n\}) = find\_rewriting(ind, \{L_1, \ldots, L_n\})$
3:   $\Phi_{\bowtie} = \times_{i=1}^{n} \Phi_i$
4:   $\Psi = \varnothing$
5:   **for** $\langle S_1, \ldots, S_m \rangle \in \Phi_{\bowtie}$ **do**
6:     $\Psi \leftarrow compute\_joinability(\{S_1, \ldots, S_m\}, \{L_1, \ldots, L_n\})$
7:   **end for**
8:   $\Psi_{sort} \leftarrow sort(\Psi)$
9:   **return** $\langle f_\alpha, \Psi_{sort,k} \rangle$
10: **end function**

In the following, we discuss the degree of joinability index and the procedure for its computation. Two sources are joinable if they have the same values for domains that are mapped to the same dimensional levels. To check this condition, the corresponding domains should be compared in order to determine how many values are shared between the sources through set containment. However, a full comparison is not practical in a Data Lake scenario. For this reason, we resort to the LSH Ensemble to provide an estimated evaluation of the joinability of two data sources. Typical use of LSH Ensemble is based on single join attribute at a time (similarity between sets), while in our case the match needs to be performed on sets of dimensional levels. Hence, we apply a combination function (e.g., a concatenation of strings) to the domains that represent the dimensional levels, in order to map them onto a single domain before applying the hashing function. To give an example, if the query requires levels *Geo.Country* and *Time.Day*, the hash will be calculated on the concatenation of domains *country_region + updated* for source S1 (a possible value is "Italy 2020-11-30"). Such "combined MinHashes" corresponding to concatenated domains for each source are precomputed during the integration and mapping step and stored in order to speed up the query answering.

Finally, as summarized by Algorithm 2 (lines 3–7), the degree of joinability is iteratively calculated by executing the LSH_Ensemble for each pair of data sources (S_i, S_{i+1}) (line 4) and considering the product of the obtained values (line 6). Given that the containment is an asymmetric index, we consider the application of LSH_Ensemble in both directions (from source i to source $i+1$ and vice versa, according to the semantics of an inner join) and check for the highest threshold t (line 5). The search of such a threshold is done through binary search.

Algorithm 2. Computing degree of joinability

1: **function** COMPUTEDEGREEOFJOINABILITYSCORE($\{S_1, \ldots, S_m\}, \{L_1, \ldots, L_m\}$)
2:    $MH \leftarrow retrieve\_combined\_MinHashes(\{S_1, \ldots, S_m\}, \{L_1, \ldots, L_m\})$
3:    joinability = 1
4:    **for** i=1..n-1 **do**
5:       get the highest t such that max of $|LSHEnsemble(MH_i, MH_{i+1})|$ and $|LSHEnsemble(MH_{i+1}, MH_i)|$ is > 0
6:       $joinability = joinability * t$
7:    **end for**
8:    **return** $\langle j, \{S_1, \ldots S_m\}\rangle$
9: **end function**

---

6   Evaluation

An evaluation of the approach on the case study is proposed here. Tests have been carried out on an Intel Core i5-1135G7, 8 cores @ 2.40 GHz, x86_64 architecture, with 8 GB RAM running Linux Fedora 34. A single-thread implementation of the approach has been used, relying on the Python library datasketch 1.5.7 [18] for the implementation of MinHash and LSH Ensemble and on pandas 1.3.3 for manipulation of data structures.

Integration and Mapping Discovery. A preliminary setup of the Knowledge Graph for the Data Lake has been performed by defining dimensional levels and measures. Members of levels have been defined programmatically from available online resources (e.g., list of countries and corresponding ISO alpha 2 and alpha 3 codes[8]). For any loaded data source, initialization includes computation of

---

[8] https://gist.github.com/tadast/8827699.

MinHashes for any domain, mapping with the dimensional levels and precomputation of the combined MinHashes for domains mapped to dimensional levels. For LSH Ensemble we set the number of hashing permutations to 256 and number of parts to 32. We report the average and the total execution time in Table 2 and some of the mappings in Table 3. Domains are processed in less than 1.6 s on average.

Table 2. Case study: execution time for MinHash generation and mapping.

		S1	S2	S3	S4	S5
Hashing computation	Avg [s]	1.654	0.050	0.005	0.009	0.075
	Total [s]	28.125	1.557	0.076	0.376	0.375
Mapping to dimensional levels	Avg [s]	<0.001	<0.001	<0.001	<0.001	<0.001
	Total [s]	0.006	0.012	0.011	0.033	0.003
Precomputation of dimensional MinHashes for querying	Total [s]	21.235	0.151	0.423	1.610	0.918

Table 3. The set of Knowledge Graph levels and measures which source domains are mapped to.

Source	K levels	K measures
S1	Time.Day, Geo.Country	Positive, Recovered, Deaths
S2	Time.Day, Geo.Province	ICU, Positive, Negative, Recovered, Deaths
S3	Time.Day, Geo.Country	Positive, Deaths
S4	Time.Day, Geo.Country	Cumulative_Positive, Cumulative_Deaths
S5	Time.Day, Geo.Country	ICU, Cumulative_ICU

Query Answering. Let us assume the user is interested in analysing measures *Positive* and *ICU_on_positives_rate* at *Geo.Country* and *Time.Day* levels. As for the first measure, the *find_rewriting* returns $(Positive, \{\{S1\}, \{S3\}\})$. In this case, no join is needed as the measure is directly available from multiple sources.

As for the second measure, the function returns $(\frac{ICU}{Positives}, \{\{S5\}, \{S1, S3\}\})$. Combination of sources are produced and two alternative solutions are combining S5 with either S1 or S3. They are checked for joinability as follows:

- S5,S1: the degree of joinability is 0.78, with a query time equal to 3.109 s;
- S5,S3: the degree of joinability is 0.31, with a query time equal to 3.283 s.

As a result, the solution (S5,S1) is preferred over (S5,S3). This is motivated by the fact that S5 and S1 include data for both years 2020 and 2021, while S3 includes data only on year 2020. Therefore, the degree of joinability of S3 is lower than that of S1, as the former shares a smaller subset of data with S5.

Discussion. The approach proposed in [16] requires, for a given query, a number of set containment evaluations increasing linearly with the number of sources. On the other hand, the present approach enables to reduce such a number to only the relevant sources (2 in the example) by performing a preliminary evaluation based on formula rewriting. In general, by considering M measures and N sources, the approach requires a number of evaluations equal to $\frac{N}{M}$, on average. If indicators are not available at the requested dimensional schema, decomposing indicators in components requires a further number of evaluations. By considering an average number s of dependencies per indicator and a number l of hierarchical levels in the formula graph, the overall number of components to check for an indicator can analytically be estimated as $(1 + \sum_{i=1}^{l} s^i)\frac{M}{N}$, e.g., for $M = 200$, $N = 10000$, $s = 3$, $l = 2$, corresponding to average formula graphs for real-world frameworks of indicators, the number of evaluations amounts to 500.

7   Conclusion

This paper has introduced a knowledge-based approach for analytic query-driven discovery in a Data Lake, which is characterized by the formal representation of indicators' formulas and efficient mechanisms for source integration and mapping discovery. Starting from a query, which is expressed ontologically as a measure of interest and relevant analysis dimensions, the framework determines the set of sources that are capable of collectively responding, by exploiting math-aware reasoning on indicator formulas. A quantitative evaluation of the result, in terms of joinability of sources, is provided through the degree of joinability index. Future work will be devoted to individuate real case studies for extensive evaluation and to extend the approach towards interesting research directions. In particular, the degree of joinability could be adapted to evaluate the completeness of a data source with respect to the Knowledge Graph concepts. This would enable to determine the scope of a source and paves the way for an efficient evaluation of the overlapping or complementarity among sources, and possible more efficient indexing approaches. Merging capabilities could also be beneficial to find unionable sources and hence to vertically integrate data providing the same measures. Finally, dynamic calculation of indicators can be envisaged for a variety of analytical tasks, including interactive data exploration or navigation [17].

References

1. Broder, A.Z .: On the resemblance and containment of documents. In: Proceedings of Compression and Complexity of SEQUENCES 1997 (Cat. No. 97TB100171), pp. 21–29. IEEE (1997)
2. Diamantini, C., Lo Giudice, P., Potena, D., Storti, E., Ursino, D.: An approach to extracting topic-guided views from the sources of a data lake. Inf. Syst. Front. **23**, 243–262 (2021)

3. Diamantini, C., Potena, D., Storti, E.: Analytics for citizens: a linked open data model for statistical data exploration. Concurr. Comput. Pract. Exp. **33**(8), e4186 (2021)

4. Diamantini, C., Potena, D., Storti, E.: A semantic data lake model for analytic query-driven discovery. In: The 23rd International Conference on Information Integration and Web Intelligence, iiWAS2021, pp. 183–186. Association for Computing Machinery, New York, NY, USA (2021)

5. Farid, M., Roatis, A., Ilyas, I.F., Hoffmann, H., Chu, X.: CLAMS: bringing quality to Data Lakes. In: Proceedings of the International Conference on Management of Data (SIGMOD/PODS 2016), pp. 2089–2092. ACM, San Francisco, CA, USA (2016)

6. Fernandez, R.C.: Seeping semantics: linking datasets using word embeddings for data discovery. In: 2018 IEEE 34th International Conference on Data Engineering (ICDE), pp. 989–1000. IEEE (2018)

7. Giebler, C., Gröger, C., Hoos, E., Schwarz, H., Mitschang, B.: Leveraging the data lake: current state and challenges. In: Ordonez, C., Song, I., Anderst-Kotsis, G., Tjoa, A.M., Khalil, I. (eds.) Big Data Analytics and Knowledge Discovery. pp, pp. 179–188. Springer International Publishing, Cham (2019). https://doi.org/10.1007/978-3-030-27520-4_13

8. Hai, R., Geisler, S., Quix, C.: Constance: an intelligent data lake system. In: Proceedings of the International Conference on Management of Data (SIGMOD 2016), pp. 2097–2100. ACM, San Francisco, CA, USA (2016)

9. Hale, T., Webster, S., Petherick, A., Phillips, T., Kira, B.: Oxford COVID-19 government response tracker. Technical report, Blavatnik School of Government (2020)

10. Indyk, P., Motwani, R.: Approximate nearest neighbors: towards removing the curse of dimensionality. In: Proceedings of the Thirtieth Annual ACM Symposium on Theory of Computing, pp. 604–613 (1998)

11. Microsoft. Covid-19 data lake. https://docs.microsoft.com/en-us/azure/open-datasets/dataset-covid-19-data-lake. Accessed 23 Feb 2022

12. Miller, R.J.: Open data integration. Proc. VLDB Endow. **11**(12), 2130–2139 (2018)

13. Nargesian, F., Zhu, E., Miller, R.J., Pu, K.Q., Arocena, P.C.: Data lake management: challenges and opportunities. Proc. VLDB Endow. **12**(12), 1986–1989 (2019)

14. Oram, A.: Managing the Data Lake. O'Reilly, Sebastopol (2015)

15. Sawadogo, P., Darmont, J.: On data lake architectures and metadata management. J. Intell. Inf. Syst. **56**(1), 97–120 (2020). https://doi.org/10.1007/s10844-020-00608-7

16. Zhu, E., Nargesian, F., Pu, K.Q., Miller, R.J.: LSH ensemble: internet-scale domain search. Proc. VLDB Endow. **9**(12), 1185–1196 (2016)

17. Zhu, E., Pu, K.Q., Nargesian, F., Miller, R.J.: Interactive navigation of open data linkages. Proc. VLDB Endow. **10**(12), 1837–1840 (2017)

18. Zhu, E., Markovtsev, V.: ekzhu/datasketch: first stable release, February 2017. https://doi.org/10.5281/zenodo.290602

Insight-Based Vocalization of OLAP Sessions

Matteo Francia, Enrico Gallinucci, Matteo Golfarelli,
and Stefano Rizzi(✉)

DISI - University of Bologna, Bologna, Italy
{m.francia,enrico.gallinucci,matteo.golfarelli,stefano.rizzi}@unibo.it

Abstract. Carrying out OLAP analyses in hands-free scenarios requires lean forms of communication between the users and the system, based for instance on natural language. In this paper we introduce VOOL, a framework specifically devised for vocalizing the insights resulting from OLAP sessions. VOOL is self-configurable, extensible, and is aware of the user's intentions expressed by OLAP operators. To avoid overwhelming the user with very long descriptions, we pursue the vocalization of selected insights automatically extracted from query results. These insights are detected by a set of modules, each returning a set of independent insights that characterize data. After describing and formalizing our approach, we evaluate it in terms of efficiency and effectiveness.

Keywords: Vocalization · OLAP · Data warehouse

1 Introduction

The democratization of data access pushes towards the adoption of OLAP (On-Line Analytical Processing) tools, which make data fruition and analysis easier by enabling "point-and-click" queries on the multidimensional cubes stored in a data warehouse. The scenarios requiring hand-free interfaces (e.g., in the field of augmented reality [9] or smart assistants [7]) make this push even more pressing and ask for the introduction of leaner forms of communication between the users and the system, based for instance on natural language. As argued in [28], this setting is not only motivated by the needs of specific user groups, such as visually-impaired ones. More in general, we are assisting to a shift of user-computer communication towards voice interfaces, which are more convenient if users are distracted or unable to access screen and keyboard. While the translation of natural language into actionable OLAP queries has recently been addressed [7], the way to the vocalization of query results has been paved only partially. The goal of this paper is to contribute to bridging this gap.

The description of the many facets shown by the cube resulting from an OLAP query can span from simple insights (e.g., min/max or Top-k) to complex ones (e.g., clusters and outliers); these, in turn, can be representative of very different amounts of facts in the cube. Additionally, according to the OLAP paradigm, data analyses come in the form of sessions, where a query q' can be

© Springer Nature Switzerland AG 2022
S. Chiusano et al. (Eds.): ADBIS 2022, LNCS 13389, pp. 193–206, 2022.
https://doi.org/10.1007/978-3-031-15740-0_15

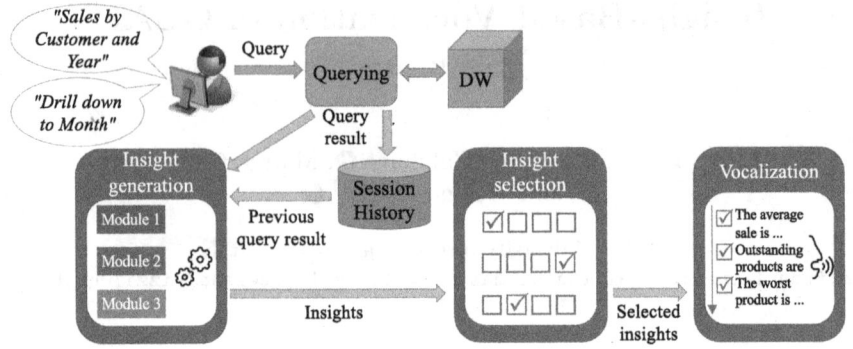

Fig. 1. Functional view of VOOL

obtained by applying an OLAP operator to the previous one, q. Hence, differently from a generic sequence of stand-alone queries, q and q' are strongly related, which enables the detection of insights based on the comparison of the results of consecutive queries. These insights should also be related to the intention expressed by the user when applying an OLAP operator; for instance, drill-down refines the previous result while slice-and-dice shifts the focus to a specific part of the result. Overall, in our vision, the desiderata for a framework for the vocalization of OLAP sessions are the following:

#1 *Intention-awareness*: it must generate vocalizations that describe the comparison of the results of subsequent queries rather than those of a single one; in generating such vocalizations, it must consider the user's intention as expressed by the OLAP operator and aggregation operator applied.

#2 *Extensibility*: it must rely on interfaces that make ad-hoc modules easily pluggable since vocalization is inherently multi-faceted.

#3 *Timeliness*: it must produce vocalizations responsively, avoiding long delays in returning results to the user.

#4 *Conciseness*: it must produce vocalizations that take a limited time not to overwhelm the user.

Following these desiderata, in this paper we present *VOOL*, a framework specifically devised for the VOcalization of OLap sessions. A functional view of VOOL is sketched in Fig. 1 (the querying component is grayed out since it is out of the scope of this paper and has been extensively discussed in [7]). Given the result of either a completely-specified query (e.g., "Sales by Customer and Year") or an OLAP operator that refined the previous query (e.g., "Drill down to Month"), the insight generator executes concurrent modules, each returning a set of independent insights that characterize this result. Out of all the insights returned, the insight selector applies an optimization algorithm to return only the most relevant insights given a limited budget (e.g., related to the duration of vocalization); these insights are then sorted into a comprehensive description that is vocalized to the user.

The remainder of the paper is organized as follows. Section 2 discusses the related work. Section 3 provides an overview of VOOL by sketching its functional architecture and the vocalization steps. Section 4 formalizes the necessary background. Section 5 describes the vocalization process, while Sect. 6 details one of the modules we implemented to support the VOOL framework. Finally, Sect. 7 evaluates the approach by means of a set of tests, draws the conclusions, and envisions the directions for evolving VOOL.

2   Related Work

The first research area that intersects with our contribution is *exploratory data analysis*, a knowledge discovery process in which users explore datasets through sessions that concatenate a sequence of operations. In this context, two interesting research directions are *recommendation* and *insight extraction*. As to recommendation, many studies focus on learning users' preferences and profiling data to give recommendations on the exploration path [25,27]. This is orthogonal to our approach since our goal is not to suggest to the user how to build a session, but rather to return concise insights on the data resulting from a user-defined session. As to insight extraction, OLAP comes with well-known operators to explore multidimensional cubes [22]. Additional operators have been recently classified as [13] *coverage* (returning insights that cover tuples with certain values [4,12]), *information* (returning insights providing information about the distribution of measure values [10]), and *contrast* (returning insights occurring with some values but not the others [8,24]). These operators are complementary to VOOL, since they are potential modules to be plugged into VOOL (as we have already done for [4,8,10,29]). Cinecubes [11], the contribution closest to VOOL, compares the results of a query to results obtained over sibling values or drill-downs to produce insight. With respect to Cinecubes, VOOL allows the description and vocalization of a user-defined OLAP session and also leverages the user's intentions to drive insight extraction.

Another research area closely related to our work is that of *conversational systems*. Natural language interfaces to operational databases enable users to specify complex queries without previous training on formal programming languages (such as SQL) and software [1]. Some examples of approaches that translate natural language into formal SQL/OLAP queries are [7,17,23]. In hand-free scenarios, some emphasis has been given on the one hand to providing effective summarizations of query results, which enables the creation of concise analytic reports [6,9]; on the other hand, some vocalization approaches have been proposed. In [26], the authors translate a database subset into a narrative that synthesizes the contents of the subset following a set of rules and templates. In [5], the authors leverage the provenance of tuples in the query result, detailing not only the results but also their explanations. Finally, a couple of works are placed in the context of multidimensional data and OLAP. In [28], the authors sample the database to evaluate alternative speech fragments; OLAP queries are not fully evaluated and sampling focuses on result aspects that are relevant for voice output. In [21] an end-to-end dialog system is introduced, but the vocalization approach is limited (when too many rows should be returned, only the count of rows is returned).

Overall, in the light of the above-mentioned contributions, it appears that the road to full-fledged conversation-driven OLAP is not paved yet, since end-to-end conversational frameworks are not provided in the domain of analytic sessions over multidimensional data. The closest contribution to VOOL is [28]; however, differently from VOOL, that approach only copes with stand-alone queries, so the vocalization does not take into account the comparison of the sequential query results emerging from OLAP sessions; besides, it is not extensible and does not provide a dynamic interest-based vocalization of the insights.

3   Overview

The interaction with VOOL takes place as follows. (i) A user issues an *initial* query (typically, the first one in a session), whose result is computed; (ii) a set of vocalization insights (i.e., descriptions of insights) are extracted out of the query result; (iii) the most interesting insights are vocalized. Every time the user issues a new query by applying an OLAP operator (obtaining a *refined query*), this process is repeated; the difference is that the insights extracted may describe not only the result of the last query, but also its comparison with the results of the previous query.

Vocalization of Initial Queries. The result returned by the *Querying* component is sent to the insight generation step, in which a set of modules analyzes the query result to produce different types of insights. Each insight is characterized by a natural language description, an interestingness, a coverage (i.e., the number of tuples covered by the description), and the cost necessary for its vocalization (e.g., the number of words of its natural language description). An example of natural language description for an insight produced by a Top-k module is "The facts with highest Quantity are Beer with 80, Wine with 70, and Cola with 30". Since the number of insights can be arbitrarily high (a module can return any number of insights and there is no limit to the number of modules), the insight selection step determines the insights eligible for vocalization in such a way that the total time necessary for vocalization does not overcome a given time budget and the total interestingness is as high as possible. Finally, the selected insights are vocalized from the most general (i.e., those with high coverage) to the most specific (those with low coverage).

Vocalization of Refined Queries. The results of the current query and of the previous one are sent to the insight generation step. In this case, both cubes are used to extract the insights entailing the comparison and description of consecutive results. For instance, in the sales domain, after drilling down sales from product category to product, a user might be interested in outstanding products that were previously hidden within average-performing categories. After insight generation, vocalization proceeds as for an initial query.

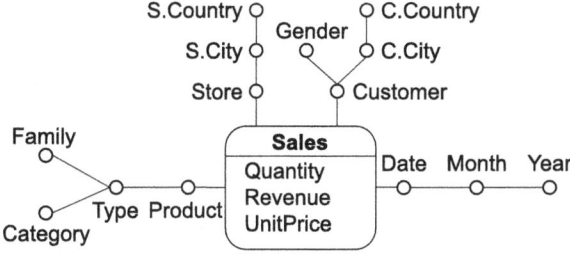

Fig. 2. (Simplified) DFM representation of the Sales cube schema

4 Formal Background

A *cube* is the multidimensional representation of a business phenomenon relevant for decision making, and is defined through the following steps.

Definition 1 (Hierarchy and Cube Schema). *A* hierarchy *is a couple* $h = (L, \succeq)$ *where (i) L is a set of categorical* levels, *each level l being coupled with a domain of* members, $Dom(l)$; *(ii) $\succeq$ is a* roll-up *partial order of L. A* cube schema *is a couple $C = (H, M)$ where (i) H is a set of hierarchies; (ii) M is a set of numerical measures, each coupled with an aggregation operator $op(m) \in \{\mathsf{sum}, \mathsf{avg}, \mathsf{min}, \mathsf{max}\}$.*

Example 1. As a working example we will use cube schema $\mathsf{Sales} = (H, M)$, whose conceptual representation according to the DFM [14] is shown in Fig. 2:

$$H = \{h_{\mathsf{Date}}, h_{\mathsf{Customer}}, h_{\mathsf{Store}}, h_{\mathsf{Product}}\}$$
$$M = \{\mathsf{Quantity}, \mathsf{Revenue}, \mathsf{UnitPrice}\}$$
$$\mathsf{Date} \succeq \mathsf{Month} \succeq \mathsf{Year}, \mathsf{Store} \succeq \mathsf{S.City} \succeq \mathsf{S.Country}, \ldots$$

We have $op(\mathsf{Quantity}) = op(\mathsf{Revenue}) = \mathsf{sum}$ and $op(\mathsf{UnitPrice}) = \mathsf{min}$. □

Aggregation is the basic mechanism to query cubes, and it is captured by the following definition of group-by. As normally done when working with the multidimensional model, if a hierarchy h does not appear in a group-by it is implicitly assumed that a complete aggregation is done along h.

Definition 2 (Group-by and Coordinate). *Given cube schema $C = (H, M)$, a* group-by *of C is a tuple G of levels. A* coordinate *of group-by G is a tuple of members, one for each level of G.*

Definition 3 (Base Cube). *Let G_0 be the finest group-by. A* base cube *over C is a partial function C_0 that maps the coordinates of G_0 to a numerical value for each measure m in M.*

Each coordinate γ that participates in C_0, with its associated tuple of measure values, is called a *fact* of C_0. The value taken by measure m in the fact corresponding to γ is denoted $\gamma.m$. With a slight abuse of notation, we will also consider a cube as the set of the coordinates corresponding to its facts.

Example 2. Three group-by's of Sales are $G_0 = \langle \mathsf{Date}, \mathsf{Customer}, \mathsf{Store}, \mathsf{Product} \rangle$, $G_1 = \langle \mathsf{Month}, \mathsf{C.City}, \mathsf{Gender} \rangle$, and $G_2 = \langle \mathsf{Year} \rangle$, where $G_0 \succeq_H G_1 \succeq_H G_2$. Coordinates of the three group-by's are, respectively, $\gamma_0 = \langle 2021\text{-}04\text{-}15, \mathsf{Rossi}, \mathsf{BigMart}, \mathsf{Beer} \rangle$, $\gamma_1 = \langle 2021\text{-}04, \mathsf{Rome}, \mathsf{Male} \rangle$, and $\gamma_2 = \langle 2021 \rangle$.   □

Definition 4 ((Cube) query). *Given a base cube C_0 over schema $\mathcal{C}$, a query over C_0 is a quadruple $q = (C_0, G_q, P_q, M_q)$ where (i) G_q is a group-by of $\mathcal{C}$; (ii) P_q is a (possibly empty) set of selection predicates each expressed over one level of H; (iii) $M_q \subseteq M$.*

Example 3. A query over Sales is the one returning the total quantity sold by product, which can be formalized as $q = (C_0, G_q, P_q, M_q)$ where $G_q = \{\mathsf{Product}\}$, $P_q = \varnothing$ (i.e., no selection predicate is applied), and $M_q = \{\mathsf{Quantity}\}$.   □

An *OLAP session* is a sequence of queries; the first query in a session is completely specified, while each of the others is obtained as a refinement by applying an *OLAP operator* to the result of the previous one. A formal definition of the OLAP operators we will consider in this work can be found in [7]; here we just give an intuition:

- *Roll-up* aggregates data (e.g., from Product to Type).
- *Drill-down* disaggregates data (e.g., from Type to Product).
- *Slice-and-dice* filters data based on a predicate (e.g., Product = 'Beer').

5   The Vocalization Process

As already stated, the VOOL framework includes three main stages, namely, *Insight generation*, *Insight selection*, and *Vocalization*; all of these are described in the following subsections.

5.1   Insight Generation

At this stage, a set of *modules* (e.g., the Top-k function or a clustering function) are executed to extract *insights* (e.g., the top-3 facts or a pair of clusters) describing query results. An insight consists of a set of *components*, each describing either a single fact (e.g., a fact belonging to the top-3 facts) or a group of facts (e.g., a cluster).

Definition 5 (Module). *Given two cubes C and C' resulting from two consecutive queries in an OLAP session (with $C = \varnothing$ when vocalizing an initial query), a module is a function $F(C, C') = S^F$, where S^F is a set of insights.*

The executability of a module is subject to the fulfillment of specific conditions, which may concern the applied OLAP operators, the measures involved in the query result, and the aggregation operator used in the query. Note that, consistently with desideratum #2 (Extensibility), this definition allows the application of any function capable to extract insights from one or two cubes.

$C' = q(\text{Sales}, \{\text{Product}\}, \varnothing, \{\text{Quantity}\})$

Product	Quantity
Beer	80
Wine	70
Cola	30
Bagel	8
Pizza	6
Bread	5

Fig. 3. The cube C' resulting from the (initial) query in Example 3, which represents the Quantity sold by Product

Definition 6 (Insight). *An insight $s \in F(C, C')$ is a set of components; each component $v \in s$ describes a set of facts of C', denoted as $Desc(v)$. Insight s is characterized as follows:*

(i) $NL(s)$ is the natural language description of s.

(ii) $int(s)$ is the interestingness of the insight, i.e., its estimated relevance to the decision-making process, defined as

$$int(s) = \sum_{v \in s} int(v)$$

where $int(v) \in (0, 1]$ is the interestingness of component v.

(iii) $cov(s) \in (0, 1]$ is the fraction of cube facts covered by the insight, called coverage:

$$cov(s) = \frac{|\bigcup_{v \in s} v|}{|C'|}$$

(iv) $cost(s) \in \mathbb{N}$ is the cost related to the vocalization of s, measured as the number of words in $NL(s)$.

The natural language descriptions of insights, $NL(s)$, are generated from pre-defined module-specific grammars. The interestingness of insight components, $int(v)$, is also specific of each module; an example of how $int(v)$ is defined for one module will be provided in Sect. 6. Intuitively, an insight with high coverage is more general, one with small coverage is more specific.

Definition 7 (Insight space). *Let $\mathcal{F}$ be the set of all modules. Given two consecutive cubes C and C' in an OLAP session (possibly with $C = \varnothing$), their insight space is the set of the sets of insights produced by all modules:*

$$\mathcal{S} = \{F(C, C'); F \in \mathcal{F}\} = \{\{s_1^F, \ldots, s_n^F\}; F \in \mathcal{F}\}$$

To enable concurrent and efficient insight generation and selection (as shown later), we make two assumptions on insights and modules:

1. Each insight s is *self-contained*, i.e., $NL(s)$ contains all the information necessary for vocalization and is a self-standing sentence. As a consequence, insights can be vocalized independently of each other.

Table 1. Examples of insights describing the query result in Fig. 3

Module	Insight	NL	int	cov	$cost$
Statistics	s^P	The average Quantity is 33.2	0.0	1.0	5
Top-k	s_1^T	The fact with highest Quantity is Beer with 80	0.4	0.2	9
	s_3^T	The three facts with highest Quantity are Beer with 80, Wine with 70, and Cola with 30	1.0	0.5	17
Clustering	s_1^C	Facts can be grouped into 2 clusters, the largest one has 4 facts and 12 as average Quantity	0.8	0.7	18
	s_2^C	Facts can be grouped into 2 clusters, the largest one has 4 facts and 12 as average Quantity, the second one has 2 facts and 75 as average Quantity	1.6	1.0	29
Assess	s_1^A	When compared to the previous query, the Quantity of Pizza is 6, tantamount to the average Quantity of Food that is 6.3	1.0	0.2	22

2. The insights generated from the same module F are *incremental*, i.e., they can be arranged into a sequence where the description of one insight extends the previous one by including one more component. In the following we will assume that the resulting inclusion (total) ordering is reflected in the ordering of indices: $S^F = \{s_1^F, \ldots, s_n^F\}$, with $cov(s_{i+1}^F) \geq cov(s_i^F)$, $int(s_{i+1}^F) \geq int(s_i^F)$, and $cost(s_{i+1}^F) \geq cost(s_i^F)$ for $1 \leq i \leq n-1$.

Example 4. Given the query result from Fig. 3, examples of insights produced by different modules are shown in Table 1. Note that, from the informative point of view, an insight may be an extension of another insight because it includes additional components (e.g., s_3^T extends s_1^T with two additional components corresponding to two facts, namely, Wine and Cola, while s_2^C extends s_1^C with an additional component corresponding to a cluster including two facts). □

5.2 Insight Selection

The insight space $\mathcal{S}$ can be very large, so a selection must be done on the insights to be vocalized. The goal of this step is to find the set of insights $\overline{S} \subseteq \mathcal{S}$ such that (i) the total interestingness is maximum and (ii) the total cost does not exceed a given time budget t_{voc} (see desideratum #4, Conciseness). Expressing t_{voc} in seconds makes budget definition intuitive for users. However, the insight cost has been defined as the number of words in its textual description, so as to decouple it from its vocalization (the optimal speech rate may depend on the target audience). Transforming t_{voc} into a maximum number of words is straightforward; for instance, 180 is the average number of words per minute for English speakers/readers [3].

The one formulated above is clearly an optimization problem, with two additional issues to be considered: non-redundancy and right-time response.

Non-redundancy. While, by assumption, different modules produce insights with different semantics, the insights from the same module have overlapping content (since they are built incrementally). As a consequence, given a module F and its output S^F, at most one insight $s_z^F \in S^F$ should be selected. Insight selection can then be formulated as a multiple-choice knapsack problem (MCKP), a generalization of the ordinary knapsack problem. In the MCKP, the set of items (S) is partitioned into classes (the S^F's) and the binary choice of taking or not an item is replaced by the selection of at most one item out of each class [16].

Right-Time Response. S is incrementally populated since the modules entail different complexities and execution times and are executed in a *bag-of-task* fashion (see desideratum #3, Timeliness). On the other hand, to preserve the interactivity of OLAP session, vocalization should begin shortly after query execution, without waiting until all the modules have completed their execution. Thus, insight selection is started after a fixed time t_{gen}, and the insights added to S after t_{gen} are not included in the selection process.

5.3   Vocalization

Vocalization starts with a preamble that describes the query (e.g., "The query result shows the sum of quantity grouped by product"). The preamble is a preliminary description which acts as a context for the subsequent insights. Note that, if the time necessary to vocalize the preamble is greater than t_{gen}, the user will not perceive any pause in the vocalization. After the preamble, the insights in $\overline{S}$ are vocalized. Specifically, they are sorted by descending coverage *cov* (i.e., from the most general to most specific), then their natural language descriptions NL's are concatenated and vocalized.

6   A Closer Look at the VOOL Modules

The core set of modules currently implemented is summarized below:

- *Statistics* returns general statistics on the overall result (e.g., the average value of the Quantity measure and its skewness).
- *Bottom-k/Top-k* [20], applied to a single measure, returns the worst/best performing facts (e.g., sales with lowest/highest Quantity).
- *Outliers* [19] returns the facts whose measure values deviate from the data distribution (e.g., anomalous sales).
- *Clustering* [18] returns groups of facts that maximize intra-group similarity and minimize inter-group similarity (e.g., facts with similar Quantity).
- *Correlation* returns the degree of Pearson correlation between pairs of measures (e.g., how Quantity and Revenue correlate).
- *Slicing variance*, applied to a single measure, returns the degree of correlation between the values of a measure in the cubes before and after the application of a slice-and-dice operator (e.g., how quantities by product change after applying selection predicate StoreCity='Rome').
- *Aggregation variance*, applied to a single measure, returns the facts with the highest variation in the values of that measure after a roll-up or drill-down operator (similarly to [4]; e.g., after a roll-up from Product to Category, returns the categories showing the highest variation in products' Quantity).

Note that some of these modules are inspired from well-known approaches (e.g., [4,8,10,29]); in some of these cases we just had to devise a textual description of the insight and/or adapt the returned measures of interestingness/relevance. As already mentioned, this set can smoothly be extended with modules that follow the requirements expressed in Sect. 1.

In the remainder of this section we describe an end-to-end implementation of the Top-k module, which operates on both initial and refined queries. For simplicity, we will drop the superscript denoting the module from the notation of insights. Let $q = (C_0, G_q, P_q, M_q)$ be an initial query, with $M_q = \{m\}$, and C' the resulting cube. The goal of the Top-k module is to describe the three facts in C' having the highest values of m, namely, $\{\gamma_1, \gamma_2, \gamma_3\}$ (we will assume that $\gamma_1.m \geq \gamma_2.m \geq \gamma_3.m \geq \ldots$). Three insights including from one to three components are returned:

$$s_1 = \{\{\gamma_1\}\}, \ s_2 = \{\{\gamma_1\}, \{\gamma_2\}\}, \ s_3 = \{\{\gamma_1\}, \{\gamma_2\}, \{\gamma_3\}\}$$

These insights are characterized as follows:

$$NL(s_k) = \begin{cases} \text{``The fact with highest } m \text{ is } \gamma_1 \text{ with } \gamma_1.m\text{''}, & \text{if } k = 1; \\ \text{``The two facts with highest } m \text{ are } \gamma_1 \text{ with } \gamma_1.m \text{ and} \\ \gamma_2 \text{ with } \gamma_2.m\text{''}, & \text{if } k = 2; \\ \text{``The three facts with highest } m \text{ are } \gamma_1 \text{ with } \gamma_1.m, \\ \gamma_2 \text{ with } \gamma_2.m, \ \gamma_3 \text{ with } \gamma_3.m\text{''}, & \text{if } k = 3; \end{cases}$$

$$cov(s_k) = \frac{k}{|C'|}$$

As to the interestingness, for each component $v_k = \{\gamma_k\}$ it is

$$int(v_k) = \frac{\gamma_k.m - \gamma_{\overline{k}}.m}{\sum_{i=1}^{\overline{k}} (\gamma_i.m - \gamma_{\overline{k}}.m)}$$

where $\overline{k} > 3$. While the coverage formula is obvious, the interestingness of s_k corresponds to the percentage of m that is retained by the Top-k tuples (e.g., the total Quantity retained by the Top-3 products with respect to the overall units sold by the Top-$\overline{k}$ facts). The reason for limiting the denominator to the Top-$\overline{k}$ facts rather than summing on all the query results is to avoid that a *long tail* of several low values makes $int()$ meaningless. Conversely, by considering only the highest non-top values (in our implementation we set $\overline{k} = 6$) the interestingness function properly expresses how high are the Top-3 as compared to the next ones. Finally, the reason why all measure values are shifted by $\gamma_{\overline{k}}.m$ is to cope with the case of negative values (e.g., if the measure expresses a temperature).

As to refined queries, while NL and cov remain unchanged, the interestingness of a component changes depending on the result of the previous query. Given two consecutive cubes C and C', a fact in C' is considered to be interesting (in the sense of *peculiar*) if its measures deviate significantly from those in the corresponding fact(s) of C [8]. This is based on the idea of prior belief [2]: specifically,

$C' = q(\text{Sales}, \{\text{Product}\}, \emptyset, \{\text{Quantity}\})$

Product	Quantity
Beer	80
Wine	70
Cola	30
Bagel	8
Pizza	6
Bread	5

$C = q(\text{Sales}, \{\text{Category}\}, \emptyset, \{\text{Quantity}\})$

Category	Quantity
Beverages	180
Food	19

Fig. 4. Example of corresponding facts (gray lines) between the results of two consecutive queries, C and C', the latter being obtained by drilling down the former from Category to Product

the interestingness is defined as the difference of belief for corresponding facts in the cubes before and after the application of an OLAP operator. For instance, after drilling down from Category to Product, the more the Quantity of Beer deviates from the Quantity of Beverages, the higher its peculiarity; in other words, a user is less likely to expect a product with outstanding sales coming from a category with middling sales. Measuring interestingness in this way requires to define, for each fact in C', the "corresponding fact(s)" in C. To this end we use, as in [8], a *proxy function* $proxy_C(\gamma)$ (with $\gamma \in C'$) that models a one-to-many (many-to-one) mapping in case of drill-down (roll-up), and a one-to-one mapping in case of slice-and-dice or addition/removal of a measure (see Fig. 4 for an example). Intuitively, if the OLAP operator changes the group-by, the corresponding fact(s) of C are determined via the roll-up order; if the operator changes the selection predicate, the corresponding facts of C are one-to-one mapped to the facts of C'; if the operator changes the measure, the corresponding facts are the empty set. For the formal definition of proxy and peculiarity $pec()$, we refer the reader to [8]. Finally, the interestingness of component $v_k = \{\gamma_k\}$ describing the results of a refined query is defined as for initial queries, but weighing measure values on fact peculiarity:

$$int(v_k) = \frac{(\gamma_k.m - \gamma_{\overline{k}}.m) \cdot pec(\gamma_k)}{\sum_{i=1}^{k}(\gamma_i.m - \gamma_{\overline{k}}.m) \cdot pec(\gamma_i)}$$

Example 5. As already shown in Table 1, if C' is the cube in Fig. 3 resulting from an initial query, examples of insights are

$s_1^{\mathrm{T}} = (NL = $ "The fact with higher Quantity is Beer with 80",

$\quad int = 0.44, cov = 0.20, cost = 9)$

$s_3^{\mathrm{T}} = (NL = $ "The three facts with higher Quantity are Beer with 80,

$\quad$ Wine with 70, and Cola with 30",

$\quad int = 0.98, cov = 0.50, cost = 17)$

On the other hand, if C' is the result of a drill-down from Category to Product as in Fig. 4, the interestingness changes as follows:

$$s_1^T.int = \frac{(80-5) \cdot 0.21}{64.33} = 0.24$$

$$s_3^T.int = \frac{(80-5) \cdot 0.21 + (70-5) \cdot 0.36 + (30-5) \cdot 1.0}{64.33} = 0.98$$

Intuitively, following the prior belief principle, since Beer is the top product of the top-selling category, Beer is less interesting than Cola (which is the worst-selling beverage).   □

7 Evaluation and Conclusion

In this paper we have presented VOOL, an approach for vocalizing selected insights out of the results of an OLAP session. To test the approach we have implemented a prototype using Java and Python; the necessary mining models are imported from the Scikit-Learn library and the insights are vocalized through the text-to-speech Google APIs.

To evaluate the efficiency of VOOL, we made some experiments against the Foodmart cube [15] to understand how the performance of each module scales with respect to the cardinality of the query result. To this end we executed 10 OLAP sessions, each involving 3 OLAP steps; different combinations of modules were tested, and all the modules were invoked in at least one session. The tests were run on an Intel(R) Core(TM)i7-6700 CPU@3.40 GHz with 8 GB RAM. The tests were repeated 10 times and the average results are reported. Figure 5 shows the scalability of each module against query results with increasing cardinalities (up to 10^4). We emphasize that, since our work focuses on the vocalization of OLAP sessions —and not on the generic mining of multidimensional cubes—, 10^4 is large enough to be considered unrealistic for OLAP, since the results analyzed by users are usually constrained by the visualization and interaction metaphors adopted [9]. Noticeably, for query results including 10^4 facts, the computation of all the modules requires less than 1 s. The only exception is Clustering, which requires 7 s on average for query results with cardinality 10^4.

To assess the effectiveness of VOOL, we made some preliminary tests with 10 users, mainly master students in data science, with basic or advanced knowledge of business intelligence and data warehousing. The users were briefly introduced to the vocalization problem and to VOOL, then they were assigned three OLAP sessions with different analysis goals (e.g., "As a shop owner, you are analyzing the sum of quantity sold in each product department") and the query results were vocalized. On a scale from 1 (very poor) to 5 (very high), the average results show that both the user experience and the description of query results are deemed to be good (4.2 ± 0.6 and 3.8 ± 0.9, respectively). The insights with highest/lowest appreciation are Aggregation variance and Statistics, respectively (the latter sometimes is too simple to describe the whole result); the users asked to refine some of the proposed modules and suggested extensions with new ones.

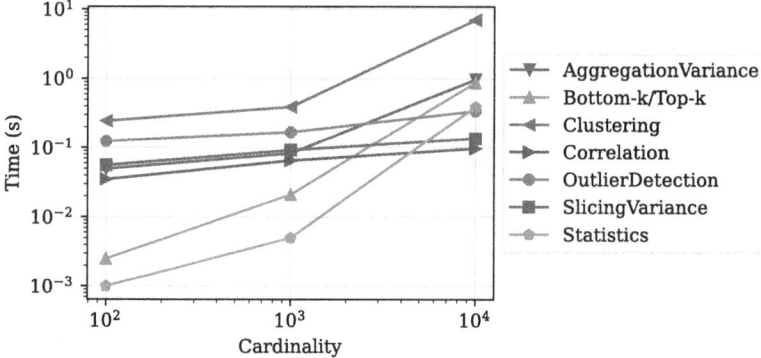

Fig. 5. Performance scalability of the modules

Overall, these preliminary results confirm the effectiveness and efficiency of VOOL. Besides refining and extending the modules, other directions that can be envisioned for future research are: (i) handling the redundancy of insights over single queries (since multiple modules can vocalize the same tuples, the interestingness of an insight should also depend on those previously selected) and sessions (vocalizing the same insight twice or more reduces its interestingness); (ii) introducing a "tell me more" interaction, where users can ask for further details as well as insights retrieved after the time budget; and (iii) conducting additional tests to assess the correlation between the insights vocalized and the users' intentions.

References

1. Affolter, K., Stockinger, K., Bernstein, A.: A comparative survey of recent natural language interfaces for databases. VLDB J. **28**(5), 793–819 (2019). https://doi.org/10.1007/s00778-019-00567-8
2. Bie, T.: Subjective interestingness in exploratory data mining. In: Tucker, A., Höppner, F., Siebes, A., Swift, S. (eds.) IDA 2013. LNCS, vol. 8207, pp. 19–31. Springer, Heidelberg (2013). https://doi.org/10.1007/978-3-642-41398-8_3
3. Brysbaert, M.: How many words do we read per minute? A review and meta-analysis of reading rate. J. Mem. Lang. **109**, 104047 (2019)
4. Das, M., Amer-Yahia, S., Das, G., Yu, C.: MRI: meaningful interpretations of collaborative ratings. Proc. VLDB Endow. **4**(11), 1063–1074 (2011)
5. Deutch, D., Frost, N., Gilad, A.: Explaining natural language query results. VLDB J. **29**(1), 485–508 (2020)
6. El, O.B., Milo, T., Somech, A.: Towards autonomous, hands-free data exploration. In: Proceedings of CIDR (2020)
7. Francia, M., Gallinucci, E., Golfarelli, M.: COOL: a framework for conversational OLAP. Inf. Syst. **104**, 101752 (2022)
8. Francia, M., Golfarelli, M., Marcel, P., Rizzi, S., Vassiliadis, P.: Assess queries for interactive analysis of data cubes. In: Proceedings of EDBT, pp. 121–132 (2021)
9. Francia, M., Golfarelli, M., Rizzi, S.: A-BI$^+$: a framework for augmented business intelligence. Inf. Syst. **92**, 101520 (2020)

10. Francia, M., Marcel, P., Peralta, V., Rizzi, S.: Enhancing cubes with models to describe multidimensional data. Inf. Syst. Front. **24**(1), 31–48 (2022)
11. Gkesoulis, D., Vassiliadis, P., Manousis, P.: CineCubes: aiding data workers gain insights from OLAP queries. Inf. Syst. **53**, 60–86 (2015)
12. Golab, L., Karloff, H.J., Korn, F., Srivastava, D.: Data auditor: exploring data quality and semantics using pattern tableaux. Proc. VLDB Endow. **3**(2), 1641–1644 (2010)
13. Golab, L., Srivastava, D.: Exploring data using patterns: a survey and open problems. In: Proceedings of DOLAP@EDBT/ICDT, pp. 116–120 (2021)
14. Golfarelli, M., Maio, D., Rizzi, S.: The dimensional fact model: a conceptual model for data warehouses. Int. J. Cooper. Inf. Syst. **7**(2–3), 215–247 (1998)
15. Hyde, J.: Foodmart. https://github.com/julianhyde/foodmart-data-mysql. Accessed 18 Jan 2021
16. Kellerer, H., Pferschy, U., Pisinger, D.: The multiple-choice knapsack problem. In: Knapsack Problems, pp. 317–347. Springer, Heidelberg (2004). https://doi.org/10.1007/978-3-540-24777-7_11
17. Li, F., Jagadish, H.V.: Understanding natural language queries over relational databases. SIGMOD Rec. **45**(1), 6–13 (2016)
18. Likas, A., Vlassis, N., Verbeek, J.J.: The global k-means clustering algorithm. Pattern Recogn. **36**(2), 451–461 (2003)
19. Liu, F.T., Ting, K.M., Zhou, Z.: Isolation forest. In: Proceedings of ICDM, pp. 413–422 (2008)
20. Luo, Z.W., Ling, T.W., Ang, C.H., Lee, S.Y., Cui, B.: Range top/bottom k queries in OLAP sparse data cubes. In: Mayr, H.C., Lazansky, J., Quirchmayr, G., Vogel, P. (eds.) DEXA 2001. LNCS, vol. 2113, pp. 678–687. Springer, Heidelberg (2001). https://doi.org/10.1007/3-540-44759-8_66
21. Lyons, G., Tran, V., Binnig, C., Çetintemel, U., Kraska, T.: Making the case for query-by-voice with echoquery. In: Proceedings of SIGMOD, pp. 2129–2132 (2016)
22. Romero, O., Abelló, A.: On the need of a reference algebra for OLAP. In: Song, I.Y., Eder, J., Nguyen, T.M. (eds.) DaWaK 2007. LNCS, vol. 4654, pp. 99–110. Springer, Heidelberg (2007). https://doi.org/10.1007/978-3-540-74553-2_10
23. Saha, D., Floratou, A., Sankaranarayanan, K., Minhas, U.F., Mittal, A.R., Özcan, F.: ATHENA: an ontology-driven system for natural language querying over relational data stores. PVLDB **9**(12), 1209–1220 (2016)
24. Sarawagi, S.: Explaining differences in multidimensional aggregates. In: Proceedings of VLDB, pp. 42–53 (1999)
25. Sarawagi, S.: User-adaptive exploration of multidimensional data. In: Proceedings of VLDB, pp. 307–316 (2000)
26. Simitsis, A., Koutrika, G., Alexandrakis, Y., Ioannidis, Y.E.: Synthesizing structured text from logical database subsets. In: Proceedings of EDBT, pp. 428–439 (2008)
27. Song, L., Gan, J., Bao, Z., Ruan, B., Jagadish, H.V., Sellis, T.: Incremental preference adjustment: a graph-theoretical approach. VLDB J. **29**(6), 1475–1500 (2020). https://doi.org/10.1007/s00778-020-00623-8
28. Trummer, I., Wang, Y., Mahankali, S.: A holistic approach for query evaluation and result vocalization in voice-based OLAP. In: Proceedings of SIGMOD, pp. 936–953 (2019)
29. Zgraggen, E., Zhao, Z., Zeleznik, R.C., Kraska, T.: Investigating the effect of the multiple comparisons problem in visual analysis. In: Proceedings of CHI, p. 479 (2018)

Advanced Querying

Querying Temporal Anomalies in Healthcare Information Systems and Beyond

Christina Khnaisser[1]($\boxtimes$)(iD), Hind Hamrouni[2](iD), David B. Blumenthal[3](iD),
Anton Dignös[2](iD), and Johann Gamper[2](iD)

[1] Université de Sherbrooke, Sherbrooke, Canada
`christina.khnaisser@usherbrooke.ca`
[2] Free University of Bozen-Bolzano, Bozen, Italy
`{hind.hamrouniEpBenkhaled,anton.dignoes,johann.gamper}@unibz.it`
[3] Friedrich-Alexander-Universität Erlangen-Nürnberg, Erlangen, Germany
`david.b.blumenthal@fau.de`

Abstract. Finding anomalies in temporal relational databases is a difficult and challenging task, in particular if data is integrated from different sources. The problem is especially pressing in healthcare information systems, where temporal anomalies can pinpoint critical events such as erroneous drug administration or prescription. In this paper, we define three different temporal anomalies, which we call temporal redundancy, contradiction, and incompleteness. We define two different operators for each of these anomalies: the retrieval operator to retrieve all tuples of a relation that cause anomalous behaviour, and the labelling operator to annotate a temporal relation with additional information that marks normal and anomalous tuples. Finally, we present and evaluate different implementation techniques for the two operators for relational database systems.

1 Introduction

In many application domains, checking and maintaining the integrity of relational databases over time has become increasingly important. In particular, detecting and avoiding temporal inconsistencies enables better data quality and enhances the accuracy of downstream temporal data analysis [10]. Independently of the concrete application domain, the following data inconsistencies need to be addressed when working with temporal data [6, 18]: *redundancy* (redundant data), *contradiction* (contradicting data), and *incompleteness* (incomplete data). In this paper, these inconsistencies are called temporal anomalies.

As a concrete use case, we consider healthcare information systems. Electronic prescription systems recently emerged in many countries to enhance access to drug prescriptions and administration. These systems help practitioners in identifying when a new drug administration is started, when the doses are changed, or when a prescription is stopped, etc. However, errors occur when feeding data into healthcare information systems. Especially when data are integrated from multiple data sources, temporal

Supported by Ministère de l'Économie et de l'Innovation – Québec and by the Autonomous Province of Bozen-Bolzano with research call "Research Südtirol/Alto Adige 2019" (project Enabling Industrial-Strength, Open-Source Temporal Query Processing – ISTeP).

S. Chiusano et al. (Eds.): ADBIS 2022, LNCS 13389, pp. 209–222, 2022.
https://doi.org/10.1007/978-3-031-15740-0_16

anomalies often occur [11]. For instance, if a patient gets medical services from different healthcare providers for several diseases, he/she might be prescribed the same drug by each provider, potentially with different doses [23]. Also, the use of more than one pharmacy can lead to anomalies because systems often do not communicate with each other [19]. Data inconsistencies include, among others, redundant prescriptions or contradicting dosages.

Temporal anomalies in relational databases may be indicative of poor quality of the temporal data (e.g., due to errors that occurred when feeding the data into the system) or inconsistent behaviour of the real-world system modelled by the database. Especially in healthcare information systems, detecting temporal anomalies is hence critical (1) to avoid misinterpretation in temporal data analysis and (2) to flag potentially life-threatening events such as the administration of a drug that has not been prescribed.

Example 1. Consider the simplified real-world example in Fig. 1. It shows data about drug prescriptions and administrations that were integrated from different databases. The relation **Prescription** stores drug prescriptions, where Pat is the patient ID, Name is the name of the prescribed drug product, Dose is the prescribed dose, and T is the time period during which the drug prescription is valid. The relation **Administration** stores the drugs that actually were administered together with the actual amount and the time period.

Prescription

	Pat	Name	Dose	T
d_1	P1	Ibuprofen	600mg	[2,10)
d_2	P1	Ibuprofen	600mg	[2,6)
d_3	P1	Ibuprofen	600mg	[4,12)
d_4	P1	Cardizem	60mg	[8,10)
d_5	P2	Amlodipine	5mg	[1,10)
d_6	P2	Morphine	5mg	[6,8)
d_7	P3	Clopidogrel	75mg	[1,12)
d_8	P3	Clopidogrel	85mg	[10,14)
d_9	P3	Clopidogrel	95mg	[8,11)

Administration

	Pat	Name	Dose	T
a_1	P1	Ibuprofen	600mg	[2,12)
a_2	P1	Cardizem	60mg	[8,13)
a_3	P2	Morphine	8mg	[6,8)
a_4	P2	Amlodipine	5mg	[1,4)
a_5	P2	Amlodipine	10mg	[4,10)
a_6	P3	Clopidogrel	75mg	[3,12)

Fig. 1. Temporal relations **Prescription** and **Administration**.

In this example, we can spot different types of temporal anomalies. For instance, the tuples d_1, d_2, and d_3 in the **Prescription** table indicate that the same drug and dose is prescribed to the same patient over overlapping time periods (*temporal redundancy*). In the same table, the tuples d_7 and d_8 report a contradicting dose of a drug prescribed to a patient at the same time (*temporal contradiction*). There are also anomalies across the two tables. For instance, tuple a_2 in table **Administration** reports the administration of Cardizem to patient P1 during the time period $[8, 13)$, but there is no prescription in the **Prescription** table that covers the full period of the administration (*temporal incompleteness*).

In this paper, we provide a formal definition of the above mentioned temporal anomalies together with two new operators that allow, respectively, to retrieve the anomalous tuples from a dataset or to label the tuples in a dataset with an indication whether they are anomalous or not. These two operators support data scientists in checking the quality of the data and identifying potential anomalies or inconsistencies. While the running example is from the medical domain, where the detection of anomalous data is particularly important, the solutions are general and can be applied in other domains as well. Moreover, for all three types of anomalies, we present implementation techniques of the two operators for relational database management systems.

To summarize, the technical contributions are as follows:

- We formally define three types of temporal anomalies in temporal databases, named *temporal redundancy*, *temporal contradiction*, and *temporal incompleteness*.
- We provide two operators, termed *anomaly retrieval* and *anomaly labelling*, for the processing of these anomalies and show how they can be implemented efficiently using temporal aggregation in SQL.
- We perform extensive experiments on real-world medical data to show the feasibility of the different anomaly operations.

The rest of the paper is organized as follows. Section 2 presents related work. Section 3 introduces preliminary concepts and the definition of the temporal anomalies, followed by their implementation in Sect. 4. Section 5 discusses experimental results. Section 6 concludes the paper and points to future work.

2   Related Work

Studies on various aspects of temporal databases started around the 1980 and are still an active research area. A large number of temporal models and languages have been proposed since then [1,20]. The incorporation of temporal data definition and manipulation features in the SQL:2011 standard [16] supports the association of data with time periods, representing application time or system time, as well as basic temporal queries, i.e., predicates over time periods and time-slice queries. However, there is little support for temporal data integrity. Hence, temporal data integrity constraints can be defined, detected and repaired only at the application level. Date et al. [6] identified four potential problems that need to be addressed when working with relations that contain intervals: redundancy, contradiction, circumlocution and denseness. We based our definition of the temporal anomalies on these problems because they can be intuitively generalized for many application domains without modifications of the RDBMS [15]. In addition to defining the anomalies, we provide two operators that allow data scientists to identify such anomalies and that can be implemented in existing RDBMSs.

Lorentzos et al. [17] define *unfold/fold* operators with an SQL extension for interval data [18] to facilitate the usage of relational operators when querying interval relations. The unfold operator transforms interval data to point data by replacing each tuple with a set of tuples with point data. The fold operator performs the inverse operation and merges identical tuples having consecutive points into tuples with intervals. The operators can be applied to any interval relation before and after any relational operator.

We will use unfold/fold as one technique to implement our temporal operators to query for anomalies.

Böhlen et al. [2] propose a *coalescing* operation for eliminating redundancies. During the operation, tuples with identical non-temporal attribute values are coalesced (merged) if their temporal attributes meet or overlap. However, the study shows that implementations with various algorithms based on nested-loop, explicit partitioning, explicit sorting, temporal sorting, temporal partitioning, and combined explicit/temporal sorting are quite expensive. A more recent study presents a new efficient implementation of coalescing using window functions [24] that were introduced in SQL:2003. The proposed algorithm performs a single scan with linear scalability in terms of the database size. The intended use of these works on coalescing is to eliminate redundancies, which is different from our setup, where we want to identify tuples that cause redundancies since this might be an indication of poor data quality.

Combi et al. [4] define keys and attribute temporal constraints at a conceptual level. Among these constraints they mention snapshot-reducible keys to avoid redundancy, time-invariant keys and attribute constraints to avoid contradiction, and time-invariant relationships to avoid temporal gaps after a starting point in time. The snapshot-reducible constraint ensures that at any point in time the key attribute uniquely identifies an entity. The time-invariant constraint ensures that the same entity cannot have two different values for the key or a non-key attribute for different points in time. The time-invariant relationship constraint and its variant can be seen as a special case of inconsistency since they ensure that the involved entities can start at any point in time during the existence of the involved entities, but it has to hold for all the remaining time. In [5] temporal constraints are studied from a modelling and reasoning perspective with a particular focus on the medical domain. The focus of our work is different. We provide an approach to query for anomalies that do not satisfy a constraint and provide an efficient implementation that can be readily used in contemporary database systems.

Finally, there have been many works in the past years on processing temporal data, particularly, focusing on temporal join [3,8,22] and temporal aggregation [14,21]. While many works on temporal aggregation that we use in this paper are based on main memory algorithms, and thus require an external implementation, some works provide in-database solutions [7] or solutions based on plain SQL [9]. In this paper, we focus on solutions that are readily available in contemporary database systems, and we adopt and extend the window-function based approach for temporal aggregation from [9].

3   Temporal Anomalies

3.1   Preliminaries

We assume a linearly ordered, discrete time domain, Ω^T. A time interval is a set of contiguous time points, and $T = [t_s, t_e)$ denotes the closed-open interval of points from t_s inclusive to t_e exclusive. We use $|t| = t_e - t_s$ to denote the duration of the time interval T and $T \cap T'$ to denote the set of time points shared by two intervals T and T', which, if not empty, is itself an interval. The schema of a temporal relation is given by $R = (A_1, \ldots, A_m, T)$, where $\mathbf{A} = A_1, \ldots, A_m$ are the non-temporal attributes with domains Ω_i and T is the time interval attribute with domain $\Omega^T \times \Omega^T$ representing, for instance, the tuple's valid time. A temporal relation **r** with schema R is a finite set of tuples, where each tuple has a value in the appropriate domain for each attribute in the schema. We use $r.A_i$ to denote the value of attribute A_i in tuple r, and $r.T = [r.t_s, r.t_e)$ to refer to its time interval. We assume the following (non-temporal) relational algebra operators: selection σ, generalized projection π, union $\cup$, renaming of attributes and relations $/$, and aggregation ϑ. Further we assume SQL's duplicate elimination with keyword DISTINCT within aggregation functions to produce the aggregation function over distinct values. We use the temporal relational aggregation operator ϑ^T [1,9], for which each snapshot in the result corresponds to the result of applying non-temporal aggregation to the corresponding snapshot of the input, and is defined as follows:

Definition 1 (Temporal Aggregation). *Let* **r** *be a temporal relation with non-temporal attributes* **A** *and time interval attribute T. Further, let* **C** $\subset$ **A**, $agg(A_i)$ *be an aggregation function over an attribute* $A_i \in$ **A**, *and* $\tau_t(\mathbf{r})$ *be the timeslice operator [12] that extracts from a temporal relation* **r** *the tuples projected to the attributes* **A** *that are valid at time point t. The result of temporal aggregation is defined using non-temporal aggregation as follows:*

$$\forall t \in \Omega^T : \tau_t\big(\mathbf{C}\vartheta^T_{agg(A_i)}(\mathbf{r})\big) = \mathbf{C}\vartheta_{agg(A_i)}\big(\tau_t(\mathbf{r})\big)$$

3.2 Definition of Temporal Anomalies

Temporal Redundancy. A temporal redundancy occurs when more than one tuple in a relation exist that share the same values for a given set of attributes at a time point t.

Definition 2 (Temporal redundancy). *Let* **r** *be a temporal relation with non-temporal attributes* **A** *and temporal attribute T. A tuple* $r \in$ **r** *causes a temporal redundancy at a time point t iff*

$$\exists r' \in \mathbf{r} : r \neq r' \wedge r.\mathbf{A} = r'.\mathbf{A} \wedge t \in r.T \wedge t \in r'.T$$

Consider the example in Fig. 1. The **Prescription** table contains redundant prescriptions with respect to patient, drug name, and dose in the tuples d_1, d_2 and d_3: the prescription Ibuprofen 600 mg for patient P1 is prescribed 2 times during $[2, 4)$, 2 times during $[6, 10)$, and 3 times during $[4, 6)$.

Temporal Contradiction. A temporal contradiction occurs when more than one tuple in a relation exist that share, at some time point t, the same values for a given subset of the attributes, but have different values for the remaining attributes.

Definition 3 (Temporal contradiction). *Let* **r** *be a temporal relation with non-temporal attributes* **A** *and temporal attribute T. Further, let* **C** $\subset$ **A** *and* **C'** $=$ **C**/**A**. *A tuple* $r \in$ **r** *causes a temporal contradiction for attributes* **C** *at time point t iff*

$$\exists r' \in \mathbf{r} : r.\mathbf{A} = r'.\mathbf{A} \wedge r.\mathbf{C'} \neq r'.\mathbf{C'} \wedge t \in r.T \wedge t \in r'.T$$

Consider the example in Fig. 1. The tuples d_7, d_8, and d_9 in the **Prescription** table contain contradicting dosage information: the prescriptions of Clopidogrel for patient P1 are {75 mg, 95 mg} during $[8, 10)$, {75 mg, 85 mg, 95 mg} during $[10, 11)$, and {75 mg, 85 mg} during $[11, 12)$.

Temporal Incompleteness. A temporal incompleteness between two relations occurs when for a tuple of one relation no tuple in the other relation exists that shares the same values for a given set of attributes at a time point t.

Definition 4 (Temporal incompleteness). *Let* $\mathbf{r}$ *and* $\mathbf{s}$ *be two temporal relation with non-temporal attributes* $\mathbf{A}_1$ *and* $\mathbf{A}_2$, *respectively, and temporal attribute* T. *Further, let* $\mathbf{C} \subseteq \mathbf{A}_1 \wedge \mathbf{C} \subseteq \mathbf{A}_2$. *A tuple* $x \in \mathbf{r} \vee x \in \mathbf{s}$ *causes a temporal incompleteness for attributes* $\mathbf{C}$ *at time point* t *iff*

$$t \in x.T \wedge (\nexists s \in \mathbf{s} : x.\mathbf{C} = s.\mathbf{C} \wedge t \in s.T \vee \nexists r \in \mathbf{r} : x.xxC = r.\mathbf{C} \wedge t \in r.T)$$

Consider the example in Fig. 1. The **Prescription** and **Administration** tables are not complete with respect to the drug name and the patient for the tuples d_4 with a_2, d_7, d_8 with a_6. There is a gap in **Prescription** during $[10, 13)$ for the medication Cardizem for P1. Similarly, there is a gap in the Administration during $[1, 3)$ and $[12, 14)$ for the medication Clopidogrel for P3.

3.3 Anomaly Labelling and Retrieval

To support data scientists in detecting anomalies, we propose two different operators to highlight and extract anomalies in a dataset. First, *anomaly labelling* is a relational query that extends a relation with a labelling attribute that highlights inconsistent tuples with a label. This kind of operation is very helpful when anomalies need to be resolved upstream by a domain expert to improve the quality of the available data. Second, *anomaly retrieval* is a relational query that extracts only portions of the data that are affected by an anomaly (including their label). This kind of operation is very helpful when anomalies need to resolved downstream, so that the affected portions of the data can be quickly identified.

In the following, we use temporal aggregation to define the two operators for all three temporal anomalies.

Temporal Redundancy. For the redundancy labelling, the relation is extended with the number of occurrences of tuples. The count aggregation function is used to compute the number of occurrences of a tuple during a time interval. For a tuple, an occurrence value of 1 represents that the tuple is defined only once in the relation (i.e., there is no redundancy).

Definition 5 (Temporal redundancy labelling and retrieval). *Let* $\mathbf{r}$ *be a temporal relation with non-temporal attributes* $\mathbf{A}$ *and temporal attribute* T. *We define the following two operators for temporal redundancy:*

- *Temporal redundancy labelling:*

$$\mathbf{A}\,\vartheta^T_{count(*)/Count}(\mathbf{r})$$

- *Temporal redundancy retrieval:*

$$\sigma_{Count>1}\left(\mathbf{A}\,\vartheta^T_{count(*)/Count}(\mathbf{r})\right)$$

The result of the retrieval operation for temporal redundancies is simply the restriction of the labelling result to those tuples that have an occurrence value strictly greater than 1.

Example 2. Consider the relation **Prescription** from Example 1. To label or retrieve temporal redundancies we use the queries from Definition 5. We have **r** = **Prescription** and **A** = $(Pat, Name, Dose)$. The result for temporal redundancy labelling is shown in Fig. 2a, and the result for retrieval is shown in Fig. 2b.

Pat	Name	Dose	T	Count
P1	Ibuprofen	600mg	[2,4)	2
P1	Ibuprofen	600mg	[4,6)	3
P1	Ibuprofen	600mg	[6,10)	2
P1	Ibuprofen	600mg	[10,12)	1
P1	Cardizem	60mg	[8,10)	1
P2	Amlodipine	5mg	[1,10)	1
P2	Morphine	5mg	[6,8)	1
P3	Clopidogrel	75mg	[1,12)	1
P3	Clopidogrel	85mg	[10,14)	1
P3	Clopidogrel	95mg	[8,11)	1

(a) Redundancy labelling

Pat	Name	Dose	T	Count
P1	Ibuprofen	600mg	[2,4)	2
P1	Ibuprofen	600mg	[4,6)	3
P1	Ibuprofen	600mg	[6,10)	2

(b) Redundancy retrieval

Fig. 2. Temporal redundancy labelling and retrieval for the **Prescription** table using **A** = $(Pat, Name, Dose)$.

Temporal Contradiction. To provide a meaningful label for the temporal contradiction, we use the `array_agg` aggregation function[1], which returns an array of all values in the aggregation. More specifically, to identify contradictions we use `array_agg` in combination with `DISTINCT` to get an array of distinct (contradicting) values.

Definition 6 (Temporal contradiction labelling and retrieval). *Let* r *be a temporal relation with non-temporal attributes* **A** *and temporal attribute* T*. Further, let* **C** $\subset$ **A** *and* **C'** = **C**/**A**. *We define the following two operators for temporal contradiction:*

– *Temporal contradiction labelling:*

$$\pi_{\mathbf{C},T,c/Contradiction}\left(\mathbf{C}\vartheta^T_{array\_agg(\texttt{DISTINCT}\,(\mathbf{C'}))/c}(\mathbf{r})\right)$$

– *Temporal contradiction retrieval:*

$$\pi_{\mathbf{C},T,c/Contradiction}\left(\sigma_{|c|>1}\left(\mathbf{C}\vartheta^T_{array\_agg(\texttt{DISTINCT}\,(\mathbf{C'}))/c}(\mathbf{r})\right)\right)$$

The result of the retrieval operation for temporal contradictions is the restriction of the labelling result to those tuples for which the array contains more than one element.

---

[1] See: https://www.postgresql.org/docs/current/functions-aggregate.html for PostgreSQL or https://docs.microsoft.com/en-us/u-sql/functions/aggregate/array-agg for MS SQLServer.

Example 3. Consider the relation **Prescription** from Example 1. To find contradictions in the dose for each patient and drug, we use $\mathbf{C} = (Pat, Name)$. The result for temporal contradiction labelling in Fig. 3a shows the contradicted doses (*Contradiction* attribute) for each combination of patient and drug during a time interval. The tuples in gray have more than one contradicting dose in the label attribute. The result for the retrieval is shown in Fig. 3b and includes only tuples with contradictions in the dose.

Pat	Name	T	Contradiction
P1	Ibuprofen	[2,10)	600mg
P1	Ibuprofen	[2,6)	600mg
P1	Ibuprofen	[4,12)	600mg
P1	Cardizem	[8,10)	60mg
P2	Amlodipine	[1,10)	5mg
P2	Morphine	[6,8)	5mg
P3	Clopidogrel	[1,8)	75mg
P3	Clopidogrel	[8,10)	75mg, 95mg
P3	Clopidogrel	[10,11)	75mg, 85mg, 95mg
P3	Clopidogrel	[11,12)	75mg, 85mg
P3	Clopidogrel	[12,14)	85mg

Pat	Name	T	Contradiction
P3	Clopidogrel	[8,10)	75mg, 95mg
P3	Clopidogrel	[10,11)	75mg, 85mg, 95mg
P3	Clopidogrel	[11,12)	75mg, 85mg

(a) Contradiction labelling   (b) Contradiction retrieval

Fig. 3. Temporal contradiction labelling and retrieval for the **Prescription** table using $\mathbf{C} = (Pat, Name)$.

Temporal Incompleteness. For the incompleteness labelling, the relation is extended with an additional column that indicates, for each tuple, the relations in which the tuple is temporally covered. In the following definition, the label *Src* returns the array of distinct relations that contain the tuple during a time interval. If the array contains both input relations, there is no temporal gap, i.e., the tuple is temporally covered by one or more tuples in both relations. If, instead, the array contains only one element, this element indicates the relation in which the tuple is present; hence the other relation contains a temporal gap.

Definition 7 (Temporal incompleteness labelling and retrieval). *Let* $\mathbf{r}$ *and* $\mathbf{s}$ *be two temporal relation with non-temporal attributes* $\mathbf{A}_1$ *and* $\mathbf{A}_2$ *respectively, and a temporal attribute T. Further, let* $\mathbf{C} \subseteq \mathbf{A}_1 \wedge \mathbf{C} \subseteq \mathbf{A}_2$. *We define the following two operators for temporal incompleteness:*

- *Temporal incompleteness labelling:*

$$\pi_{\mathbf{C},T,Src}\left(\mathbf{C}\vartheta^T_{array\_agg(\text{DISTINCT } Src)/Src}\left(\pi_{\mathbf{C},T,'\mathbf{r}'/Src}(\mathbf{r}) \cup \pi_{\mathbf{C},T,'\mathbf{s}'/Src}(\mathbf{s})\right)\right)$$

- *Temporal incompleteness retrieval:*

$$\pi_{\mathbf{C},T,Src}\left(\sigma_{|Src|<2}\left(\mathbf{C}\vartheta^T_{array\_agg(\text{DISTINCT } Src)/Src}\left(\pi_{\mathbf{C},T,'\mathbf{r}'/Src}(\mathbf{r}) \cup \pi_{\mathbf{C},T,'\mathbf{s}'/Src}(\mathbf{s})\right)\right)\right)$$

The result of the retrieval operation for temporal incompleteness is the restriction of the labelling result to those tuples for which the array contains only one element.

Example 4. Consider the relations **Prescription** and **Administration** from Example 1. To find for each patient incomplete prescriptions or administrations of a drug, we use $\mathbf{C} = (Pat, Name)$. The result of temporal incompleteness labelling shown in Fig. 4a. The last column shows the tables in which the tuple is covered. The tuples highlighted in gray are incomplete since the label contains only one of the two relations. For instance, the third tuple records the administration of Cardizem to patient P1 over the time period [10, 13), which is not covered by a corresponding entry in the **Prescription** table. The result of the retrieval operation that only returns the incomplete tuples is shown in Fig. 4b.

Pat	Name	T	Src
P1	Ibuprofen	[2,12)	presc, admin
P1	Cardizem	[8,10)	presc, admin
P1	Cardizem	[10,13)	admin
P2	Amlodipine	[1,10)	presc, admin
P2	Morphine	[6,8)	presc, admin
P3	Clopidogrel	[1,3)	presc
P3	Clopidogrel	[3,12)	presc, admin
P3	Clopidogrel	[12,14)	presc

(a) Incompleteness labelling

Pat	Name	T	Src
P1	Cardizem	[10,13)	admin
P3	Clopidogrel	[1,3)	presc
P3	Clopidogrel	[12,14)	presc

(b) Incompleteness retrieval

Fig. 4. Temporal incompleteness labeling and retrieval for the **Prescription** and **Administration** tables using $\mathbf{C} = (Pat, Name)$.

4 SQL Implementations

In this section, we present three different implementations for the retrieval and labelling operators of the three anomalies.

4.1 Unfold/Fold

The implementation based on unfold/fold [18] works in a three step fashion. First, time intervals are transformed into time points of the base granularity (unfold). For this we can use user-defined functions in contemporary database systems. Second, a non-temporal operator is performed on the time points, e.g., non-temporal aggregation with the time point as an additional grouping attribute. Third, the result is transformed back to an interval-based relation (fold), for which we can again use user-defined functions.

For temporal redundancy, the labelling operation is performed by simply using a `count` aggregation in the second step of the approach. For the retrieval operation, the tuples are restricted to those with a count value greater than 1 before the fold operation. For temporal contradiction, the approach is very similar to temporal redundancy, with the exception that, instead of a count, the array_agg aggregation over distinct values is used (cf. Definition 6). For temporal incompleteness, we unfold both relations and add the relation name as a label. We then union the two intermediate results and aggregate using the array_agg aggregation function over the distinct label. Similar to the other two anomalies, also in this case the query is very similar to the definition of the operators (cf. Definition 7).

4.2   Unfold/Fold Join Filtered

The main issue of the unfold/fold approach is the very large intermediate results that need to be processed. To tackle this problem, we use a semi-join based technique that filters the input before it is unfolded. This filtering technique works only for temporal redundancy and contradiction. Also, it can only be applied for the retrieval operation, but not for the labelling, because the semi-join removes input tuples that do not contribute to an anomaly. After the filtering step, the approach is the same as unfold/fold above.

For temporal redundancy, before performing the unfold operation, we add a row number to the relation, followed by a self semi-join on the attributes of interest, overlapping intervals, and different row numbers. In such a way, for temporal redundancy we restrict the input of unfold to tuples that definitely produce a redundancy. For temporal contradiction, we use a similar approach that omits the row number and additionally constrains the join to contradictions, i.e., different values.

4.3   Window Function

For this implementation approach, we use the temporal aggregation technique from [9], which is based on the SQL window functions and is very similar to the coalescing approach from [24]. This approach is able to compute an accumulative aggregation function, e.g., count of sum, using two window functions over the union of start and end points of a relation. It is currently the most efficient approach for temporal aggregation applicable in contemporary database systems.

For temporal redundancy, we can use this approach and simply calculate the accumulative aggregation function count as a label. For the retrieval operation, we restrict the tuples to those with a label that is greater than 1. For temporal contradiction, we need to adjust the approach, because the array_agg aggregation function over distinct values cannot be computed incrementally. In this case, we first calculate the count aggregation to produce the final intervals. Then, we join the result of the aggregation with the original relation on the attributes of interest and on overlapping time intervals. Finally, we compute the array_agg aggregation function on distinct values over the result of the join, thereby grouping on the attributes of interest and the final intervals produced by the aggregation. For the retrieval of the contradictions, we additionally restrict the aggregation result before the join to only those tuples with a count larger than 1 (since we can only have a contradiction if at least two tuples with the same attribute values exist at the same time). For temporal incompleteness, the approach is similar as for contradiction. We first use window functions to produce a count aggregation over the union of the two input relations using the attributes of interest. This will produce the final intervals for which at least one tuple in either relation exists. Then we join the aggregation result with the union of the two input relations extended with an attribute containing the relation name on the attributes of interest and overlapping intervals. Finally, we use the `array_agg` aggregation function over distinct relation names to produce the final result for the incompleteness labelling operator. For the retrieval operator we restrict the result to tuples that only contain one relation name in the label.

5   Experimental Evaluation

5.1   Setup and Dataset

The experiments were run on an Intel(R) Xeon(R) CPU E5-2667 v3 @ 3.20 GHz machine with 94 GB main memory running Ubuntu 64-bit (version 16.04.7 LTS). As a database system we use PostgreSQL version 14, where all configuration parameters are kept to the default values. We do not use parallelism to avoid interference with other processes on the same server.

In the experiments, we use the real-world MIMIC-IV (Medical Information Mart for Intensive Care) [13] dataset, which is a large medical dataset comprising patients' health data in critical care units between 2008 and 2019. More specifically, we use the two tables *Prescriptions* and *Pharmacy*. The Prescriptions table records 13 280 145 prescribed medications for 226 305 patients. The durations of the prescriptions range from 1 to 1 192 days with an average of 3.3 days. The Pharmacy table records 10 911 112 filled medications that were prescribed to 219 367 patient. The durations range from 1 to 75 981 days with an average of 3.5 days, and contain 9 205 different medications. While the timestamps in the two tables are recorded at a granularity of minutes, we use days as the main granularity.

5.2   Results

In the first experiment, we evaluate the runtime of the different implementations for temporal redundancy. We use the Prescription table and the attributes subject_id, drug, dose_val_rx, and [startdate, enddate) from the MIMIC dataset as an input and vary the number of tuples from 2 to 10 million. The results for the retrieval and labelling operators are shown in Fig. 5. We observe that the implementation based on unfold/fold is very slow due to the blow-up of the intermediate result into individual time points caused by the unfold operation before the aggregation. The join-based implementation for the retrieval operation is able to reduce the size of the intermediate result by filtering before the unfold and fold. This yields a substantial improvement in the performance, but it is not applicable for the labelling operation. The approach based on window functions is by far the most efficient one for both operations.

In the second experiment, we evaluate the approaches for temporal contradiction on the Prescription table from the MIMIC dataset. For this case, we use the subject_id and drug to find contradictions in the doses. The results for retrieval and labelling for various number of input tuples are shown in Fig. 6. We observe a similar behaviour as

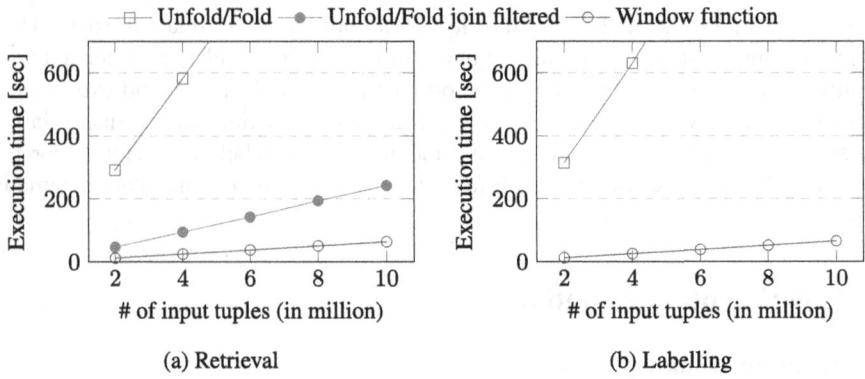

Fig. 5. Runtime for temporal redundancy on the Prescription table.

for the redundancy, but the join filtering implementation is less effective compared to redundancy. This is because less attributes are used, and thus more tuples share the same values. The approach based on window functions is again the most efficient approach.

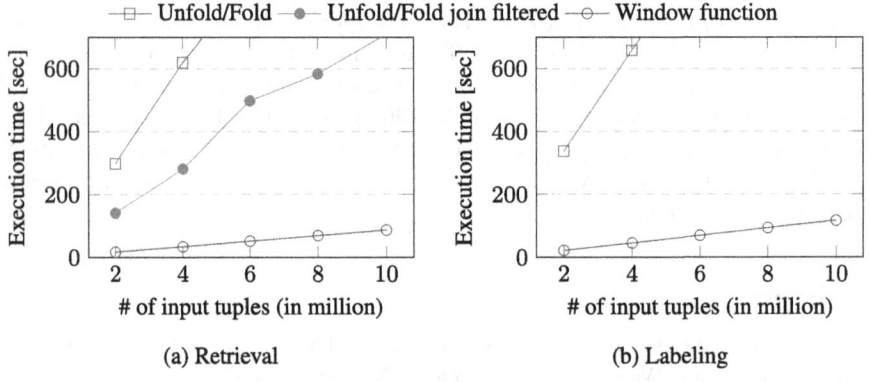

Fig. 6. Runtime for temporal contradiction on the Prescription table.

Finally, we evaluate the approaches for temporal incompleteness. For this case we use the attributes subject_id, pharmacy_id (used to link a prescription to its administration), and drug, which are in common between the Prescription and Pharmacy tables. The results for retrieval and labelling for various number of input tuples are shown in Fig. 7. Note that the join based filtering for unfold/fold has a very large runtime, thus we omit it from the result. Again we can see that unfold/fold suffers form the large intermediate results that are unioned and then aggregated. The approach based on window functions provides much better results.

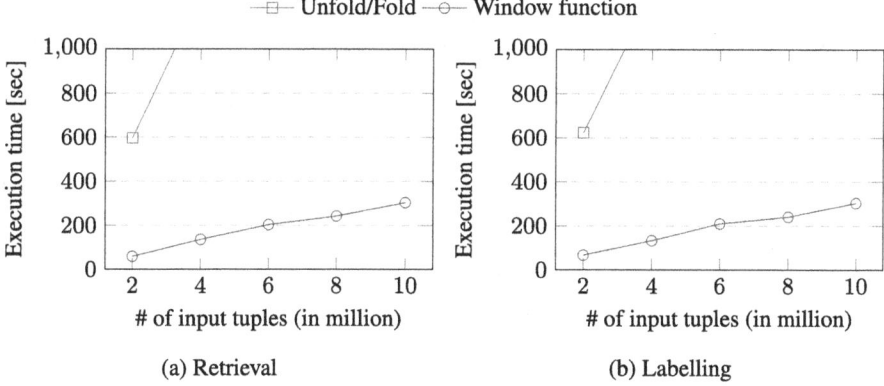

Fig. 7. Runtime for temporal incompleteness.

6 Conclusion and Future Work

Temporal anomalies may be used to verify data inconsistencies by detecting or enforcing data integrity over time. Standard SQL can be used to store temporal data, but enforcing integrity and querying is still a challenge, especially in healthcare databases with a large number of entities. Thus, providing a generic implementation of these anomalies will help database designers in writing efficient queries to detect temporal anomalies or even automate their detection for various application domain. This paper presents temporal queries and efficient implementations to help developers in detecting common temporal anomalies.

In future work, we would like to (1) extend the temporal anomalies operations to cover bitemporal models, cross-relational temporal contradictions, and semantic temporal conflicts, (2) develop a federated version for privacy-preserving anomaly mining, and (3) define a temporal repair system using healthcare expert preferences.

References

1. Böhlen, M.H., Dignös, A., Gamper, J., Jensen, C.S.: Temporal data management – an overview. In: Zimányi, E. (ed.) eBISS 2017. LNBIP, vol. 324, pp. 51–83. Springer, Cham (2018). https://doi.org/10.1007/978-3-319-96655-7_3
2. Böhlen, M.H., Snodgrass, R.T., Soo, M.D.: Coalescing in temporal databases. In: VLDB, pp. 180–191. Morgan Kaufmann (1996)
3. Bouros, P., Mamoulis, N., Tsitsigkos, D., Terrovitis, M.: In-memory interval joins. VLDB J. **30**(4), 667–691 (2021). https://doi.org/10.1007/s00778-020-00639-0
4. Combi, C., Degani, S., Jensen, C.S.: Capturing temporal constraints in temporal ER models. In: Li, Q., Spaccapietra, S., Yu, E., Olivé, A. (eds.) ER 2008. LNCS, vol. 5231, pp. 397–411. Springer, Heidelberg (2008). https://doi.org/10.1007/978-3-540-87877-3_29
5. Combi, C., Keravnou-Papailiou, E., Shahar, Y.: Temporal Information Systems in Medicine, 1st edn. Springer, New York (2010). https://doi.org/10.1007/978-1-4419-6543-1
6. Date, C.J., Darwen, H., Lorentzos, N.A.: Time and Relational Theory: Temporal Databases in the Relational Model and SQL. Morgan Kaufmann (2014)

7. Dignös, A., Böhlen, M.H., Gamper, J., Jensen, C.S.: Extending the kernel of a relational DBMS with comprehensive support for sequenced temporal queries. ACM Trans. Database Syst. **41**(4), 26:1–26:46 (2016)

8. Dignös, A., Böhlen, M.H., Gamper, J., Jensen, C.S., Moser, P.: Leveraging range joins for the computation of overlap joins. VLDB J. **31**(1), 75–99 (2021). https://doi.org/10.1007/s00778-021-00692-3

9. Dignös, A., Glavic, B., Niu, X., Gamper, J., Böhlen, M.H.: Snapshot semantics for temporal multiset relations. Proc. VLDB Endow. **12**(6), 639–652 (2019)

10. Dong, X.L., Kementsietsidis, A., Tan, W.: A time machine for information: looking back to look forward. SIGMOD Rec. **45**(2), 23–32 (2016)

11. Ethier, J.F., Goyer, F., Fabry, P., Barton, A.: The prescription of drug ontology 2.0 (PDRO): more than the sum of its parts. Int. J. Environ. Res. Public Health **18**(22), 12025 (2021)

12. Jensen, C.S., Snodgrass, R.T.: Timeslice operator. In: Liu, L., Özsu, M.T. (eds.) Encyclopedia of Database Systems, pp. 3120–3121. Springer, Boston (2009). https://doi.org/10.1007/978-0-387-39940-9_1426

13. Johnson, A., Bulgarelli, L., Pollard, T., Horng, S., Celi, L.A., Mark, R.: Mimic-iv (2020). https://doi.org/10.13026/A3WN-HQ05. https://physionet.org/content/mimiciv/0.4/

14. Kaufmann, M., et al.: Timeline index: a unified data structure for processing queries on temporal data in SAP HANA. In: SIGMOD Conference, pp. 1173–1184 (2013)

15. Khnaisser, C., Lavoie, L., Burgun, A., Ethier, J.-F.: Past indeterminacy in data warehouse design. In: Benslimane, D., Damiani, E., Grosky, W.I., Hameurlain, A., Sheth, A., Wagner, R.R. (eds.) DEXA 2017. LNCS, vol. 10439, pp. 90–100. Springer, Cham (2017). https://doi.org/10.1007/978-3-319-64471-4_9

16. Kulkarni, K.G., Michels, J.: Temporal features in SQL: 2011. SIGMOD Rec. **41**(3), 34–43 (2012)

17. Lorentzos, N.A., Johnson, R.G.: Extending relational algebra to manipulate temporal data. Inf. Syst. **13**(3), 289–296 (1988)

18. Lorentzos, N.A., Mitsopoulos, Y.G.: SQL extension for interval data. IEEE Trans. Knowl. Data Eng. **9**(3), 480–499 (1997)

19. Lyson, H.C., et al.: A qualitative analysis of outpatient medication use in community settings: observed safety vulnerabilities and recommendations for improved patient safety. J. Patient Saf. **17**(4), e335–e342 (2019)

20. Özsoyoglu, G., Snodgrass, R.T.: Temporal and real-time databases: a survey. IEEE Trans. Knowl. Data Eng. **7**(4), 513–532 (1995)

21. Piatov, D., Helmer, S.: Sweeping-based temporal aggregation. In: Gertz, M., et al. (eds.) SSTD 2017. LNCS, vol. 10411, pp. 125–144. Springer, Cham (2017). https://doi.org/10.1007/978-3-319-64367-0_7

22. Piatov, D., Helmer, S., Dignös, A.: An interval join optimized for modern hardware. In: ICDE, pp. 1098–1109. IEEE Computer Society (2016)

23. Won, S.-M., Kim, M.-H., Kim, J.-M.: Administration management system design for smart phone applications in use of QR code. In: Park, J.J.J.H., Ng, J.K.-Y., Jeong, H.Y., Waluyo, B. (eds.) Multimedia and Ubiquitous Engineering. LNEE, vol. 240, pp. 585–592. Springer, Dordrecht (2013). https://doi.org/10.1007/978-94-007-6738-6_71

24. Zhou, X., Wang, F., Zaniolo, C.: Efficient temporal coalescing query support in relational database systems. In: Bressan, S., Küng, J., Wagner, R. (eds.) DEXA 2006. LNCS, vol. 4080, pp. 676–686. Springer, Heidelberg (2006). https://doi.org/10.1007/11827405_66

Soft Spatial Querying on JSON Data Sets

Paolo Fosci[(⊠)] and Giuseppe Psaila[(⊠)]

University of Bergamo, Viale Marconi 5, 24044 Dalmine, (BG), Italy
{paolo.fosci,giuseppe.psaila}@unibg.it
http://www.unibg.it

Abstract. *JSON* (JavaScript Object Notation) has become popular for exchanging data sets over the Internet. Many data sets are "geo-tagged", since they represent spatial entities. As an effect, spatial analysts have to perform spatial queries on *JSON* data sets. While working with large data sets, crisp (on/off) spatial relations could be marginally effective; instead, soft relations and "soft spatial querying" could be the right tools, because they reveal the extent of a given spatial relation. In this paper, we present the recent evolution of *J-CO-QL$^+$*, the query language of the *J-CO* Framework (under development at University of Bergamo, Italy) towards soft spatial querying on geo-tagged *JSON* data sets.

Keywords: Geo-tagged JSON documents · Soft spatial querying ·
Fuzzy operators and fuzzy spatial relations

1 Introduction

The advent of Open Data portals has pushed forward the distribution of geographical information: public authorities publish authoritative data sets that describe territories. In this context, *JSON* (JavaScript Object Notation) has become very popular to represent data sets over the Internet. In our opinion, this popularity is due to its simple syntax, which makes *JSON* data sets easy to process. Consequently, often analysts and data engineers have to work with *JSON* data sets, possibly describing "geo-tagged" data, i.e., data that are associated with geometrical descriptions of real-world entities on the Earth surface.

In this regard, *GeoJSON* is playing an important role, since it is a standard format that relies on *JSON* as hosting syntax to represent "geographical information layers". Many data sets concerning spatial data are provided as *GeoJSON* documents, on which spatial analysts have to perform "spatial queries".

When spatial querying is performed on large data sets, often spatial relations between spatial entities cannot be on/off (or crisp), because what it matters is the "extent" to which the relation is met. Furthermore, the query itself could be explained in a vague or tolerant way, so as to catch unexpected situations. Therefore, we are moving towards "soft spatial querying".

At University of Bergamo (Italy), we are devising the *J-CO* Framework [19]: its goal is to provide analysts with tools to flexibly acquire, integrate and query

© Springer Nature Switzerland AG 2022
S. Chiusano et al. (Eds.): ADBIS 2022, LNCS 13389, pp. 223–237, 2022.
https://doi.org/10.1007/978-3-031-15740-0_17

JSON data sets, in a way that is independent of the platform that provides and stores data sets. In the original design of the *J-CO-QL* query language [3], we were focused on the native support for spatial querying on geo-tagged *JSON* data sets. Later, we realized that soft querying could be a valuable tool for analysts: we started extending *J-CO-QL* towards evaluating membership degrees of *JSON* documents to fuzzy sets in order to perform soft querying (see [11]). We understood that, to fully introduce soft-querying capabilities, it was necessary to revise the whole language, that has become *J-CO-QL$^+$*: constructs have been unified, re-arranged and made more intuitive and flexible, for better supporting soft querying.

The contribution of this paper is to present the recent constructs to perform soft spatial querying, through the unified `JOIN` statement (until [11,19] there were two distinct `JOIN` statements). The new `JOIN` statement, which pairs documents coming from two data sets, now supports fuzzy sets; furthermore, soft spatial relations can be now evaluated, thus enabling soft spatial querying.

The paper is organized as follows. Section 2 introduces the relevant background for the paper. Section 3 provides the main contribution of the paper. Finally, Sect. 4 draws the conclusions and highlights possible future works.

2   Background

In this section, we briefly present the background of our work, i.e., basic notions on fuzzy sets (Sect. 2.1), the literature on soft querying databases (Sect. 2.2) and a short introduction to the *J-CO* Framework (Sect. 2.3).

2.1   Brief Introduction to Fuzzy Sets

The notion of Fuzzy Set (and related theory) was introduced by Zadeh in [20]; here, we report a brief introduction.

Let us denote by U a non-empty universe, either finite or infinite.

Definition 1. *A fuzzy set $A \subseteq U$ is a mapping $A : U \to [0,1]$. The value $A(x)$ is referred to as the membership degree of the item $x \in U$ to the A fuzzy set.*

We can say that a fuzzy set A in U is characterized by a membership function $A(x)$ that associates each item $x \in U$ with a real number in the range $[0,1]$; this value denotes the degree with which x is a member of A, also called "membership degree". Specifically, if $A(x) = 0$, this means that x does not belong at all to A; if $0 < A(x) < 1$, this means that x belongs to A only partially, with an extent that is the membership degree; $A(x) = 1$ means that x fully belongs to A.

As a example, consider the universe U of scientific papers. The fuzzy set called *RelevantPapers* denotes those papers that are relevant for the research conducted by a researcher named Jane: she judges the relevance of a paper x, by expressing its membership degree; clearly, $RelevantPapers(x) = 1$ means that the x paper is fully relevant; given two papers x_1 and x_2, if $RelevantPapers(x_1) = 0.8$ and $RelevantPapers(x_1) > RelevantPapers(x_2)$, this means that x_1 is more relevant than x_2, even though x_1 is not fully relevant to Jane's research (its membership degree is 0.8).

Some properties can be easily introduced.

A fuzzy set is empty if and only if its membership function is identically zero for each $x \in U$.

Two fuzzy sets A and B are equal, denoted as $A = B$, if and only if $A(x) = B(x)$ for all $x \in U$.

The classical set-oriented operations can be extended to fuzzy sets.

Definition 2. *Consider two fuzzy sets A and B in U.*

Considering the items $x \in U$ in the intersection $I = A \cap B$, they have the membership function $I(x) = min(A(x), B(x))$.

Considering the items $x \in U$ in the union $S = A \cup B$, they have the membership function $S(x) = max(A(x), B(x))$.

Considering the items $x \in U$ in the complement of A, i.e., $C_A = U - A$, they have the membership function $C_A(x) = 1 - A(x)$.

Soft conditions linguistically express the fact that an item belongs to fuzzy sets. Consequently, the AND operator is mapped to the fuzzy intersection, the OR operator is mapped to the fuzzy union, the NOT operator is mapped to the fuzzy complement.

For example, considering again Jane's research, we could think about a second fuzzy set called *HighlyCitedPapers*, which denotes the level of citations; so, Jane could look for papers x such that

$$HighlyCitedPapers(x) \text{ AND } RelevantPapers(x),$$

which is an example of soft condition. The resulting membership degree denotes the relevance of a paper x w.r.t. the condition; so it can be used to rank results.

In our work, we consider the universe U of *JSON* documents. Thus, given a document $d \in U$, the membership degree $A(d)$ denotes the extent with which d belongs to the A fuzzy set.

2.2   Related Work

Soft querying on data sets is not a novel topic: in the past, it was deeply investigated on top of relational databases. The first approach was to work on an already existing database; under this hypothesis, the query language should support soft querying, but the underlying database model cannot change; these query languages extend the concept of "selection condition", as argued by [2], so as to formulate "soft queries" in which selection conditions can rely on vague predicates; the goal is to obtain queries that are tolerant to thresholds, which is a typical problem in classical crisp querying; for example, given a non-soft predicate income >=100000 to select people whose annual income is no less than 100000 Euros, a person with 99999 Euros as annual income would not be selected. By adopting fuzzy sets, it is possible to express the concept of WantedIncome in a soft way, so that 99999 Euros is considered not fully wanted, i.e., has a membership degree slightly less than 1. Since relational databases were considered, many extensions of SQL towards soft querying were proposed: they preserve the underlying classical relational data model but provide capabilities of soft querying table rows through fuzzy-set theory. Since it is impossible to be exhaustive here, we mention *SQLf* (see [8,9]) and the attempt to implement it (in [13]); the second proposal we mention is *FQUERY for Access* (see [15] and [21]), which was designed to work within Microsoft Access; finally, we mention the proposal called *SoftSQL* (see [4–6]), which also covers the definition of user-defined "linguistic predicates" through a dedicated statement, to be used in soft selection conditions in the SELECT statement.

Removing the hypothesis of working on top of an existing relational database, the data model can be extended towards "fuzzy-relational databases" (see [7] and [17]) for representing uncertain and imprecise data directly within the database, by means of "fuzzy values". Here, we mention *FSQL*, presented in [12,14], which is the most remarkable proposal, in our opinion.

The recent advent of the notion of *JSON* document stores (NoSQL databases specifically designed to store *JSON* documents) seems stimulating novel research on soft querying, this time on *JSON* data sets and stores.

The work [1] proposed *fMQL*, an extension of *MQL* (the *MongoDB* Query Language). The authors worked under the hypothesis that *JSON* documents are previously labeled with "fuzzy labels".

Recently, [16] proposed an extension of the *MongoDB* data model towards a fuzzy *JSON* document store, supporting fuzzy values in single fields.

2.3   The *J-CO* Framework

The *J-CO* Framework [3,18,19] is a tool able to retrieve, integrate and query collections of *JSON* documents either stored in *JSON* document stores or directly provided by Web sources. The core of the framework is *J-CO-QL$^+$*, a novel declarative query language specifically designed to query heterogeneous *JSON* data sets, by natively dealing with spatial relations between geo-tagged documents. The *J-CO* Framework is composed of several software tools; the interested reader can refer to [19] for a complete description. Here, we just mention the *J-CO-QL$^+$ Engine*, which actually executes *J-CO-QL$^+$* queries/scripts; in particular, it is worth noticing that this is independent of any *JSON* document store, while it is able to retrieve data from and store data to several stores (but it processes data independently of specific processing capabilities provided by specific *JSON* stores).

Here, we briefly introduce the data and execution models.

Data Model. *J-CO-QL$^+$* works on collections of standard *JSON* documents. Two special root-level fields, named ~fuzzysets and ~geometry, play a specific role. The ~fuzzysets field works as a map $fsn \rightarrow md$, where fsn is a fuzzy-set name and md is the corresponding membership degree (see Fig. 3c); this way, the degrees of membership to multiple fuzzy sets of a document can be simultaneously represented. The ~geometry field represents spatial geometries [3,19], by relying on the GeometryCollection format of *GeoJSON*.

Execution Model. We define the concept of *query-process state* as a tuple $s = \langle tc, IR, DBS, FO \rangle$. Its members have the following meaning: $s.tc$ is called *temporary collection* of documents; $s.IR$ (*Intermediate Results*) is a local database where the process can temporarily store collections; $s.DBS$ is a set of database descriptors, used to connect to *JSON* databases; $s.FO$ is the set of *fuzzy operators* (see Sect. 3.2) defined throughout the query. All members in the initial state s_0 are empty sets.

A query (or script) is a sequence of instructions. An instruction i_j works on the $s_{(j-1)}$ query-process state and generates a new query-process state s_j, where one member can be changed. In particular, consider the temporary collection tc: i_j receives the $s_{(j-1)}.tc$ temporary collection as input and may generate a new $s_j.tc$ temporary collection as output.

3   Case Study and *J-CO-QL$^+$* Script

We now present the main contribution of the paper. Section 3.1 introduces the case study. Then, Sect. 3.2 presents how *J-CO-QL$^+$* deals with fuzzy concepts. Section 3.3 shows how to pre-process *GeoJSON* documents for soft querying. Finally, Sect. 3.4 shows how to perform soft spatial queries in *J-CO-QL$^+$*.

3.1   Case Study

Suppose that an analyst has to face the following problem.

Problem 1. *Consider a European Union (EU) region R and its provinces P(R), whose geographical description is provided. Consider also a registry of Gross Domestic Product (GDP(p)) about P(R), in which GDP in 2017 and 2018 are reported for a EU province p. Finally, consider a GeoJSON document that describes the set L of lakes that are either partially or totally in the territory of R. We want to look for provinces $p \in P(R)$ such that there is a lake $l \in L$ partially falling in p's territory (but in a significant way), such that p had a significant increase in GDP from 2017 to 2018.*

The goal of the analyst is to discover if there is a relation between the fact a significant part of a lake falls into the territory of a given province p and the province increase in GDP.

In the rest of the paper, we refer to the following territory and data sets:

- R is the Italian region named Lombardia, where our University is located.
- EU NUTS describe administrative regions in the EU. They were downloaded from the Eurostat website[1] as a shape file. The data set encompasses many

---

[1]   https://ec.europa.eu/eurostat/web/gisco/geodata/reference-data/administrative-units-statistical-units/nuts.

administrative levels: Level 0 corresponds to countries; Level 1 corresponds to macro-regions; Level 2 corresponds to regions; finally, Level 3 corresponds to provinces. The shape file was converted to geo-tagged *JSON* documents where the geometry is compliant with *GeoJSON* through QGIS, from the 2016 EU NUTS, we filtered the 12 Level-3 ones (provinces) in Lombardy.

- Registry data about provinces were downloaded as an Excel file from the Eurostat *data browser*[2]. Only rows concerning the 12 provinces in Lombardy were selected and converted into *JSON* documents.
- Data about lakes were downloaded from Regione Lombardia Open Data portal[3]. They were provided as shape files and then converted to *GeoJSON* through QGIS. The *GeoJSON* document encompasses 5469 lakes.

Listing 1 *J-CO-QL$^+$* script, part 1.

```
1. CREATE FUZZY OPERATOR IncreasingGDP
      PARAMETERS    gdp0 TYPE Float, gdp1 TYPE Float
      PRECONDITION  gdp0 > 0 AND gdp1 > 0
      EVALUATE      100*(gdp1-gdp0) / gdp0
      POLYLINE      [ (0, 0.0), (1, 0.8), (3, 1.0)];

2. CREATE FUZZY OPERATOR InterestingLake
      PARAMETERS    ownership TYPE Float
      PRECONDITION  ownership IN_RANGE [0, 1]
      EVALUATE      ownership
      POLYLINE      [ (0, 0), (0.25, 0), (0.3, 1), (0.8, 1), (0.85, 0), (1, 0) ];

3. USE DB AbdisDb ON SERVER MongoDB 'http://127.0.0.1:27017';

4. GET COLLECTION GeojsonLakes@AbdisDb;

5. EXPAND
      UNPACK WITH ARRAY .features
      ARRAY .features TO .LakeGeoData;

6. FILTER
      CASE WHERE WITH .LakeGeoData.item
        GENERATE
          SETTING GEOMETRY .LakeGeoData.item.geometry
          BUILD {
            .properties: .LakeGeoData.item.properties
          };

7. SAVE AS Lakes;
```

In Problem 1, two concepts are naturally imprecise. The first one is "Increasing GDP": when a GDP is increasing in a significant way? Small variations of GDP are absolutely normal, both in a positive and in a negative way; so, it is necessary to define this imprecise concept in a possibly tolerant way.

The second imprecise concept is a "significant portion" of lake in the territory of a province p. When is a lake portion significant, in such a way it is not marginal and the lake is not completely contained in the territory of the province p?

The *J-CO-QL$^+$* script to address Problem 1 is presented in Listings 1 and 2. We will present it in the remainder of this section.

---

[2] https://ec.europa.eu/eurostat/databrowser/explore/all/all_themes?lang=en.

[3] https://www.dati.lombardia.it/Territorio/Lago-10000-CT10/qm9t-uzst.

Listing 2 *J-CO-QL⁺* script, part 2.

```
 8. JOIN OF COLLECTIONS GeoNuts@AbdisDb AS G, NutsInfo@AbdisDb AS I
       SET GEOMETRY LEFT
       CASE WHERE WITH  .G.properties.NUTS_ID,    .G.properties.CNTR_CODE,
                        .G.properties.LEVL_CODE,  .I.NUTSCode
         AND .G.properties.NUTS_ID    = .I.NUTSCode
         AND .G.properties.CNTR_CODE  = "IT"
         AND .G.properties.LEVL_CODE  = 3
       GENERATE
         CHECK FOR FUZZY SET ImprovingWealth
           USING IncreasingGDP(TO_FLOAT(.I.GDP2017), TO_FLOAT(.I.GDP2018))
         BUILD {
           .nutsName: .G.properties.NUTS_NAME,
           .nutsId: .G.properties.NUTS_ID,
           .gdp2017: TO_FLOAT(.I.GDP2017),
           .gdp2018: TO_FLOAT(.I.GDP2018)  };

 9. JOIN OF COLLECTIONS temporary AS N, Lakes AS L
       ON GEOMETRY INTERSECT
       SET GEOMETRY INTERSECTION
       SET FUZZY SETS LEFT ALL, HOW_INCLUDE (RIGHT) AS OwnedLake
       CASE WHERE KNOWN FUZZY SETS ImprovingWealth, OwnedLake
            AND WITH .L.properties
       GENERATE
         CHECK FOR
         FUZZY SET WishedLakes
             USING InterestingLake (MEMBERSHIP_OF (OwnedLake)),
         FUZZY SET Wanted
             USING ImprovingWealth AND WishedLakes
         ALPHACUT 0.7 ON Wanted
         BUILD {
           .nutsName  : .N.nutsName,
           .nutsId    : .N.nutsId,
           .gdp2017   : .N.gdp2017,
           .gdp2018   : .N.gdp2018,
           .lake      : .L.properties.NOME_LG,
           .rank      : MEMBERSHIP_OF (Wanted)     }
       DEFUZZIFY;

10.    SAVE AS RankedNuts@AbdisDb;
```

3.2 Definition of Fuzzy Concepts

The first two instructions in Listing 1 define two "fuzzy operators". In *J-CO-QL⁺*, a fuzzy operator is an operator that evaluates the membership degree of a *JSON* document to a fuzzy property or relation. Remember that we have to define two concepts, i.e., "increasing GDP" and "interesting lake"; the two fuzzy operators defined on lines 1 and 2 correspond to these two concepts.

Increasing GDP. The instruction on line 1 of Listing 1 defines the fuzzy operator called `IncreasingGDP`. Its goal is to evaluate if a province had a relevant increase of GDP in two consecutive years. Details are presented hereafter.

- The `PARAMETERS` clause specifies the formal parameters of the operator. Specifically, two floating-point parameters are defined, called `gdp0` and `gdp1`.
- The `PRECONDITION` clause specifies a condition that must be met by actual parameters, before evaluating the operator (otherwise an exception is raised and the evaluation is stopped). Specifically, GDPs must be greater than zero.

- The EVALUATE clause specifies a mathematical expression evaluated on the parameters: the value it returns is used as x-axis value of the membership function specified in the POLYLINE clause. Specifically, the expression computes the percentage of variation of the two GDPs.
- The POLYLINE clause specifies the membership function used to obtain the membership degree. It is defined as a sequence of points, where x-coordinates are free, while y-coordinates are in the range $[0, 1]$. Between two consecutive points, the function is a segment connecting the two points. For x values less than the leftmost point, its y-coordinate is returned as membership degree; similarly, for x values greater than the rightmost point.

  Figure 1a depicts the polyline. Notice that the membership degree increases from 0 to 0.8 up to 1% of GDP increase, becoming significantly interesting after 1% (degree greater than 0.8). From 1% to 3%, it slowly reaches 1: any further increase is considered equally of interest.

InterestingLake. The instruction on line 2 defines the second fuzzy operator, named InterestingLake. Its goal is to evaluate if a lake is interesting (for the analysis) on the basis of the ownership degree of its area by the province.

The operator receives one single formal parameter: it is called ownership, since it is the degree with which the province owns the area of a lake. The expression in the EVALUATE clause does not make any computation on the value of the parameter, which is used as it is as x-coordinate on the membership function.

The POLYLINE clause defines the membership function, as depicted in Figure 1b. The trapezoidal shape excludes those lakes whose area marginally falls into a province, because the membership degree is 0 up to an ownership degree of 25%. It is 1 in the range $[0.3, 0.8]$. After 0.85 it is 0 again (this way, we exclude those lakes that are substantially fully contained in the territory of a province).

Notice that by means of the two fuzzy operators, we are able to quantify the relevance of GDP increase in our context, as well as the relevance of a lake (based on the percentage of its area owned by a province). Nevertheless, intermediate membership degrees allow for being tolerant and consider borderline situations with respect to thresholds (0.3 and 0.8, for the InterestingLake operator).

3.3 Processing a GeoJSON Document

In Listing 1, lines from 3 to 6 acquire and process a *GeoJSON* document that describes lakes, stored in a *MongoDB* database called AdbisDb. This part of the script is not strictly connected with soft querying: hereafter, we provide a short introduction (the interested reader can refer to our previous works [10,19]).

- The USE DB instruction on line 3 connects to the *MongoDB* server and to the desired database.
- The GET COLLECTION instruction on line 4 retrieves the content of the collection named GeojsonLakes, stored in the AdbisDb database; the collection

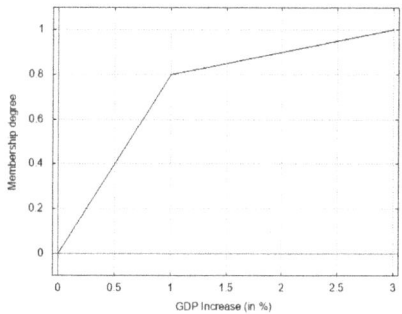

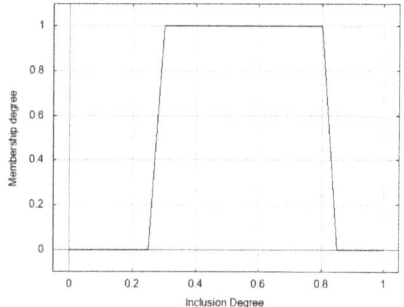

(a) Membership function of the **IncreasingGDP** fuzzy operator.

(b) Membership function of the **InterestingLake** fuzzy operator.

Fig. 1. Membership functions of the fuzzy operators.

contains one single *GeoJSON* document describing lakes; an excerpt of this document is reported in Figure 2a: notice that it is a single document, the root-level field of which, called **features**, is an array of documents; each nested document describes a "feature", i.e., a spatial entity with "properties" (the **properties** field) and "geometry" (the **geometry** field).

- The **EXPAND** instruction on line 5 unnests documents in the **features** array: the instruction generates a new collection of *JSON* documents, which contains as many documents as the ones contained in the **features** array. In details, each unnested document contains all fields in the source document, except the **features** array field; in its place, the **LakeGeoData** field is added, which contains two fields named **item** and **position**, where the former one contains the item actually unnested from within the original array, while the latter denotes the position occupied by the unnested item in the array.
- The **FILTER** instruction on line 6 restructures the documents in the temporary collection generated by line 5. Specifically, the **SETTING GEOMETRY** clause takes the **geometry** field nested within the **LakeGeoData** field as the geo-tagging of the document (as an effect, the **~geometry** field is added to the document). Then, the **BUILD** block restructures the document, putting the nested **properties** field at the top level. An example of documents contained in the new temporary collection is reported in Figure 2b.
- The instruction on line 7 saves the temporary collection into the *IR* (Intermediate Results) database, so as to reuse it later during the query process.

3.4 Soft Spatial Querying

The second part of the *J-CO-QL⁺* script (reported in Listing 2) actually performs a soft query on the data. Provided that line 10 actually saves the final temporary collection into the **AdbisDb** database, the key instructions are on lines 8 and 9.

```
{
  "type"        : "FeatureCollection",
  "features" : [
    {
      "type"        : "Feature",
      "properties" : {
        "OBJECTID"  : 4206,
        "AREA"      : 361162914.20957,
        "PERIMETER" : 171513.61094,
        "EID"       : 224,
        "STRATO_CTR" : "LG",
        "COD_LG"    : "LG224",
        "NOME_LG"   : "Garda (Lago di)"
      },
      "geometry"  : {
        "type"        : "Polygon",
        "coordinates" : [
          [ [ 10.55890, 45.53928 ],
            ...,                [ ..., ...] ] ]
      }
    },
    {...}, ..., {...} ]
}
```

```
{
  "properties" : {
    "OBJECTID"  : 4206,
    "AREA"      : 361162914.20957,
    "PERIMETER" : 171513.61094,
    "EID"       : 224,
    "STRATO_CTR" : "LG",
    "COD_LG"    : "LG224",
    "NOME_LG"   : "Garda (Lago di)"
  },
  "~geometry"  : {
    "type"        : "Polygon",
    "coordinates" : [
      [ [ 10.55890, 45.53928 ],
        ...,                [ ..., ...] ] ]
  }
}
```

(a) Excerpt of the *GeoJSON* document in the **GeojsonLakes** collection.
(b) **Example** of document in collection Lakes after line 6.

Fig. 2. Sample documents concerning lakes.

Processing NUTS. The first step is to process data about EU NUTS for the 12 considered provinces, stored within the **GeoNuts** collection in the **AdbisDb** database. Figure 3a reports a sample document: they follow the same structure of *GeoJSON* features; nevertheless, they were previously pre-processed, so they already respect the *J-CO-QL$^+$* data model (see the **~geometry** field).

The **NUTSInfo** collection (stored within the same database) actually contains registry data, about provinces. A sample document is reported in Figure 3b: notice that this is a very simple *JSON* document, which provides extra data to those provided by documents in the **GeoNUTS** collection (see Figure 3a).

The **JOIN** instruction on line 8 of Listing 2 pairs documents in the two collections, so as to extend a document g in the **GeoNUTS** collection (aliased as **G**) with the corresponding document i in the **NutsInfo** collection (aliased as **I**). The instruction behaves as follows.

- For each pair (g, i), a new document d is generated, where the **G** field contains the source g document, while the **I** field contains the source i document.
- The **SET GEOMETRY** clause denotes (if present) the geometry to assign to the d document. In this case, the left geometry is taken, so as to keep geometries of NUTS. As an effect, now d has the **~geometry** field too.
- The **CASE WHERE** clause selects those documents the analyst is interested in. Specifically, documents describing an Italian and level-3 (provinces) NUTS, that has been paired to the corresponding registry data.
- The **CHECK FOR FUZZY SET** clause evaluates the membership degree of the d document to the **ImprovingWealth** fuzzy set: this is done by means of the **USING** sub-clause, where a "soft condition" is expressed. Specifically, the

`IncreasingGDP` fuzzy operator is used (see Sect. 3.2). As a consequence, the d document is further extended with the `~fuzzysets` field.

- Finally, the `BUILD` block simplifies the document, as shown in Figure 3c (notice that GDP values are converted from strings, as in the source `NutsInfo` collection, to numerical values through the `TO_FLOAT` built-in function).

```
{
  "properties" : {
    "NUTS_ID"    : "ITC47",
    "LEVL_CODE"  : 3,
    "CNTR_CODE"  : "IT",
    "NAME_LATN"  : "Brescia",
    "NUTS_NAME"  : "Brescia",
    "MOUNT_TYPE" : 2,
    "URBN_TYPE"  : 2,
    "COAST_TYPE" : 3,
    "FID"        : "ITC47"
  },
  "~geometry"  : {
    "type"       : "Polygon",
    "coordinates": [
      [ [ 10.515752, 46.343224 ],
        ...,              [ ..., ...] ] ]
  }
}
```

(a) Example of document in the `GeoNUTS` collection.

```
{
  "NUTSCode"   : "ITC47",
  "NUTSLabel"  : "Brescia",
  "Area"       : 4786,
  "Pop2017"    : 1262.5,
  "Pop2018"    : 1265.9,
  "GDP2017"    : "42151.46",
  "GDP2018"    : "43311.54"
}
```

(b) Example of document in the `NUTSInfo` collection.

```
{
  "nutsId"    : "ITC47",
  "nutsName"  : "Brescia",
  "gdp2017"   : 42151.46,
  "gdp2018"   : 43311.54,
  "~fuzzysets" : {
    "ImprovingWealth": 0.975217040222694
  },
  "~geometry"  : {
    "type"       : "Polygon",
    "coordinates": [
      [ [ 10.515752, 46.343224 ],
        ...,              [ ..., ...] ] ]
  }
}
```

(c) Example of document after line 8.

```
{
  "nutsId"    : "ITC47",
  "nutsName"  : "Brescia",
  "gdp2017"   : 42151.46,
  "gdp2018"   : 43311.54,
  "lake"      : "Garda (Lago di)",
  "rank"      : 0.975217040222694,
  "~geometry"  : {
    "type"       : "Polygon",
    "coordinates": [
      [ [ 10.515752, 46.343224 ],
        ...,              [ ..., ...] ] ]
  }
}
```

(d) Example of document after line 9.

Fig. 3. Sample documents concerning NUTS and obtained by Listing 2

Cross-Analyzing Lakes and Provinces. We are now ready to cross-analyze lakes and provinces, by performing a soft spatial query (on line 9 of Listing 2).

- The instruction builds pairs (n, l), where n comes from the temporary collection generated by line 8 (see the `temporary` keyword), while l comes from the `Lake` collections generated by line 6 and temporarily stored within the IR database; the collections are aliased as N and L, respectively.
- The `ON GEOMETRY` clause expresses a "spatial condition" to generate pairs. Specifically, a pair (n, l) is selected if n's and l's geometries intersect.

– A new document d is generated, having the N field (containing the n document) and the L field (containing the l document).
  Based on the SET GEOMETRY clause, the ~geometry field is added too, containing the intersection of the two source geometries (i.e., the portion of lake that falls into the territory of the province).
– To complete the novel d document, it is necessary to deal with fuzzy sets. First of all, fuzzy sets whose membership was already evaluated for input documents must be considered (recall, looking at Figure 3c, that documents generated by line 8 had the membership degree to the fuzzy set called IncreasingWealth). Second, only here it is possible to exploit "fuzzy spatial relations", i.e., gradual relations concerning geometries of source documents. Both these tasks are done by the SET FUZZY SETS clause:
  • LEFT ALL specifies that all membership degrees to fuzzy sets evaluated for documents coming from the left collection (in this case, those generated by line 7) must be kept (i.e., the IncreasingWealth fuzzy set);
  • the membership degree to the OwnedLake fuzzy set is obtained by means of the HOW_INCLUDE function, which provides the degree of inclusion of the intersection in the area of the right (as specified within parentheses) geometry. As a result, d is further extended with the ~fuzzysets field, which contains two fields (i.e., IncreasingWealth and OwnedLake).
  • Alternatively, specific fuzzy-set names and the side they come from (i.e., LEFT or RIGHT) could be specified.

J-CO-QL^+ provides three functions that evaluate a fuzzy spatial relation between two source geometries:

• HOW_INCLUDE(RIGHT) (resp. HOW_INCLUDE(LEFT)) provides the degree the intersection is included in the right (respectively left) geometry;
• HOW_MEET(RIGHT) (resp. HOW_MEET(LEFT)) provides the degree the right perimeter touches the left one (resp., the left perimeter touches the right one);
• HOW_INTERSECT() provides the degree the area of the intersection is contained in the area of the united geometries.

We can now continue explaining the remainder of line 9.

– The CASE WHERE condition selects those documents so far obtained for which the membership degrees to the ImprovingWealth and OwnedLake fuzzy sets has been evaluated (see the KNOWN FUZZY SETS predicate).
– Within the GENERATE block, two FUZZY SET branches are specified in the CHECK FOR clause.
  Specifically, the first one evaluate a fuzzy set that is called WishedLakes: its membership degree denotes the degree of interest the lake has for the query. This is done by means of the InterestingLake fuzzy operator (see Sect. 3.2), which receives the membership degree to the OwnedLake fuzzy set, obtained through the MEMBERSHIP_OF function (remember that a lake is interesting only if the percentage of its area that falls into the territory of the province is approximately between 30% and 80%).

– On the basis of the last evaluated fuzzy set, the second FUZZY SET branch evaluates the membership degree to the last Wanted fuzzy set. The USING soft condition says that analyst is interested in those documents that describe a province that shows IncreasingWealth and whose territory contains WishedLakes. Notice the use of the AND logical operator, now in the fuzzy version (that returns the minimum membership degree, see Definition 2).
– The ALPHACUT clause filters documents on the basis of their membership degree to the Wanted fuzzy set: specifically, documents whose membership degree is no less than 0.7 (i.e., high satisfaction of the condition) are kept.
– The BUILD block generates the final structure of documents, flattening them. Furthermore, the rank field is added, taking the membership degree to the Wanted fuzzy set as value, because the document is "de-fuzzified" (i.e., the ~fuzzysets field is removed). Figure 3d reports a sample output document.

The reader can notice how the last USING clause expresses the soft condition in a linguistic way, if fuzzy set names are properly chosen: indeed, we are looking for provinces that have shown increasing wealth and contain a wished lake. The final membership degree is used to rank retrieved documents, so as to express the relevance of the document to the query. Even if the soft condition is not directly expressed on a fuzzy set directly obtained by means of a spatial fuzzy relation, nevertheless, the WishedLakes fuzzy set derives from the OwnedLake fuzzy set, evaluated in the SET FUZZY SETS clause when source documents are actually paired. So, this is a simple yet complete example of soft spatial querying performed on *JSON* documents through *J-CO-QL$^+$*.

To conclude this section, we highlight the novelty introduced in the presented version of the language. First of all, the general organization of sub-clauses associated to the WHERE selection condition has been significantly improved. Furthermore, in place of having two distinct JOIN statements, one for spatial join and one for non-spatial join, now a unified JOIN statement is provided; not only, apart from the fact that this statement now supports fuzzy-set evaluation, novel fuzzy spatial functions have been introduced.

4   Conclusions

The paper presented how the *J-CO-QL$^+$* language (the query language of the *J-CO* Framework) is now capable to support "soft spatial queries": evaluation of membership degree to fuzzy sets is now integrated with soft spatial relations within the redesigned JOIN statement; it is now able to deal with membership degrees already available in joined documents; membership degrees to new fuzzy sets based on spatial relations now can be added too. The *J-CO-QL$^+$* script presented in the paper addressed a problem of soft spatial querying in which authoritative data sets coming from various sources were integrated.

As a future work, the main activity will be devoted to complete the redesign of all the statements, so as to fully support soft querying. Specifically, we will address the concept of *fuzzy aggregator*, so as to define complex aggregations of membership degrees in aggregated documents. Then, we plan to address multiple models of fuzzy sets, so as to deal with them in an integrated and unified way. The *J-CO* Framework is available on a public GitHub repository[4].

References

1. Castelltort, A., Laurent, A.: Towards fuzzy querying of NoSQL document-oriented databases. In: Laurent, A., Strauss, O., Bouchon-Meunier, B., Yager, R.R. (eds.) IPMU 2014. CCIS, vol. 444, pp. 384–395. Springer, Cham (2014). https://doi.org/10.1007/978-3-319-08852-5_40
2. Blair, D.C.: Information retrieval, 2nd edn. c.j. van rijsbergen. london: Butterworths; 1979: 208 pp. price: $32.50. J. Am. Soc. Inf. Sci. **30**(6), 374–375 (1979). https://doi.org/10.1002/asi.4630300621
3. Bordogna, G., Ciriello, D.E., Psaila, G.: A flexible framework to cross-analyze heterogeneous multi-source geo-referenced information: The J-CO-QL proposal and its implementation. In: Proceedings of the International Conference on Web Intelligence, pp. 499–508 (2017)
4. Bordogna, G., Psaila, G.: Modeling soft conditions with unequal importance in fuzzy databases based on the vector p-norm. In: IPMU Conference, Malaga (2008)
5. Bordogna, G., Psaila, G.: Soft aggregation in flexible databases querying based on the vector p-norm. Int. J. Uncert. Fuzzi. Knowl. Based Syst. **17**(supp01), 25–40 (2009)
6. Bordogna, G., Psaila, G.: Customizable flexible querying in classical relational databases. In: Galindo, J. (ed.) Handbook of Research on Fuzzy Information Processing in Databases, pp. 191–217. IGI Global (2008)
7. Bosc, P., Prade, H.: An introduction to the fuzzy set and possibility theory-based treatment of flexible queries and uncertain or imprecise databases. In: Uncertainty Management in Information Systems, pp. 285–324. Springer, New York (1997). https://doi.org/10.1007/978-1-4615-6245-0
8. Bosc, P., Pivert, O.: SQLF: a relational database language for fuzzy querying. IEEE trans. Fuzzy Syst. **3**(1), 1–17 (1995)
9. Bosc, P., Pivert, O.: SQLF: query functionality on top of a regular relational database management system. In: Knowledge Management in Fuzzy Databases, pp. 171–190. Springer, Heidelberg (2000). https://doi.org/10.1007/978-3-7908-1865-9
10. Fosci, P., Marrara, S., Psaila, G.: Soft querying Geojson documents within the J-Co framework. In: 16th International Conference on Web Information Systems and Technologies (WEBIST 2020), pp. 253–265 (2020)
11. Fosci, P., Psaila, G.: Towards flexible retrieval, integration and analysis of JSON data sets through fuzzy sets: a case study. Information **12**(7), 258 (2021)
12. Galindo, J.: New characteristics in FSQL, a fuzzy SQL for fuzzy databases. WSEAS Trans. Inf. Sci. Appl. **2**(2), 161–169 (2005)
13. Galindo, J., Medina, J.M., Pons, O., Cubero, J.C.: A server for fuzzy SQL queries. In: Andreasen, T., Christiansen, H., Larsen, H.L. (eds.) FQAS 1998. LNCS, vol. 1495, pp. 164–174. Springer, Heidelberg (1998). https://doi.org/10.1007/BFb0055999
14. Galindo, J., Urrutia, A., Piattini, M.: Fuzzy Databases: Modeling, Design, and Implementation. IGI Global (2006)

---

[4] Github repository - https://github.com/JcoProjectTeam/JcoProjectPage.

15. Kacprzyk, J., Zadrożny, S.: Fquery for access: Fuzzy querying for a windows-based DBMS. In: Bosc, P., Kacprzyk, J. (eds.) Fuzziness in Database Management Systems. Studies in Fuzziness, vol. 5. Physica, Heidelberg (1995). https://doi.org/10.1007/978-3-7908-1897-0_18

16. Medina, J.M., Blanco, I.J., Pons, O.: A fuzzy database engine for mongoDB. Int. J. Intell. Syst. **37**, 5691–5724 (2022)

17. Medina, J.M., Pons, O., Vila, M.A.: Gefred: a generalized model of fuzzy relational databases. Inf. Sci. **76**(1), 87–109 (1994)

18. Psaila, G., Fosci, P.: Toward an analyist-oriented polystore framework for processing JSON geo-data. In: International Conferences on Applied Computing 2018, Budapest; Hungary, 21–23 October 2018. pp. 213–222. IADIS (2018)

19. Psaila, G., Fosci, P.: J-CO: a platform-independent framework for managing geo-referenced JSON data sets. Electronics **10**(5), 621 (2021)

20. Zadeh, L.A.: Fuzzy sets. Inf. Control **8**(3), 338–353 (1965)

21. Zadrozny, S., Kacprzyk, J.: Fquery for access: towards human consistent querying user interface. In: ACM Symposium on Applied Computing, pp. 532–536 (1996)

Maximum Range-Sum for Dynamically Occurring Objects with Decaying Weights

Ashraf Tahmasbi[(✉)] and Goce Trajcevski

Iowa State University, Ames, IA 50011, USA
`{tahmasbi,gocet25}@iastate.edu`

Abstract. This work addresses a novel variant of the Maximum Range-Sum (MaxRS) query for settings in which spatial point objects occur dynamically and, upon occurrence, their significance (i.e., weight) decays over time. The objective of the original MaxRS query is to find a location to place (the centroid of) a fixed-size spatial rectangle so that the sum of the weights of the point objects in its interior is maximized. The unique aspect of the problem studied in this paper, which we call DDW-MaxRS (Dynamic and Decaying Weights MaxRS), is that the placement of its solution can vary over time due to the joint impact of the arrival of new objects and the change of the corresponding weights of the existing objects over time. To improve the efficiency of the DDW-MaxRS problem processing, we propose a memory-efficient approximate algorithmic solution that will naturally infuse uncertainty in the answer. We formally analyze the error bounds' properties and provide experimental results to quantify the effectiveness of the proposed approach.

Keywords: Maximum range-sum · Dynamic objects · Approximation

1 Introduction

The advances in GPS-equipped mobile devices and networking technologies have enabled generation of large volumes of location-aware data [7,20]. Often, in addition to location and time, such data includes various semantic (numerical and/or categorical) attributes, which are of interest in many applications relying on location-based services (LBS), such as social networks, emergency response, recommendations, etc. [18]. These settings, in turn, have brought about novel problems and challenges in query processing and analytics, the efficient management of which requires methodologies that extend the traditional spatial/Spatio-temporal and Moving Objects Database (MOD) approaches [8,25].

One such problem (re)addressed in the recent years is the Maximum Range-Sum (MaxRS) query, also known as the Maximum-enclosing Rectangle Problem [26], described as follows: given a set of (weighted) spatial point-objects O and a rectangle R with fixed dimensions ($a \times b$), determine the location for placing R such that the sum (of weights) of the objects in R's interior is maximized.

Research supported by the NSF SWIFT grant 2030249.

S. Chiusano et al. (Eds.): ADBIS 2022, LNCS 13389, pp. 238–252, 2022.
https://doi.org/10.1007/978-3-031-15740-0_18

Figure 1 shows the positions of a set of input points and their respective weights. The shaded region indicates the position of the query rectangle such that the sum of the weights of the (red-colored) objects in its interior is maximum among all the other possible placements in the horizontal plane.

What motivates this work is that in certain domains, in addition to the arrival/occurrence of new points, one may face a situation in which the weights of the previously inserted objects need not be constant over time (e.g., it may decay). For example, when monitoring a CO concentration through participatory sensing [3], mobile citizens periodically report the value of the measurement from their current location, based on incentives [11]. One of the objectives of managing carbon footprint is to avoid uneven distribution in urban environments [14,19]. A traffic management system may want to enforce reduced driving through an area of a given size (limited by the constraint not to cause too severe congestion) for some time. As time evolves, when re-evaluating the critical area to reduce traffic, one must consider that the value of ppm (particles per million) measured in a previous location of the mobile volunteer may have declined over time.

A complimentary scenario comes from burglary crime management. To determine the placement of a fixed-size region that will cover a maximum number of ongoing burglaries [17] – upon reporting a new one, one needs to consider that the typical "stay-time" of burglars for each previously reported one in a location is between 8–12 min [13]. Similarly, when estimating the (placement of a) fixed size region in which there is the highest likelihood of getting infected by a virus due to the concentration of infected persons, one needs to consider that the capacity of transmitting the infection decays over time [21].

While at their core, such problems resemble the settings of the MaxRS problem, they also accentuate the joint impact of two factors: (i) dynamic appearances of objects at given locations, with varying weight/importance (e.g., the severity of symptoms); and (ii) variability over time of the objects' weights.

None of the available variants of the MaxRS problem can be straightforwardly applied to address the settings described above. Towards that, the main contributions of this work are threefold: (i) We take a first step towards for-

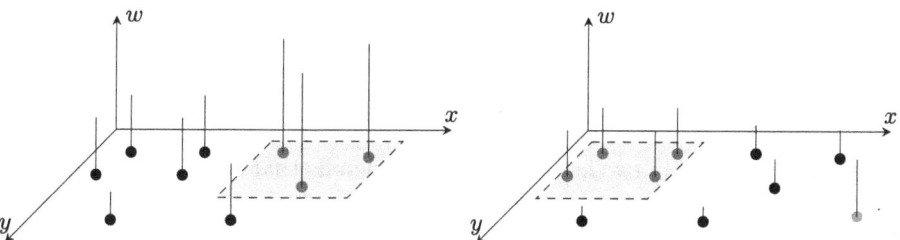

Fig. 1. Location of MaxRS at time t; weights of each object are shown as its third dimension over the z axis.

Fig. 2. MaxRS at time $t + 1$; decayed weights changed the answer and a new WD object arrived (shown in orange). (Color figure online)

malizing a novel *Dynamic and Decaying Weights MaxRS* (DDW-MaxRS) query; (ii) We identify and formally prove properties that ensure desired bounds for a novel approximate solution and introduce a corresponding algorithm; and (iii) We present experiments showing the effectiveness of the proposed approach.

We introduce the notation and formalize the problem in Sect. 2, followed by the algorithmic solutions in Sect. 3. We discuss generalizations of the parameters in Sect. 4 and present experimental observations in Sect. 5. Related literature is presented in Sect. 6 and the concluding remarks in Sect. 7.

2   Preliminaries

We consider a time-varying set of objects $O_{n_t}(t) = \{o_1, o_2, \ldots, o_{n_t}\}$, each of which is associated with a location in 2D Euclidean space, within a suitable coordinate system. Each object o_i is a triplet $((x_i, y_i), w_i(t), t_i)$, where i is its unique identifier, t_i is the time step in which this object arrived, (x_i, y_i) represents its location, and $w_i(t)$ is a scalar indicating its (relative) importance – e.g., the capability for spreading infection over time – defined as:

$$w_i(t) = \begin{cases} w_i \times \lambda_i^{(t-t_i)}, & \text{if } t \geq t_i \\ 0, & \text{otherwise} \end{cases} \tag{1}$$

where w_i indicates its initial importance/weight, and $\lambda_i < 1$ captures the decay over time – i.e., the diminishing of the importance of a particular object. In the rest of this paper, we assume that at each time step t, at most, one new object o_i is generated in the regions of interest. For brevity, we will refer to such objects as WD (weighted and decaying) objects in the rest of this paper. For a given axis-parallel rectangle R, we use $C(l_R, R)$ to denote the planar region covered by R when its center is placed in location l_R.

The Dynamic and Decaying Weights variant of the MaxRS over a time interval $\Delta(t) = [t_s, t_e]$ (DDW-MaxRS$((O_{n_s}(t)), R, (t_s, t_e)))$ is defined as follows:

Definition 1. *Given a (starting) set $O_{n_s}(t_s)$ of WD objects $O_{n_s}(t) = \{o_1, o_2, \ldots, o_{n_s}\}$, and a query rectangle R, the answer to* DDW-MAXRS$((O_{n_s}(t)), R, (t_s, t_e))$*, denoted $A_{DDW\text{-}MAXRS}$ $(O_{n_s}(t), R, (t_s, t_e))$ is a sequence of locations $L_R = \{l_R(t_s), \ldots, l_R(t_e)\}$ for placing the centroid of R, such that at each $t \in [t_s, t_e]$, $\sum_{\{o_i \in (O_{n_t}(t) \cap C(l_R, R))\}} w_i(t)$ is maximal.*

We are interested in two key aspects: (1) trade-offs that arise when sacrificing the accuracy of the answer for the benefit of the efficiency of the processing algorithm; and (2) avoiding a complete re-computation of the $A_{DDW\text{-}MAXRS}$ $(O_{n_s}(t), R, (t_s, t_e))$ at every time instant (i.e., naively invoking the traditional Max-RS algorithm for each $O_{n_t}(t)$). For (1), we consider a variant similar to [22], introducing a factor of ε to quantify the error compared with the exact answer. We have the following definition (updating the corresponding definition of $(1-\varepsilon)$-Approximate MaxRS from [22]) for an approximate answer at a time instant t:

Definition 2. *Given a set $O_n(t)$ of n WD objects $O_n(t) = \{o_1, o_2, ..., o_n\}$, the answer to $(1 - \varepsilon)$ Approximate DDW-MaxRS query at any time t, denoted A_ε DDW-MAXRS $(O_{n_t}(t), R)$, is a position $l_R(t)$ for placing the center of R such that the covered wight of R is a $(1 - \varepsilon)$-approximate answer, i.e.,*

$$(1 - \varepsilon) \times m^* \leq m \leq m^* \tag{2}$$

where $m = \sum_{\{o_i \in (O_n(t) \cap C(l_R, R))\}} w_i(t)$ is the weight returned by A_ε DDW-MAXRS $(O_n(t), R)$, and m^* is the weight returned by A DDW-MAXRS $(O_n(t), R)$.

However, in our case, answers may vary over time. For illustration, consider Fig. 2, showing a future time instant of the scenario in Fig. 1. It shows the change of the (location of the) answer due to the different decays of the weights over time. In addition, it shows an arrival of a new object (bottom-right) which, however, does not affect the answer.

3   Approximate Solution to *DDW-MaxRS*

We note that a naïve way to calculate the answer to DDW-MaxRS would be to invoke the traditional MaxRS solution (cf. [6]) at each time step, incorporating a new WD object, and weight modifications of the older object due to the decay over time. The time complexity of the exact solution, in this case is $O(n_t \log n_t)$ for each time instant t, where n_t is the (progressively increasing) number of WD objects at time t. To avoid such re-computations every time instant, one can rely on the approach to maintain the MaxRS in dynamic settings [2].

In the rest of this Section, we focus on a specific case of the DDW-MaxRS problem, where we assume a fixed decay – i.e., $\lambda_i = \lambda$ for all objects/points (we will later relax this assumption in Sect. 4).

To explain the intuition behind the approximate solution, let S denote the 2D space of all the possible locations for all the objects. Assuming a query rectangle R of size $[a \times b]$, we partition this space into cells of the same size $[a/j \times b/j]$ for some value j. Without loss of generality (i.e., with small "boundary zones"), we assume that S is fully covered by a 2D array of a whole number of cells. Upon arrival, each new object o_i, based on its location, is mapped to a cell. For each cell $c_{u,v}$, we store a list of points mapped to it. Given this discretization of S, upon an arrival of a new object o_m at some time instant t_m, we introduce its *extended dual rectangle* $R^e_{o_m}$ of a size $[(2j + 1)a/j \times (2j + 1)b/j]$ cells, centered in the center of the cell to which o_m belongs.

As we will formally demonstrate below, $R^e_{o_m}$ enables a more efficient calculation of the A_ε DDW-MAXRS $(O_{n_M}(t_m), R)$ based on the solution at the previous time instant $t_m - 1$. The rationale is that if there is a change in the solution since $t_m - 1$, it must be due to o_m and a subset of the older points located in the cells that are intersecting $R^e_{o_m}$.

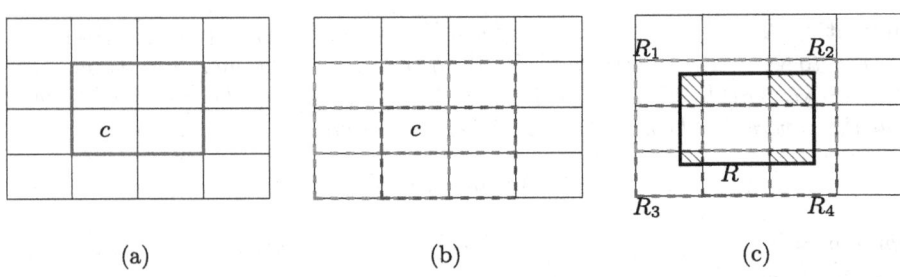

Fig. 3. For $j = 2$: (a) extended cell $ExtCell(c)$ (b) All extended cells containing c, (c) Naive approach provides at least a $\frac{1}{4}$-approximate solution to DDW-MaxRS.

Algorithm 1: Grid-based Approximate Solution

---

    input: ExtCList is the sorted list of ExtCells based on their weights, p^* the
             solution to A_ε DDW-MaxRS$(O_n(t), R)$ with weight w^*
    // Target rectangle R is of size $a \times b$

1 **if** *new arrival* o_i **then**
2   Find the cell c that o_i maps into
3       $CellSet(c) \leftarrow CellSet(c) \cup o_i$
4       $W(c, t) \leftarrow W(c, t) + w_i(t)$
5   **for** $ExtCell(d)$ *containing* c **do** // There are j^2 of such extended cells
6           $W(ExtCell(d), t) \leftarrow W(ExtCell(d), t) + w_i(t)$
7   Update the position of $ExtCell(d)$ in ExtCList to keep it sorted
                // this step takes $O(\log N)$, where N is the total number
                of cells
8   **end**
9       $p^* \leftarrow$ center of ExtCList$[-1]$ // ExtCList$[-1]$ is the ExtCell with the
           maximum weight
10     $w^* \leftarrow W(ExtCellList[-1], t)$
11 **end**
12 **else**
13     $w^* \leftarrow \lambda \times w^*$
14 **end**

    Recall that we aim for an approximate solution(s) which, at each time instant, provides a faster calculation of the answer(s) at the price of sacrificing accuracy in terms of the covered weight. We quantify this error by ε as defined in Definition 2.

    For each cell c, we store a list of points mapped to it (OList(c)). At any time t, the weight of a cell c is computed as: $W(c, t) = \sum_{\{o_i \in \text{OList}(c)\}} w_i(t)$. We also define an extended cell ExtCell(c) as the set of cells (CellSet(c)) in the grid of $j \times j$ cells with cell c as its most left-bottom cell (shown in Fig. 3a). The weight of an extended cell ExtCell(c) is computed as: $W(ExtCell(c), t) = \sum_{\{cell \in CellSet(c)\}} W(cell, t)$.

    We propose an approximate solutions based on the idea of diving the space S into cells, and the pseudo code is shown in Algorithm 1.

3.1   Properties of the Approximate Solution

To return an approximate solution to the DDW-MaxRS problem, we do as follows: (i) At the initial time t_0, we keep a list of extended cells and order them based on their weights (we will refer to this list as ExtCList)). We then return the extended cell with the highest weight as the approximate solution. (ii) Let ExtCell(c^*) be our approximate solution to DDW-MaxRS at time $t - 1$ (for any $t > t_0$). We will not change the approximate solution if no new point arrives at time step t. We only update its weight. However, if a new point arrives, we first find the cell c this point maps into it. Note that there are j^2 extended cells containing c (shown in Fig. 3b for the choice of $j = 2$). We update the weight of c and the associated extended cells and their position in the ExtCList so that our list remains sorted. We then return the extended cell with the maximum weight as the solution to our DDW-MaxRS. The critical thing to remember is that the relative order of all the other extended cells remains the same.

Lemma 1. *At any time $t > t_0$, upon arrival of a new point o, the relative order of extended cells, that o does not map into, remains intact.*

The proof is rather straightforward, based on the fact that the weight of each extended cell that the new point is not mapped to decay by a factor of $\lambda < 1$. This algorithm provides *at least* $\frac{1}{4}$-approximate solution (cf. Lemma 2).

Lemma 2. *At any time t, the grid-based algorithm returns at least $\frac{1}{4}$-approximate solution to the DDW-MaxRS problem.*

Proof. For simplicity, we discuss $j = 2$, but the proof for larger values of j is similar. Let m be the weight of the extended cell returned by the algorithm at time t. Also, let the exact solution to the DDW-MaxRS problem, at time t, be as shown in Fig. 3c, which means it does not fall within any of our defined extended cells but rather overlaps with multiple of them (R_1, R_2, R_3, and R_4).

The proof follows the fact that the weight of points covered in the intersection of the exact solution R with each of them is bounded by m, and hence, the total weight covered by R cannot be more than $4m$. Note that this is a *tight bound* $\square$.

Intuitively, larger values for j imply a finer grid, and one may expect that it will improve the accuracy. However, having a very large j could result in cells containing only 1 (or 0) point, and there will be no benefit in dividing the space into cells. There is a trade-off in the choice of j between the accuracy of results and the computational efficiency. We study the effect of this parameter on the results in Sect. 5.

3.2   Memory-Efficient Approximate Solution

When mapping WD objects to grids, not all cells will include a point. A significant number of cells will be either empty or receive objects infrequently. It does not makes sense to store such objects or keep track of these sparse cells. This sub-section introduces an approach to clear sparse cells without hurting the accuracy. To this end, let us formally define a sparse cell:

Definition 3. *A cell c is called θ-sparse at time t if:*

$$W(c,t) \leq \theta \tag{3}$$

According to Definition 3, we call a cell θ-sparse at time t if its weight is less than θ, where $\theta > 0$ is a threshold we set. Later, We will discuss how to choose this threshold. Lemma 3 investigates lower and upper bounds on the sum of all WD objects' weights present in the system at any time t:

Lemma 3. *Let t_l be the last time step in which a new point arrived. At any time step $t \geq t_l$, we have:*

$$(n_0 + 1)\lambda^t < \sum_{\{o_i \in O_{n_t}\}} w_i(t) < (n_0 - 1)\lambda^t + \frac{1}{1 - \lambda} \tag{4}$$

where n_0 is the number of points in the system at time $t = 0$.

Proof. The proof is straightforward: given n_0 WD objects were present in the system at time t_0, the total weight is at least $n_0\lambda^t + \lambda^{t-t_l}$. On the other hand, in the worse case, we had one new WD object added at each time step from $t = 0$ to $t = t_l$, and the total weight of those objects is bounded by $\frac{1}{1-\lambda}$. □

In the rest of this section, we assume $n_0 = 1$ for simplicity, but the results are generalizable to $n_0 > 1$.

Let an algorithm clear all the θ-sparse cells (for some predetermined θ). One question that arises is how does it affect our grid? How many nonempty cells will remain after clearing such cells? Lemma 4 answers this question:

Lemma 4. *At any time t, the number of nonempty cells after deleting the θ-sparse cells will never exceed $L = \log_\lambda \theta(1 - \lambda)$.*

Proof. The proof is similar to the proof of Theorem 4.4 in [23]. Let t_c be the last time instant in which cell c received a new point. Therefore, we have:

$$W(c,t) = \lambda^{t-t_c} \times W(c, t_c) \tag{5}$$

Consider the case when $t - t_c > L = \log_\lambda \theta(1 - \lambda)$, then we have:

$$W(c,t) \overset{(i)}{=} \lambda^{t-t_c} W(c,t_c) < \lambda^L W(c,t_c) \overset{(ii)}{<} \frac{\lambda^L}{1 - \lambda} = \frac{\theta(1 - \lambda)}{1 - \lambda} = \theta \tag{6}$$

Where (i) holds because of (5) and (ii) is true because, according to Lemma 3, the sum of all weights at time t is less than $1/(1 - \lambda)$ (for $n_0 = 1$); therefore, the weight of each cell is also less than this upper bound. 6 concludes that any cell c that has not received any points within the last L time steps will be θ-sparse and erased at time t. Therefore, the number of non-empty cells is at most L. □

Our proposed memory-efficient solution is quite similar to Algorithm 1 with the main difference that θ-sparse cells (and consequently empty ExtCells) will be removed after line 4. As a consequence of this change, we need to address two fundamental questions:

Q1: *For any cell c, how much does the new weight differ from the actual weight of c if it was never cleared before? Lemma 5 answers this question.*

Lemma 5. *Suppose the last time a cell c was cleared because of being θ-sparse was t_d. If at current time t the weight of c is $\hat{W}(c,t)$, then we have:*

$$W(c,t) \leq \hat{W}(c,t) + (\frac{\theta}{1-\lambda^L})\lambda^{t-t_d} \tag{7}$$

where $L = \log_\lambda \theta(1-\lambda)$ and $W(c,t)$ is the actual weight of c if it was never deleted before.

Proof. The proof follows a similar approach as the proof of Theorem 4.3 in [23]. Suppose cell c has been previously deleted at time steps $t_1, t_2, ..., t_d$. Therefore, we knew that: $\hat{W}(c,t_i) \leq \theta$ for $i = 1, 2, ..., d$. Thus, if we choose θ to be a constant (not a function of t_is) and if we never cleared all these previous weights, the actual weight would be:

$$W(c,t) - \hat{W}(c,t) = \sum_{i=1}^{d} \hat{W}(c,t_i)\lambda^{t-t_i} \leq \sum_{i=1}^{d} \theta \times \lambda^{t-t_i} = \theta\lambda^{t-t_d} \times \sum_{i=1}^{d} \theta\lambda^{t_d-t_i}$$

On the other hand, according to Lemma 4, for each $i = 2, ..., d$, we have $t_i - t_{i-1} \geq L + 1$, where $L = \log_\lambda \theta(1-\lambda)$. Therefore $W(c,t) - \hat{W}(c,t) \leq \theta\lambda^{t-t_d}\left(\frac{1}{1-\lambda^L}\right)$. $\square$

Q2: *How does the error in the weights of cells (as the result of clearing sparse cells) affect our approximate solution ? Lemma 6 addresses this question.*

Lemma 6. *At any time t, the proposed memory-efficient approximate solution with the choice of $\theta \leq \frac{1}{(1-\lambda)}\left(1 - \frac{\lambda}{\alpha}\right)$, returns at least $\frac{1}{4(1+\alpha)}$-approximate solution to a DDW-MaxRS query, where $\alpha > \lambda$.*

Proof. Let c denote the cell with the maximum weight at time t. According to Lemma 5, we have $W(c,t) \leq \hat{W}(c,t) + \frac{\theta\lambda}{1-\lambda^L}$, where $L = \log_\lambda \theta(1-\lambda)$. Now, similar to the proof of Lemma 2, we can argue that the weight covered by R (the true answer of DDW-MaxRS at time t) is bounded as follows:

$$W(R,t) \leq 4 \times W(c,t) \leq 4 \times \left(\hat{W}(c,t) + \frac{\theta}{1-\lambda^L}\right) \leq 4 \times \left(1 + \frac{\lambda}{1-\lambda^L}\right)\hat{W}(c,t)$$

For the choice of $\theta \leq \frac{1}{(1-\lambda)}\left(1 - \frac{\lambda}{\alpha}\right)$, we have $\lambda^L = \theta(1-\lambda) \leq (1-\frac{\lambda}{\alpha})$. Therefore, $(1 + \frac{\lambda}{1-\lambda^L}) \leq (1+\alpha)$. $\square$

Upon arrival of a new point, the time complexity of the update is $O(j^2 \log L)$, where $L = \log_\lambda \theta(1-\lambda)$.

4   Generalization

So far, we have used certain assumptions on the settings of the DDW problem
– i.e., having an equal initial weight for all objects, the same decaying factor,
and a specific exponential decaying function. This section discusses how these
assumptions can be relaxed and the consequences, i.e., what needs to be changed
in the proposed solutions to make them suitable for more generalized variant(s).

Generalization to Different Initial Weights: The presented results can be
generalized to cases where initial weights are of an interval value $[w_{min}, w_{max}]$
(i.e., $w_{min} \leq w_i \leq w_{max}$). Lemma 1 and 2 are directly applicable to such WD
objects . Consequently, the approximate solution presented in Algorithm 1 can
be applied and used for this generalized version. In addition, if we define $\theta' =
w_{max} \times \theta$, Lemmas 4, 5, and 6 will hold for this generalized version (i.e., by
replacing θ with θ'). Thus, the proposed memory-efficient approximate solution
can also be used to answer this particular generalized variant.

Generalization to Different Decaying Factors: If the initial weights for
all WD objects are equal (i.e., $w_i = 1$), but their decaying factor is different
($\lambda_{min} \leq \lambda_i \leq \lambda_{max}$), Lemma 1 does *not* hold anymore, and the relative order of
cells changes over time. One possible modification to Algorithm 1 is to check all
the non-empty cells upon arrival of a new object and find the one with maximum
weight. This would result in $O(m_t)$ time complexity at each time step t, where
m_t is the number of the non-empty extended cells at time t ($m_t \leq j^2 \times n_t$).
A complimentary observation is that one could ("naïvely") consider an upper
bound, say, λ_{max} in the statements of the corresponding Lemmas. This would
retain the properties in terms of the respective inequalities (possibly affecting
the discrepancy between the values on the left-hand sides and right-hand sides).
We note that while we provide experimental observations about the impact of λ
in Sect. 5, a more detailed investigation is left for our future work.

Generalization to a Broader Class of Decaying Functions: Let us consider
a class of functions F such that $\forall f \in F : f(t) \leq \epsilon(t) \times f(t-1)$, where $\forall t : \epsilon(t) \leq
\epsilon(t-1) < 1$. If the decaying function of the weight of each WD object is from
F, then Lemma 1 and 2 hold and apply directly to this more generalized case.
As a result, Algorithm 1 can be used to provide an approximation solution to
a DDW-MaxRS query. In addition, similar results can be proven by replacing λ
with $\epsilon(1)$ in Lemmas 3–6.

5   Experimental Study

Since there are no existing solutions to the DDW-MaxRS problem, we implemented an adapted (i.e., no external memory) version of the grid-based solution
from [6] as the baseline. We compared it against the Memory Efficient Approximate Solution (MEAS for short)[1] from Sect. 3.2. The comparison was done in

---

[1] Implementation   and   the   datasets   used   are   publicly   available   at
https://github.com/Ashraf-T/DDW-MaxRS.

two aspects: (1) *Efficiency* (in terms of *speed up* compared to the baseline) and (2) *Accuracy* of the results in terms of $(1 - \epsilon)$.

Dataset: We use both synthetic and real-world datasets in our experiments. In synthetic datasets, objects are created using uniform and Gaussian distributions in a 2D space with a (relative) reference coordinate system . The real dataset we used is based on the check-ins in NYC collected for about ten months (from 12 April 2012 to 16 February 2013) [24]. We extracted the unique GPS coordinates (Latitude and Longitude) and mapped them to a Cartesian coordinate system. This transformation is done using pyproj, cartographic projections, and coordinate transformations library in Python[2] We converted the GPS coordinates from EPSG:4326 to a regional system EPSG:2263.

Table 1. Parameters

Parameter Name and Symbol	Possible Values	Default value
Object distribution	Uniform, Gaussian, Real	Uniform
No. of initial objects N_0	500, 1000, 5000, 10000, 50000	5000
Size of R	1.0×1.0, 1.5×1.5, 2.0×2.0, 2.5×2.5	1.0×1.0
Decay Factor (λ)	0.9, 0.8, 0.7, 0.6	0.9
Ratio of R to cells (j)	1, 2, 4, 8	2
Sparsity threshold (θ)	0.01, 0.05, 0.1, 0.2, 0.5	0.1

Parameters: The list of the studied parameters and their values is provided in Table 1. We note that the overall area of interest (i.e., the possible range for the locations of the objects) was bound by a rectangle of size (100×100). Unless otherwise indicated, when investigating the impact of a specific parameter, we keep the rest of them fixed to their respective default values, which are indicated in the rightmost column of Table 1. We scaled the x-y coordinates for the real dataset to be mapped to a rectangle of size (100×100). Also, for the experiments studying the impact of the number of initial objects, if available, we sampled as many data points uniformly at random from the dataset.

Environment: All the algorithms were implemented in Python 3.7. The experiments were executed on a machine with an Intel 2.2 GHz core i7 processor and 8 GB of RAM. We measured the median of the processing time and accuracy for the reported results over five runs.

Results: We now present the results in the two categories mentioned above.

(1) *Efficiency*: The results are presented in Fig. 4, which shows the values corresponding to the ratio of the processing time of A_ε DDW-MaxRS relative to the processing time taken by the exact solution. The colored bars indicate the different datasets, and for each of them, the reported processing time is obtained

---

[2] cf. https://github.com/pyproj4/pyproj.

by averaging the processing time over 100-time steps through five iterations. We considered the impact of different parameters:

– N_0 - the initial number of WD objects present in the system: MEAS significantly reduces the processing time (about 45%) when there are a large number of initial objects. The processing time for the exact solution does not change much over time because its time complexity is a function of n_t (number of objects present at time t), which increases by (at most) one at each time step. Although the impact of the initial number in our settings is quite high, even in such cases, our proposed solution yields efficiency benefits. The processing time for the approximate solution also does not vary much over time, as its worst-case complexity is in order of $O(logN)$, where N is the number of extended cells in MEAS and does not change over time. Note that there is no bar corresponding to the real dataset with $N_0 = 50000$ since the number of unique x-y coordinates in the used dataset was not enough to sample these many records.

– R: As one would expect, the larger $|R|$, the faster the execution of the approximate algorithm. For the default values of the other parameters, in Fig. 4-b, we show the impact of the size of R on the speed up relative to executing the exact algorithm. We observe that similar trends hold in both approximate solutions, except for the results of the memory-efficient solution for the real dataset.

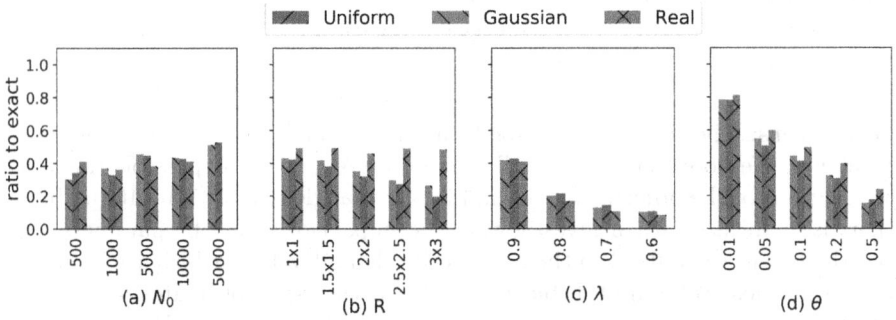

Fig. 4. Efficiency impact of (a) N_0, (b) $|R|$, (c) λ, and (d) θ

– λ: Larger decaying factor (λ) values diminish the weights of points by a greater factor after each time step. Therefore, MEAS will remove more cells as their total weights fall below the threshold and achieve a faster processing time than the exact solution and the first approximate solution in which the time complexity depends on the number of objects. Figure 4-c illustrates these results.

– θ: This parameter determines the cut-off line for the weight of sparse cells. MEAS clears more cells for the larger values of θ and achieves a higher speed up in its processing time. Figure 4-d shows the results for various choices of θ.

(2) *Accuracy*: This part presents the impact of the following four parameters: N_0, j, $|R|$ and θ. The results are shown in Fig. 5 only for the real-world dataset (the results for uniform and Gaussian distributions were similar).

– N_0: The effect of initial objects on the solution of DDW-MaxRS lasts longer for larger values of N_0. This means it may take a while before the solution of DDW-MaxRS changes due to the arrival of new objects. Figure 5-a shows the results of this experiment for MEAS. Again, no result is shown when $N_0 = 50K$ since there are not enough unique x-y coordinates in the real dataset to sample $50K$ of them.

– j - the ratio of the size of R to the size of the cells: Intuitively, a higher accuracy should be achieved for larger choices of j ; because the intersection of the optimal solution and the corner cells of an extended dual rectangle (illustrated by the hatch pattern in Fig. 3c) decrease by having smaller cells and consequently, the chance of objects being located in those areas also decreases. Figure 5-b confirms this expectation. Also worth mentioning that time complexity of the approximate solution is dominated by $log(N)$, where N is the number of extended cells, and the increase in the number of cells as the result of increasing j does not significantly affect the processing time.

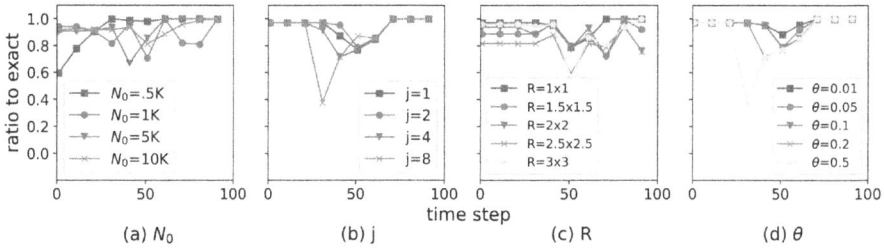

Fig. 5. Accuracy impact of (a) N_0, (b) j, (c) $|R|$, and (d) θ on MEAS for real-world dataset over time.

–R: One would expect that the larger size of the query rectangle decreases the accuracy of the approximate solution because there is a higher chance of having scenarios similar to the one shown in Fig. 3c. The reported results in Fig. 5-c support this expectation.

– θ: Choosing larger values of θ expectedly hurts the accuracy of the returned solution by MEAS. Setting larger values to θ will prune and erode more cells as sparse cells. Figure 5-d suggests that the memory-efficient solution performs pretty robust when θ is less than 0.5.

6   Related Work

The MaxRS problem has a long history, during which different variants have been addressed. It was first tackled by researchers in the computational geometry community, where a technique that finds connected components and a maximum clique of an intersection graph of rectangles in the plane was proposed in [10]. A solution based on the plane sweep strategy was presented in [15], where the input points/objects were "dualized" into rectangles, centered at the locations of the input points and with dimensions equivalent to the query rectangle R. To obtain the regions with the highest number of intersecting (dual) rectangles, an interval tree was used, updated by a sweep line technique, with events at the input points (i.e., centroids of the respective dualized rectangles). The algorithm for possible locations for placing the (center of the) query rectangle yielding the maximal number of points in its interior had $O(n \log n)$ time complexity.

The next "era" of the MaxRS research was motivated by the observations that the scalability of the existing solutions may not be good for LBS application, and a scalable solution was proposed in [5]. An extension, considering imprecision vs. efficiency trade-off and providing an approximate solution, was presented in [22]. Dynamic settings, in which objects can be inserted and/or deleted, were considered in [2], based on an aggregate graph and yielding an efficient approximation. The work considered a sliding window-based model, in which the appearance of new objects implies the removal of old objects. Recently, the dynamic variant with a constraint on the minimal number of elements from different classes that must be inside the query rectangle was addressed in [9].

Other variants of the MaxRS problem have also been considered, such as constraining the locations of spatial points to the underlying road network [16, 26]; considering moving objects, and detecting the "trajectory" of the motion of the centroid of MaxRS; MaxRS [4] where rectangles do not need to be axes parallel.

A couple of recent works have addressed uncertain variants. In [1], an index structure was proposed for efficiently answering the MaxRS query in the d dimension, in which each point was associated with an existential probability. PMaxRS (cf. [12]) considered inherent location uncertainty of spatial objects, coupling candidates generation (pruning), and sampling-based approximation (refinement) to efficiently solve the problem.

While many of the above results have motivated our present work, there are some fundamental distinguishing aspects. Namely, we jointly consider the probabilistic nature of the dynamics of the arrival of new points along with the impact of the decay factor (i.e., the value of the weight) over time.

7   Conclusions and Future Work

We introduced a novel variant of the MaxRS query – the DDW-MaxRS, which, given a spatial region of interest, combines the consideration of the arrival of new objects and the decay of their weight over time. We took the first step toward providing an approximate algorithm A_ε DDW-MaxRS$(O_n(t), R)$ and formally analyzed error bounds properties compared with the exact solution. Our experiments demonstrated the impact of the various parameters and the benefits of the A_ε DDW-MaxRS$(O_n(t), R)$.

Our future work will focus on three main topics. Firstly, we will investigate data structures that can improve scalability. Secondly, will address the complementary settings of data streams – i.e., the arrival rate of the objects is fast, and the available memory is limited – and devise approximate algorithms based on effective sampling strategies. In parallel, we will explore the extension of our results to higher dimensions and the potential joint impact of different parameters (e.g., combining N_0 and $|R|$, with different distributions for N_0 and the newly arriving objects).

References

1. Agarwal, P.K., Kumar, N., Sintos, S., Suri, S.: Range-max queries on uncertain data. J. Comput. Syst. Sci. **94**, 118–134 (2018)
2. Amagata, D., Hara, T.: Monitoring MaxRS in spatial data streams. In: EDBT, pp. 317–328 (2016)
3. Boulos, M.N.K., et al.: Crowdsourcing, citizen sensing and sensor web technologies for public and environmental health surveillance and crisis management: trends, OGC standards and application examples. Int. J. Health Geograph. **10** (2011)
4. Chen, Z., Liu, Y., Wong, R.C., Xiong, J., Cheng, X., Chen, P.: Rotating MaxRS queries. Inf. Sci. **305** (2015)
5. Choi, D.W., Chung, C.W., Tao, Y.: A scalable algorithm for maximizing range sum in spatial databases. Proc. VLDB Endow. **5**(11), 1088–1099 (2012)
6. Choi, D., Chung, C., Tao, Y.: Maximizing range sum in external memory. ACM Trans. Database Syst. **39**(3), 21:1–21:44 (2014)
7. Eldawy, A., Mokbel, M.F.: The era of big spatial data: a survey. Found. Trends Databases **6**(3–4), 163–273 (2016)
8. Güting, R.H., Schneider, M.: Moving Objects Databases. Morgan Kaufmann, San Francisco (2005)
9. Hussain, M.M., Mostafiz, M.I., Mahmud, S.M.F., Trajcevski, G., Ali, M.E.: Conditional MaxRS query for evolving spatial data. Front. Big Data **3**, 20 (2020)
10. Imai, H., Asano, T.: Finding the connected components and a maximum clique of an intersection graph of rectangles in the plane. J. Algorithms **4**(4), 310–323 (1983)
11. Koutsopoulos, I.: Optimal incentive-driven design of participatory sensing systems. In: Proceedings of the IEEE INFOCOM (2013)
12. Liu, Q., Lian, X., Chen, L.: Probabilistic maximum range-sum queries on spatial database. In: Proceedings of the 27th ACM SIGSPATIAL (2019)
13. LLC, S.: Home Burglay awareness and prevention (2013). http://www.jsu.edu/police/docs/Schoolsafety.pdf
14. Mishra, A.: Data science: The key tool cities need to reduce carbon emissions (2020)
15. Nandy, S.C., Bhattacharya, B.B.: A unified algorithm for finding maximum and minimum object enclosing rectangles and cuboids. Comput. Math. Appl. **29**(8), 45–61 (1995)
16. Phan, T., Jung, H., Youn, H.Y., Kim, U.: Efficient evaluation of maximizing range sum queries in a road network. IEICE Trans. Inf. Syst. **99-D**(5), 1326–1336 (2016)
17. Prieto Curiel, R., Collignon Delmar, S., Bishop, S.R.: Measuring the distribution of crime and its concentration. J. Quant. Criminol. **34** (2018)

18. Silva, T.H., et al.: Urban computing leveraging location-based social network data: a survey. ACM Comput. Surv. **52**(1), 17:1–17:39 (2019)
19. Singru, N., Lumain, R., Roldan, C.: Reducing Carbon Emissions from Transport Projects - ADB Strategy 2020 (2010)
20. Siow, E., Tiropanis, T., Hall, W.: Analytics for the internet of things: a survey. ACM Comput. Surv. **51**(4), 74:1–74:36 (2018)
21. Subramanian, R., He, Q., Pascual, M.: Quantifying asymptomatic infection and transmission of COVID-19 in New York city using observed cases, serology, and testing capacity. In: Proceedings of the National Academy of Sciences of the United States of America, vol. 118 (2021)
22. Tao, Y., Hu, X., Choi, D.W., Chung, C.W.: Approximate MaxRS in spatial databases. In: 39th International Conference on Very Large Data Bases 2013, VLDB 2013. vol. 6, pp. 1546–1557 (2013)
23. Wan, L., Ng, W.K., Dang, X.H., Yu, P.S., Zhang, K.: Density-based clustering of data streams at multiple resolutions. ACM Trans. Knowl. Discov. Data (TKDD) **3**(3), 1–28 (2009)
24. Yang, D., Zhang, D., Zheng, V.W., Yu, Z.: Modeling user activity preference by leveraging user spatial temporal characteristics in LBSNS. IEEE Trans. Syst. Man Cybern. Syst. **45**(1), 129–142 (2015)
25. Zhang, J., Zhu, M., Papadias, D., Tao, Y., Lee, D.L.: Location-based spatial queries. In: Proceedings of the 2003 ACM SIGMOD, pp. 443–454. ACM (2003)
26. Zhou, X., Wang, W., Xu, J.: General purpose index-based method for efficient MaxRS query. In: Hartmann, S., Ma, H. (eds.) DEXA 2016. LNCS, vol. 9827, pp. 20–36. Springer, Cham (2016). https://doi.org/10.1007/978-3-319-44403-1_2

Performance

Storage Management with Multi-Version Partitioned BTrees

Christian Riegger$^{(\boxtimes)}$ ⓘ and Ilia Petrov ⓘ

Reutlingen University, Reutlingen, Germany
{christian.riegger,ilia.petrov}@reutlingen-university.de
https://www.dblab.reutlingen-university.de

Abstract. Database Management Systems and K/V-Stores operate on updatable datasets – massively exceeding the size of available main memory. Tree-based K/V storage management structures became particularly popular in storage engines. B$^+$-Trees [1,4] allow constant search performance, however write-heavy workloads yield in inefficient write patterns to secondary storage devices and poor performance characteristics. LSM-Trees [16,23] overcome this issue by horizontal partitioning fractions of data – small enough to fully reside in main memory, but require frequent maintenance to sustain search performance.

Firstly, we propose Multi-Version Partitioned BTrees (MV-PBT) as sole storage and index management structure in key-sorted storage engines like K/V-Stores. Secondly, we compare MV-PBT against LSM-Trees. The logical horizontal partitioning in MV-PBT allows leveraging recent advances in modern B$^+$-Tree techniques in a small transparent and memory resident portion of the structure. Structural properties sustain steady read performance, yielding efficient write patterns and reducing write amplification.

We integrated MV-PBT in the WiredTiger [15] KV storage engine. MV-PBT offers an up to 2× increased steady throughput in comparison to LSM-Trees and several orders of magnitude in comparison to B$^+$-Trees in a YCSB [5] workload.

Keywords: Storage engine · Storage management · Append storage

1 Introduction

High performance persistent key-sorted No-SQL storage engines became the load-bearing backbone of online data-intensive applications. Such engines exist as standalone K/V-Stores (Key/Value Stores) [7,15] as well as in integrated in DBMS storage engines [6,11,14]. Obviously, backing tree-based K/V storage management structures – i.e. B$^+$-Trees [1], LSM [16,23] and derivatives [2,11] – natively enable necessary advanced lookup operations beside equality search, e.g. key prefix or inclusive and exclusive range searches, with (nearly) constant logarithmically scaling performance characteristics. Continuous modifications require special care to preserve constant performance characteristics and

© Springer Nature Switzerland AG 2022
S. Chiusano et al. (Eds.): ADBIS 2022, LNCS 13389, pp. 255–269, 2022.
https://doi.org/10.1007/978-3-031-15740-0_19

mentioned search features. Although B$^+$-Trees offer constant search performance to data in main memory and on secondary storage devices, modifications yield in inelastic performance characteristics. LSM-Trees sacrifice properties of a single tree structure to overcome this issue by buffering modifications in a fraction of main memory, typically tree-based components, and leveraging flash-based secondary storage device characteristics on eviction and necessary background merge operations.

Flash Technology in SSD Secondary Storage Devices exhibit individual characteristics. I/O operations possibly are independent or decomposed executed in multiple structural levels of an SSD, whereas a high internal parallelism and I/O-performance is enabled [22,24,25]. However, reads perform an order of magnitude better than writes, yielding in a asymmetric I/O behavior. Whilst reads perform nearly identical for random and sequential access patterns, write I/O is preferably sequentially performed [13]. Furthermore, pages are replaced out-of-place, wherefore much slower erases and background garbage collection is necessary [3,8].

B$^+$-Trees and Derivatives achieve a constant logarithmically scalable search performance, since root-to-leaf traversal operations depend on their height – even in case of massive amounts of stored data records. Commonly used inner nodes of traversal paths allow fast access to data in leaf nodes with few successive read I/O. However, B$^+$-Trees are probably vulnerable in case of modifications. Whilst insertions, updates and deletions of records possibly facilitate steady throughput in main memory by optimized and highly scalable maintenance procedures [11,15], massive amounts of maintainable key-sorted data yield in random write I/O and high write amplification on secondary storage devices once modifications get persisted on eviction of '*dirty*' buffers. In order to preserve strict lexicographical sort order of records, maintenance operations cause cascading node splits, whereby blank space is created to accommodate additional separator keys in inner nodes and records in leaves in the designated arrangement. As a result, sub-optimally filled nodes reduce cache efficiency and contained information is written multiple times, yielding in a high write and space amplification. Furthermore, read I/O on secondary storage devices of partially filled nodes lead to high read amplification. Therefore, for massive amounts of contained data, B$^+$-Trees become write-intensive, even in case of proportionately few modifications, yielding in following **problems**:

- low benefit from main memory optimizations, since nodes are frequently evicted
- low cache efficiency and high read amplification due to partially filled nodes
- massive space and write amplification on secondary storage devices

Alternatively, LSM-Trees are Optimized for High Update Rates and obtain a sequential write pattern, since modifications are buffered in tree-based LSM components in main memory. Components get frequently switched, merged and evicted to persistent secondary storage devices. Generally, background merge operations counteract the data fragmentation and increased read and search effort, however this behavior also increases its write amplification. Several approaches in merge policies [23] and reduction of read amplification [12,18,26,27] have been introduced. Certainly, flash allows high internal parallelism and multiple reads of parallel traversal operations. Nevertheless, since components are separate structures, they effectively leverage neither caching effects on traversal nor logarithmic capacity capabilities per height of B$^+$-Trees. Moreover, creation of new components on switch procedure is not transparent to the storage engine and relies on high-level maintenance of the database schema. Finally, due to append-based record replacement technique in LSM, key uniqueness is assumed, wherefore the application in storage engines of DBMS with non-unique indexes is complicated. **Challenges in LSM** are defined as follows:

- inefficient caching behavior of decoupled components require frequent merges and yield in considerable write amplification
- hence, high internal parallelism of flash is not leveraged for read operations
- components are non-transparent for further layers of a storage engine
- non-unique indexing requires additional care

We Propose Multi-Version Partitioned BTree (MV-PBT) as sole storage and index management structure in KV-storage engines. MV-PBT is based on Partitioned BTrees (PBT) [10], an enhancement of a traditional B$^+$-Tree. (MV-) PBT relies on manipulation of an artificial leading key column of every record – the partition number; and exploiting the regular lexicographical structure of B$^+$-Trees for partition management. Recent publications introduced (MV-) PBT as a highly scalable indexing structure in DBMS with multi-version concurrency control (MVCC) and massive index update pressure [19–21]. However, this paper focus on MV-PBT as sole storage management structure in KV-storage engines. The contributions are:

- Diminishing write amplification in append-based storage management with MV-PBT by sequential write of saturated partition managed nodes
- Transparent internal partition management and atomic partition switch operations without schema maintenance requirements
- Single root node as entry point in the B$^+$-Tree structure allows to leverage logarithmic capacity and commonly cached and traversed inner nodes
- Reduction of merge-triggered write amplification and accompanying pressure on secondary storage devices by Cached Partitions
- Leveraging scalable in-memory optimizations and compression techniques of B$^+$-Tree structures for massive amounts of data in a very hot fraction
- Prototypical implementation and experimental evaluation in WiredTiger [15], which provides competitive B$^+$-Tree and LSM-Tree implementations

Outline. We present an architectural overview of MV-PBT in Sect. 2. Sections 3 and 4 focus on reduction of write amplification by data skipping and fast retrieval

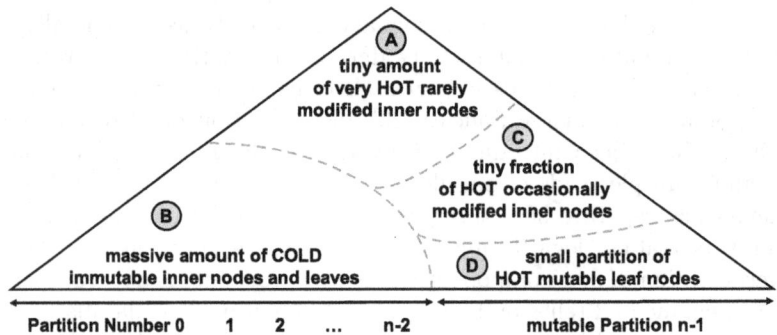

Fig. 1. Logical horizontal partitioning in MV-PBT and replacement policy of MV-PBT-Buffer yield in hot/cold separation within one single tree structure and ultimately enables a sequential write pattern of whole partitions.

in a horizontally partitioned structure and considering defragmentation only as a result of garbage collection. We evaluated the storage management structures in the homogeneous storage engine WiredTiger 10.0.1 in Sect. 5 and conclude in Sect. 6.

2   Architecture of Multi-Version Partitioned BTrees

Multi-Version Partitioned BTree (MV-PBT) as an append-based and version-aware storage and indexing structure relies on well-studied algorithms and structures of traditional B$^+$-Trees – with which they share many characteristics and areas of application. Therefore, MV-PBT is able to adopt and even leverage characteristics of advances in modern B$^+$-Tree techniques. The proposed approach facilitates straightforward horizontal partition management within one single B$^+$-Tree structure in order to keep a very hot mutable fraction of leaves in fast volatile main memory (compare Fig. 1) – the MV-PBT-Buffer including the most recent partition leaves is temporarily apart from the regular buffer replacement policy. Reaching a certain dirty memory footprint threshold initiates an atomic partition switch operation, which asynchronously finalizes in a sequential write of dense-packed cleaned data in leaves and referring inner nodes, in order to interference-freely absorb ongoing modifications. Since partitions are principally defined by the existence of associated records, they appear and vanish as simply as inserting or deleting records [10], however, auxiliary meta data structures allow a massive speed-up of operations. Append-based structures allow modifications of already persisted data by out-of-place replacement. MV-PBT enhances this behavior by additional record types, which allow internal indexing and non-uniqueness of data and enables native B$^+$-Tree-like indexing features. Moreover, maintenance of multiple record circumstances imply the adoption of multi-version capabilities by the assignment of transaction timestamps in MVCC with snapshot isolation. Low write amplification, sequential writes of dense-packed

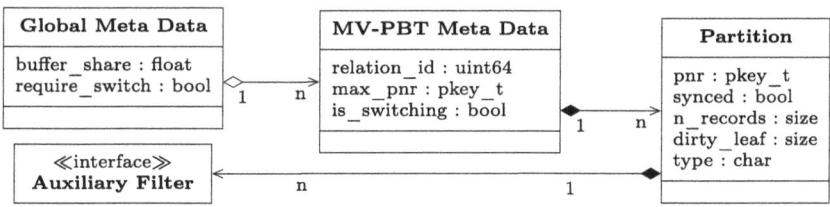

Fig. 2. Auxiliary recoverable MV-PBT data structures.

nodes, commonly utilized inner nodes with one single root as entry point, parallelized multi-partition search operations as well as multi-version indexing capabilities make MV-PBT superior as sole storage and index management structure in storage engines.

MV-PBTs Auxiliary Data Structures information is entirely contained in the B$^+$-Tree structure. For instance, the mutable most recent partition number could be identified by searching the rightmost record in the tree structure. Since cached information is frequently required and its memory footprint is very low, auxiliary data structures are cached in RAM (an excerpt is depicted in Fig. 2). MV-PBT data structures require neither locking for any atomic operation nor additional logging of modifications, since the lightweight information is completely recoverable from basic B$^+$-Tree by a scan operation. All information of horizontal partitioning is anchored within the tree structure, i.e. horizontal partitioning is transparent to further storage engine modules – contrary to schema modifications in LSM-Trees.

Multiple MV-PBT exist within a storage engine, which commonly share the MV-PBT-Buffer threshold. The MV-PBT Meta Data belongs to a specific relation in the schema. Its most recent partition number (`max_pnr`) is frequently required to determine record prefixes as well as for atomic switching operation. An MV-PBT comprises of several valid partitions, which contain a set of meta data like the number of records or specific partition type characteristics. Finally, auxiliary filter structures for point and/or range queries are referenced; e.g. fence keys, (prefix) bloom filters or hybrid point and range filters [12,18,26].

Partition Number Prefixes are prepended to each record key with the central scope of leveraging lexicographical sort capabilities of B$^+$-Trees in order to achieve a logical horizontal partitioning. Partition numbers could be of any comparable data type, e.g. 2 or 4-byte integers, and might are maintained in an artificial leading key column [10]. However, combining the partition number and the first record key attribute in a *partitioned key* type (compare Fig. 3a) enables cache efficient comparison of co-aligned attributes as evaluated in Fig. 3b. Additional storage costs are negligible due to prefix truncation techniques. *Partitioned keys* are simply allocated when setting search keys and their prefix becomes

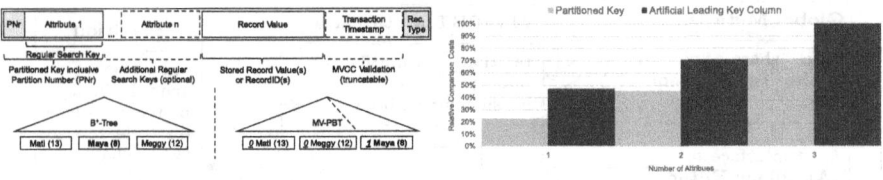

(a) Record with Partitioned Key   (b) Comparison Costs for both approaches

Fig. 3. Horizontal partition maintenance with Partitioned Keys

hidden by returning an offset in the leading key attribute in order to retain transparent horizontal partitioning.

Multi-Version Capabilities accompanying well the out-of-place replacement in MV-PBT. Multi-Version Concurrency Control (MVCC) with Snapshot Isolation (SI) are a common technique to enable high transactional parallelism in storage engines, since readers and writers are not mutually blocking as each transaction operates on a separate snapshot of data. Therefore, multiple versions records of one logical tuple are maintained in a version chain – each is valid for a different period in time. MV-PBT adopts a new-to-old ordering approach of physically materialized version records with out-of-place update scheme and one-point invalidation model [9,21] – i.e. predecessor versions remain unchanged on modification, whereas write amplification is massively reduced. Successor version records are annotated with the current transaction timestamp (which may become truncated on eviction to secondary storage devices, whenever no preceding snapshot is active) and are inserted in the most recent partition in the MV-PBT-Buffer. Thereby, it is possible to maintain multiple version records in one partition, e.g. as separate record [21] or in-memory update lists [15]. Based on the logical search succession in MV-PBT from new-to-old, transaction snapshots identify their visible version record and skip others, based on the annotated transaction timestamps. Since record data values are physically materialized in each version record, identified records are directly applicable.

Record Types in MV-PBT feature all operations over logical tuple life-cycle without modifying predecessor version records. During lifetime, it gets created, modified and deleted while it is frequently read. *Regular Records* declare the begin of the life-cycle, hence there is no predecessor version. Its transaction timestamp is applied by the inserting transaction and indicates its validation. *Replacement Records* indicate a new record value on update. Its timestamp invalidates its predecessor as well as validates itself. Replacement Records are also applied on modifications to the record key, however, invalidation requires an *Anti Record* with the predecessor key attribute values and the current transaction timestamp for invalidation. Replacement Records as well as Anti Records probably store its predecessor value for logical tuple assignment as needed in

Fig. 4. (1) After atomic partition switch, an MV-PBT consists of (A) persistent, (B) a victim and (c) a most recent partition. Internal nodes and leaves of the victim partition delay maintenance effort (e.g. split operations) by flexible page size until a reconciliation process (2.D). The (E) most recent partition consumes ongoing modifications. Finally, (3) the (F) victim partition is sequentially written to secondary storage and (G) is the only memory mapped partition.

non-uniqueness index management constraints, however, modifications to the key attributes and non-uniqueness indexing constraints with index-only visibility checks [21] allow MV-PBT to serve as sole storage and index management structure in storage engines but is out of scope in this paper. Finally, *Tombstone Records* are inserted on deletion of a logical tuple. Major difference to Anti Records is, that successor version records are impossible.

Atomic Partition Switch and Sequential Write of dense-packed leaves and referring inner nodes bring a leading edge in MV-PBT. The whole procedure consists of several partially parallelizable stages. After *(a) determination of switch requirement* by a certain dirty buffer threshold in the MV-PBT-Buffer, a *(b) valuable MV-PBT victim partition* is selected for eviction. Contrary to LSM-Trees, MV-PBT partitions become immutable and switched by *(c) atomically incrementing the most recent partition number* (`max_pnr`) in the meta data, since the required B⁺-Tree structure is already existent and logged anyways.

However, records are probably not yet in their final *(d) defragmented and dense-packed disk layout*, since structure modifications are the result of a randomly inserting workload. One approach to avoid expensive partition-internal structure modifications (e.g. node merges) is to simply re-inserting the still valid contents in their final arrangement by manipulating the partition number in a bulk load operation [21]. B⁺-Trees allow efficient split policies to support high fill factors by this operation. Finally, visibility characteristics of both partitions are swapped and the randomly grown source partition gets cropped from the tree. Another approach is to leverage modern B⁺-Tree techniques. In order to avoid

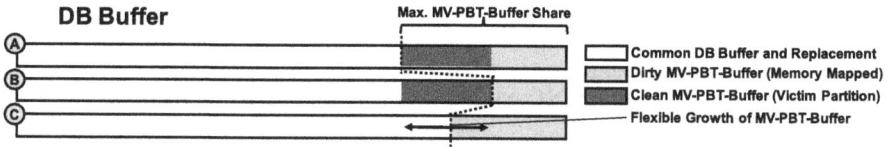

Fig. 5. Flexible MV-PBT-buffer share allows cache preserving handover of a clean victim partition from (A) the MV-PBT-buffer to (B) a common buffer replacement policy and (C) flexible growth up to a max. MV-PBT-buffer share.

structure modifications, referenced main memory nodes are allowed to flexibly grow and finally get divided and structured in the disk layout in a reconciliation process (depicted in Fig. 4).

Auxiliary *(e) filter structures* are generated as a natural by-product of defragmentation and dense-packing, since records are accessed anyways. Whenever (a fraction of) leaf nodes obtained their final layout, it is possible to *(f) perform a sequential write of leaves and referring inner nodes* by traversing the tree structure and following the sibling pointers – yielding in a bottom-up sequential write of nodes, level by level. Finally, the persisted leaves are *(g) passed to the regular replacement policy* in order to sustain a constant buffer factor and memory footprint (Fig. 5).

Basic Operations in MV-PBT are based on regular a B^+-Tree – i.e. they have logarithmic complexity. Every modifying operation is treated as an insertion of a record of a respective type. Thereby, the current transaction timestamp is set for validation in visibility checks – and one-point-invalidation of conceivable predecessors, respectively, which can be located in a preceding or the current partition. However, due to the partitioned key, each modifying operation is performed in the most recent partition in main memory. This is also valid in case of concurrent partition switch by overwriting the partition number of an insertion record key and immediate re-traversal from root. Additional constraint support is very uncommon in pure storage management since records are typically overwritten by blind insertions, however, this is facilitated by MV-PBT in preceding equality search operations.

Equality and range search operations perform root-to-leaf traversals of a (sub-)set of partitions by manipulation of the partition number in the partitioned search key. Partitions are preselected by auxiliary filter structures. Logically, partitions are searched in reverse order from the most recent to the lowest numbered one. Based on the selectivity of a query, partitions may are sequentially processed or by parallel traversals in a merge sort operation. In case of equality searches, sequential processing allow minimal read amplification, contrary, sorted range searches favorably adopt the merge sort approach, whereby multiple cursors are applied and get individually moved and returned to a higher level merge sort cursor. Thereby, record transaction timestamps are checked for visibility to a transaction snapshot. Based on a regular visibility check, invisible and invalidated records are skipped, invalidating records are remembered for exclusion of occurring predecessors (which are subsequently accessed) and matching records are returned [21].

3   Cached Partition: Stop Re-writing Valid Data

MV-PBT introduces a logical horizontal partitioning within one single tree structure in order to leverage characteristics of secondary storage devices. This data fragmentation influences the search operations in different ways. Obviously, several possible storage locations of a requested record implies additional search effort. Actually regular B^+-Trees incur increased search costs in randomly grown structures, due to diminishing cache efficiency of partially filled inner nodes. Contrary, LSM-Trees keep a read-optimized layout within each component, however, multiple entry points and referenced inner nodes are neither commonly cached nor leverage logarithmic capacity. LSM-Trees counteracting increased search effort with background merge operations, whereby write amplification of still valid data is increased.

MV-PBT preserves a read-optimized and cache-efficient layout for immutable nodes (Fig. 1B) with one commonly shared entry point and referenced inner nodes (Fig. 1A) which are subjecting to a optimal fill factor, since append based behavior of referenced data allows efficient split policies (equal to bulk loads). As outlined in Sect. 2 (*Atomic Partition Switch and sequential write*), mutable inner nodes and leaves (Fig. 1C and 1D) are a hot fraction which sustains maintenance operations of the random workload, however, modern B^+-Tree techniques allow main memory efficient delay of maintenance operations. Since the small fraction of inner nodes is commonly used, they are well cached, so that a large portion of the parallel traversal operations is performed without read latencies from secondary storage devices. Successive read I/O in multiple partitions leverage parallelism in flash persistent storage. Moreover, search performance in MV-PBT relies on data skipping by auxiliary filter structures. As a combined result, MV-PBT is able to sustain comparable search performance for higher fragmentation as in LSM-Trees.

However, variety of auxiliary filter structures imply caching and probe costs as well as massive amount of traversal operations result in high read I/O costs and shrink performance due to growing fragmentation. Instead of adversely rewriting still valid data records in a consolidated arrangement, due to asymmetry of flash and write amplification, MV-PBT introduces *Cached Partitions*. They are an internal index partition, whose records reference a preceding partition, containing the latest version record of a logical tuple in a lexicographical sort order. Several Cached Partitions may exist for a different subset of small partitions and are cyclically created while the MV-PBT evolves. Cached Partitions are the result of a background merge sort of contents in several immutable lower numbered partitions with the respective partition number as value or the contents of several preceding Cached Partitions. Background merge sort results are bulk inserted in an '*invisible*' partition while proceeding, can be paused and continue without wasting work and become finally visible by an atomic status switch.

Since a subset of partitions is fully indexed in a Cached Partition, a subsequent search operation is able to traverse the subset on the commonly cached path as needed, based on the results of the internal partition index. Cached Partitions assume responsibilities of auxiliary filter structures and allow to exclude the subset of indexed partitions from the regular logical search succession, whereby comparison costs in an internal merge sort are reduced – the effort is focused on non-indexed and Cached Partitions. Furthermore, cached index records are very space and cache efficient in the search process, since they consist of the key and one partition number (e.g. 2 or 4-byte integer) in a dense-packed arrangement.

4   Garbage Collection and Space Reclamation

Datasets and tuple values evolve over time. Storage management structures with out-of-place update approaches allow beneficial sequential write patterns and low write amplification, however, invalidated predecessor record versions remain existent on update. Search operations are able to exclude invalid version records from the result set, though visibility checking entail additional processing. Furthermore, version records which are not visible to any active transaction snapshot entail space amplification and additional storage costs.

In MV-PBT, additional search costs due to fragmentation by horizontal partitioning is well covered by Cached Partitions for insertion of new tuple version records. However, modifications to logical tuple values leave persisted obsolete version records behind, yielding in space amplification. Ideally, obsolete version records are discarded as part of the dense-packing phase on partition switch, however, many version records become invalidated after they were persisted. For the only reason of space reclamation, MV-PBT occasionally performs a garbage collection (GC) process. Similar to the creation of a Cached Partition, GC is performed by a background merge sort and bulk load operation in a not yet visible partition. Certainly, the stored record value is the regular value of the most recent record version of a tuple. As well, the GC process can throttle and continue without wasting work, since the partition is not yet accessible for querying. After the successful completion, the partition becomes visible and the records of purified preceding partitions become invalidated. Once every active search operation finished, the purified partitions are cropped from the tree structure by an efficient range truncation [15].

5   Experimental Evaluation

We present the analysis of MV-PBT as storage management structure in comparison beside the baselines LSM-Trees and B$^+$-Trees fully integrated in WiredTiger 10.0.1 (WT) [15]. LSM-Trees in WT build upon components of the provided B$^+$-Trees upon which MV-PBT is also implemented. A good comparability is achieved, since all structures commonly operate on equal code lines and B$^+$-Tree techniques, e.g.: prefix truncation, suffix truncation and snappy compression; reduced maintenance effort due to flexible page sizes; main memory page representation with sorted areas, update-lists and insertion skiplists; MVCC transaction timestamps in main memory record representation; tree-based buffer management.

Experimental Setup. We deployed WiredTiger(WT) 10.0.1 and WT with MV-PBT as storage management structure on an *Ubuntu 16.04.4 LTS* server with an eight core *Intel(R) Xeon(R) E5-1620* CPU, 2 GB RAM and an *Intel DC P3600* enterprise SSD. We used the YCSB framework [5,17] for experimental evaluation with a dataset size of approx. 50 GB, unless stated otherwise. The WT cache size is set to 100 MB and LSM-chunks as well as partitions are allowed to grow up to 20 MB. Direct IO is enabled and the OS page cache is cleaned every second in order to ensure repeatable, reliable and even conservative results.

Experiment 1: Space and Write Amplification. In Fig. 6a, B$^+$-Tree, LSM-Tree (merges are disabled for comparability) and MV-PBT are initially bulk loaded with 100 million records (key and value size are 13 and 16 bytes respectively). Prefix truncation in record keys, suffix truncation in separator keys and snappy compression allow comparable relative space requirements for all approaches. There is a clear evidence of the synergy between prefix truncation and partitioned key, since the enlarged record key by a 2 byte partition number does not result in higher space requirements. Subsequently, 5 million new records are inserted – yielding in approx. 60 new partitions/LSM-components. Due to compression techniques, the additional relative space requirement is lower than the actually added record size, with slight advantages for MV-PBT. B$^+$-Tree suffer from insertions in the read-optimized layout due to node splits – yielding in massive relative space amplification per newly inserted records. Insight: MV-PBT offers the lowest space amplification, that is between 12% and 31× better. Finally, the write amplification (Fig. 6a) is evaluated after 5 million inserts. Since almost each insertion causes escalating node splits in the read-optimized layout of a B$^+$-Tree, each insertion causes 2.76 write I/Os of half filled nodes. Sequential writes of dense-packed nodes allow LSM-Trees and MV-PBT to achieve singular writes of optimally filled nodes, yielding in much less write I/O per insertion. MV-PBT achieves a better factor due to commonly used inner nodes. Moreover, merge operations of LSM components would cause a downturn of write amplification by orders of magnitude. Insight: compared to LSM-trees, MV-PBT offers 30% less write amplification and is up to 300× better than B-Trees.

Experiment 2: Sequential Write Pattern. Figure 6b depicts a sequential write pattern with the logical block addresses (LBA) on the ordinate and evolving time on the abscissa. As a result of the partition switch operation, delayed

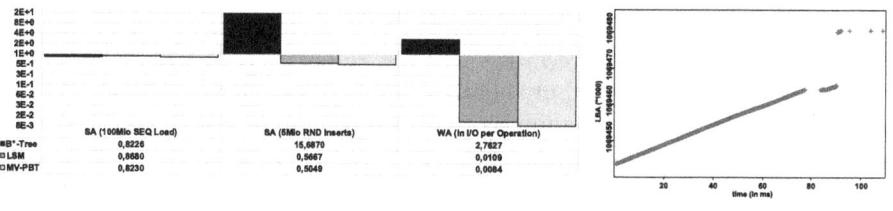

(a) Relative space and write amplification   (b) MV-PBT tree walk flush

Fig. 6. Experiments 1 and 2 evaluate the structural properties of MV-PBT.

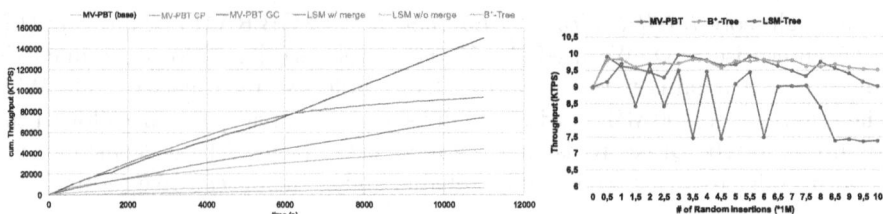

(a) Accumulated executed transactions (*1k) in YCSB Workload A with 1kB value size

(b) Intermediate State Read-Only Throughput

Fig. 7. Experiments 3 and 4 evaluate consistent performance of MV-PBT.

maintenance operations (splits) on leaves followed by inner nodes are performed in a reconciliation operation. Afterwards, leaves are identified by a tree walk and ascending written to secondary storage devices, depicted by the continuously ascending markers. Finally, the referencing levels of immutable inner nodes are sequentially written, depicted by multiple shorter continuously ascending markers. Insight: MV-PBT is able to perform advantageous sequential writes.

Experiment 3: Steady Performance by Cached Partitions and Garbage Collection. The write-heavy YCSB Workload A consists of 50% updates and reads, respectively (depicted in Fig. 7a). Write amplification in B^+-Trees yield in poor performance characteristics (7M tx). Sequential writes and low write amplification in base MV-PBT (no Cached Partition and GC) allow much higher transactional throughput, however, increasing search effort degenerates performance (44M tx), whereby LSM-Trees hold search effort down by merges (74M tx). Insight: the direct structural comparison of LSM-Trees and MV-PBT is without merges and garbage collection, whereby MV-PBT outperforms LSM (11M tx) by 4×. Enabling Cached Partitions allow MV-PBT increased read efficiency, however, memory footprint of auxiliary filter structures degenerates its capabilities over time due to effectively reduced cache (94M tx). Insight: occasional Garbage Collection in MV-PBT (every 400 Partitions) enables stable performance characteristics (151M tx), outperforming LSM-Trees by 2×.

Experiment 4: Read-Only Performance Characteristics of Intermediate Structures States. YCSB Workload C is performed several times after inserting 500k small random records for 10 min, respectively (depicted in Fig. 7b). B^+-Tree remain very stable, but slightly decrease, since the read-optimized layout breaks. LSM-Trees throughput is varying based on the number of LSM components. Insight: commonly cached inner nodes and periodically created Cached Partitions allow MV-PBT to retain comparable read performance even if 80 partitions are created after 10 million random insertions.

Experiment 5: Impact of Different Value Sizes. YCSB basic workloads (Fig. 8) are performed on small (16 bytes), medium (100 bytes) and large (1000

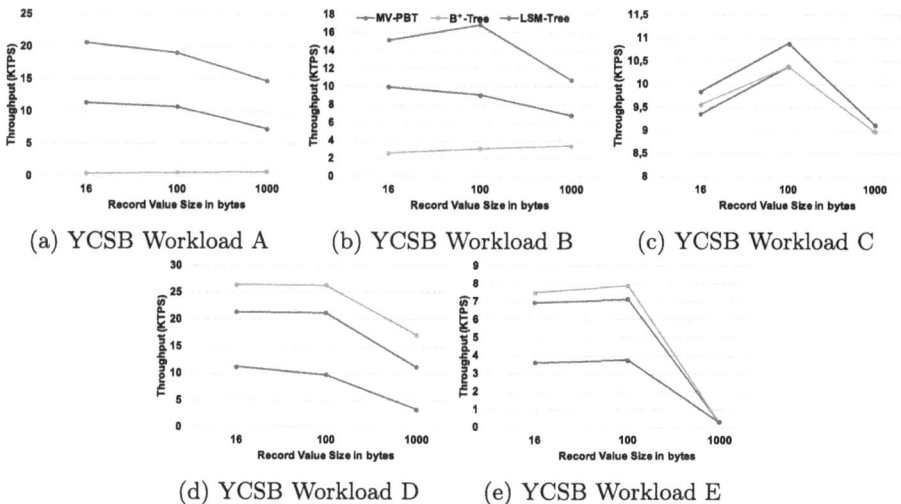

(a) YCSB Workload A   (b) YCSB Workload B   (c) YCSB Workload C

(d) YCSB Workload D   (e) YCSB Workload E

Fig. 8. Experiment 5 evaluates performance for different value sizes.

bytes) value sizes, the initial load has been adjusted to match approx. 50GB dataset size. Insight: MV-PBT outperforms its competitors in the high and medium update intensive workloads A and B, even the LSM-Tree by 2× in the workload A. The read-only workload C is performed on the read-optimized layout after load phase – comparable results prove negligible costs of partitioned key comparisons, whereas LSM-Trees are only able to retain performance for one component (compare Figs. 8c and 7b). Workload D searches for few concurrently inserted records. B$^+$-Tree benefits from well cached nodes in the traversal path due to the recent insertion. This is also valid for MV-PBT and LSM-Trees, however, concurrent insertions are not in the MVCC snapshot and cause search operations in other partitions or components, which is 2× faster in MV-PBT. Finally, MV-PBT is able to achieve comparable performance to B$^+$-Tree in the mostly scan workload E. Cached Partitions and commonly cached inner nodes enable cheap merge sort scan operations.

6 Conclusion

In this paper we present Multi-Version Partitioned BTrees (MV-PBT) as a sole storage and index management structure [21] in KV-storage engines. Logical horizontal partitioning yields beneficial appends of version records within a single tree structure. Partitions leverage properties of B$^+$-Trees by common utilization and caching of inner nodes in traversal operations, whereby constant search performance and high fragmentation are brought together. This behavior leveraged by Cached Partition in order to minimize write amplification to secondary storage devices. Contrary to LSM-Trees, merging is considered for garbage collection of obsolete version records than for sustained search performance, wherefore MV-PBT is predestinated to be applied in KV-storage engines.

References

1. Bayer, R., McCreight, E.: Organization and maintenance of large ordered indices. In: SIGFIDET 1970, New York, NY, USA (1970)
2. Bayer, R., Unterauer, K.: Prefix b-trees. ACM Trans. Database Syst. **2**, 11–26 (1977)
3. Chen, F., Koufaty, D.A., Zhang, X.: Understanding intrinsic characteristics and system implications of flash memory based solid state drives. In: SIGMETRICS (2009)
4. Comer, D.: Ubiquitous B-tree. ACM Comput. Surv. **11**, 121–137 (1979)
5. Cooper, B.F., Silberstein, A., Tam, E., Ramakrishnan, R., Sears, R.: Benchmarking cloud serving systems with YCSB. In: SoCC 2010 (2010)
6. Facebook: MyRocks a RocksDB storage engine with MYSQL (2022). http://myrocks.io
7. Facebook: RocksDB a persistent key-value store (2022). http://rocksdb.org
8. Gottstein, R.: Impact of new storage technologies on an OLTP DBMS, its architecture and algorithms. Ph.D. thesis, TU, Darmstadt (2016)
9. Gottstein, R., Petrov, I., Hardock, S., Buchmann, A.P.: SIAS-chains: snapshot isolation append storage chains. In: ADMS@VLDB (2017)
10. Graefe, G.: Sorting and indexing with partitioned b-trees (2002)
11. Levandoski, J.J., Lomet, D.B., Sengupta, S.: The BW-tree: a b-tree for new hardware platforms. In: ICDE (2013)
12. Luo, S., Chatterjee, S., Ketsetsidis, R., Dayan, N., Qin, W., Idreos, S.: Rosetta: a robust space-time optimized range filter for key-value stores. In: SIGMOD (2020)
13. Ma, D., Feng, J., Li, G.: A survey of address translation technologies for flash memories. ACM Comput. Surv. **46**, 1–39 (2014)
14. MongoDB: MongoDB: The application data platform (2022). https://www.mongodb.com
15. MongoDB-Inc.: Wiredtiger: Wiredtiger developer site (2021). https://source.wiredtiger.com/
16. O'Neil, P.E., Cheng, E., Gawlick, D., O'Neil, E.J.: The log-structured merge-tree (LSM-tree). Acta Inf. **33**, 351–385 (1996)
17. Ren, J., Kjellqvist, C., Deng, L.: Github - basicthinker/YCSB-C: Yahoo! cloud serving benchmark in c++ (2021). https://github.com/basicthinker/YCSB-C
18. Riegger, C., Bernhardt, A., Moessner, B., Petrov, I.: bloomRF: on performing range-queries with bloom-filters based on piecewise-monotone hash functions and dyadic trace-trees. CoRR (2020)
19. Riegger, C., Vinçon, T., Petrov, I.: Write-optimized indexing with partitioned b-trees. In: iiWAS 2017 (2017)
20. Riegger, C., Vinçon, T., Petrov, I.: Indexing large updatable datasets in multiversion database management systems. In: IDEAS (2019)
21. Riegger, C., Vinçon, T., Gottstein, R., Petrov, I.: MV-PBT: multi-version index for large datasets and HTAP workloads. In: EDBT (2020)
22. Ruan, X., Zong, Z., Alghamdi, M., Tian, Y., Jiang, X., Qin, X.: Improving write performance by enhancing internal parallelism of solid state drives. In: IPCCC (2012)
23. Sears, R., Ramakrishnan, R.: BLSM: a general purpose log structured merge tree. In: SIGMOD (2012)
24. Shin, I.: Verification of performance improvement of multi-plane operation in SSDS. Int. J. Appl. Eng. Res. **12**, 7254–7258 (2017)

25. Winata, Y.A., Kim, S., Shin, I.: Enhancing internal parallelism of solid-state drives while balancing write loads across dies. Electron. Lett. **51**, 1978–1980 (2015)
26. Zhang, H., et al.: Surf: practical range query filtering with fast succinct tries. In: SIGMOD 1918 (2018)
27. Zhong, W., Chen, C., Wu, X., Jiang, S.: REMIX: efficient range query for LSM-trees. In: FAST (2021)

Generalization Aware Compression of Molecular Trajectories

Md Hasan Anowar[1]([✉]) [ID], Abdullah Shamail[1]([✉]) [ID], Xiaoyu Wang[2] [ID],
Goce Trajcevski[1]([✉]) [ID], Sohail Murad[3] [ID], Cynthia J. Jameson[4] [ID],
and Ashfaq Khokhar[1] [ID]

[1] Iowa State University, Ames, IA 50011, USA
{mhanowar,ashamail,gocet25,ashfaq}@iastate.edu
[2] University of Notre Dame, Notre Dame, IN 46556, USA
xwang58@nd.edu
[3] Illinois Institute of Technology, Chicago, IL 60616, USA
murad@iit.edu
[4] University of Illinois Chicago, Chicago, IL 60607, USA
cjjames@uic.edu

Abstract. Molecular Dynamics (MD) simulation is often used to study properties of various chemical interactions in domains such as drug development when executing real experimental studies are costly and/or unsafe. Studying trajectories generated from MD simulations provides detailed atomic level location data of every atom in the experiment. The analysis of this data leads to an atomic and molecular level understanding of interactions among the constituents of the system-of-interest, however, the data is extremely large and poses formidable storage and processing challenges in the analyses and querying of associated atom level motion trajectories. We take a first step towards applying domain-specific *generalization* techniques for trajectory compression algorithms towards reducing the storage requirements and speeding up the processing of within-distance queries over MD simulation data. We demonstrate that this *generalization-aware* compression, when applied to the dataset used in this case study yields significant efficiency improvements, without sacrificing the effectiveness of within-distance queries for threshold-based detection of molecular events of interest, such as the formation of hydrogen-bonds (H-Bonds).

Keywords: Trajectory compression · Molecular dynamics simulation · Drug development · Generalization

1 Introduction

In broader terms, data compression can be perceived as any methodology which takes a dataset D with size $|D| = \beta$ as input, and produces a dataset D' with

Research was partly supported by the Eppley Foundation for Research.

S. Chiusano et al. (Eds.): ADBIS 2022, LNCS 13389, pp. 270–284, 2022.
https://doi.org/10.1007/978-3-031-15740-0_20

size $|D'| = \beta'$ as a compact representation of D, where $\beta' << \beta$. The field of data compression has a long history [18]. In the past couple of decades, owing to miniaturization of computing devices, GPS (Global Positioning Systems) and communication technologies, multiple large trajectories datasets have been generated to which compression has been extensively applied [21,26].

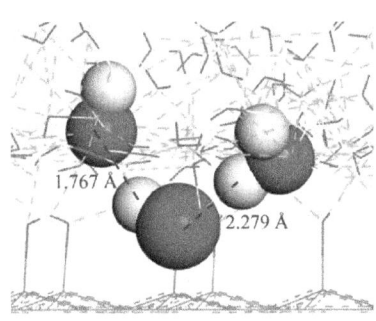

Fig. 1. Molecules and proximity

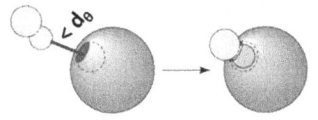

Fig. 2. Simplified proximity-based interaction

A specific domain that we focus upon in this work is the Molecular Dynamics (MD) simulation in chemistry. Due to the high costs of experiments, often times the domain scientists resort to simulation [1] which, in addition to cost reduction, decreases the failure rate and can also speed up the drug development process. However, one of the consequences of MD simulation-based studies is the large volume of generated data which is subsequently to be analyzed. Only very recently the research community investigated the compression of data corresponding to the outputs of MD simulations [9]. The objective of [9] is to have a kind of a lossless compression that will preserve the ring-polymer blends. In contrast, our work considers the MD dynamics data as a collection of trajectories of (the motion of) atoms and molecules and focuses on efficiently processing the spatio-temporal *within-distance* query over the compressed data. The rationale is that this query can indicate a potential interaction between a polymer and a drug. More specifically, given the properties of the molecules (and the forces binding the atoms), in certain cases, a structure such as Hydrogen Bond (H-Bond) may be formed between atoms from different molecules. Quite often, the precondition for such an event is that the proximity of specific atoms is less than a given threshold d_θ. An illustration is provided in Fig. 1 and Fig. 2 where Fig. 1 visualizes the overall process and Fig. 2 shows a simplified scenario of atoms from two different molecules bonding upon certain proximity. We note that there are other preconditions for bonding to occur (e.g., types of atoms within the distance threshold; angle of the planes between atomic groups in the participating molecules, etc.).

The datasets from MD simulations corresponding to molecules/atoms motion trajectories are large (exceeding 10s of GB per run). As such, the compression of these trajectories can yield significant storage savings and, in turn, speed up the processing of *within-distance* query for trajectories – which is often also a continuous one, in the sense that time-intervals before the actual proximity may be of interest. Clearly, processing any spatio-temporal query over compressed (simplified) trajectories data is subject to an uncertainty-based error. However, one can eliminate false negatives by augmenting the value of the threshold d_θ and subsequently refining the search around the locations-in-time where a proximity query may possibly be satisfied.

At the heart of the motivation for this work is the observation that incorporating certain context-awareness as part of the compression process may enable further significant speed-ups in processing *within-distance* queries for a given threshold. Studies on algorithmic approaches for trajectory data compression and the impact on the errors in queries' processing abound [21,26] and, following the parlance in cartography, we refer to this as a *generalization*. Specifically, generalization means manipulating either the input or the output of the compression (or both), for the purpose of retaining certain semantic features and, possibly, affecting the compression algorithm itself. Two classical examples are:

(1) When a map scale is lower than a certain ratio, buildings may need to be "artificially" enlarged after the compression, so that they are still visible to the human eye [20,24].
(2) When a polyline representing a river is compressed, an object (e.g., city) may turn out to be on a different (than the original) river bank [17].

Maintaining such semantic/topological consistencies can be done either with post-processing (as in (1) above), or by pre-processing of the input (e.g., adding convex hull) and additional conditions in the algorithms (as in (2) above).

In this work, we take a first step towards applying compression to trajectories of atoms from molecules participating in a chemical interaction. More specifically, we capitalize on the semantics of the interactions involved in the MD simulations for the case of flavanone drug (which is the use case for this work) to pre-process the trajectories data. This, in turn, yields even smaller input for the compression algorithms, providing increased efficiency speed-ups when processing the query of interest – in our case, the *within-distance* query, for a given threshold d_θ. Our main contributions are:

- We present domain-specific approaches to preprocess the trajectories of the molecules from the MD simulations and obtain their compact representation.
- We use these as a generalization-based application of traditional trajectory compression algorithms, whereby instead of compressing trajectories for each individual atom, we compress the trajectories of molecular "representatives". We introduce a naïve approach for the compression which works well for the specific use case in this paper, and apply two existing trajectory compression algorithms to compare the results.
- We present detailed experimental evaluations to quantify the benefits of the proposed generalization-based trajectory compression in terms of storage savings and efficiency gains in terms of processing time of the *within-distance* query.

In the rest of this paper, we present the background and problem definition in Sect. 2, and the proposed generalizations along with application of trajectory compression techniques in Sect. 3. Section 4 elaborates the data set, result analysis, and experimental evaluation of the proposed methodology, and Sect. 5 gives concluding remarks and outlines directions for future work.

2   Background

We now describe the chemical events in MD trajectories, some classical approaches for trajectory compression and related works, and then discuss the problem in greater detail.

Preliminaries

The MD trajectories used in this study are from our previous work, where we studied the chiral drug separations on the polymer surface [22,23]. The significance of this trajectory data is that it can be used to detect important chemical phenomena, like the formation of a H-Bond.

H-Bonds are special type of interactions which happen due to attractive forces between a Hydrogen (H) atom attached to a highly electro-negative atom, e.g. Nitrogen (N), or Oxygen (O) and another heavy atom which provides a lone pair of electrons [15], possibly N, O, or Fluorine (F) atoms. In H-Bond formation (illustrated in Fig. 3), the highly electro-negative atom bearing an H acts as the donor and the other electro-negative atom with lone pair electrons acts as the acceptor [11]. Formation of H-Bond is of particular interest in wide range of research like [3,8,25]. H-Bonds provide significant information in biophysics as well - as they can be used to determine the properties of biological molecules [12]. In MD simulations, if a pair of acceptor and donor atoms is within a certain structural threshold, denoted by d_θ, the instance is considered to be a formation of H-Bonds [13]. The donor-acceptor distance can vary in the range of $d_\theta \in [2.2, 4.0]$ Å [10]. In addition, there is an angle threshold for H-Bond formation (which is not considered in this work, but is left as a future work).

Traditional Compression Algorithms: Among the various compression algorithms, two popularly used ones which we consider in this work are Ramer Douglas Peucker [7] (RDP), and Scan-Pick-Move (SPM) [19]. We note that there are many other compression algorithms (e.g., the optimal algorithm, available for three or higher dimensions [2], generalizing the 2D version in [5]) which we defer for future work.

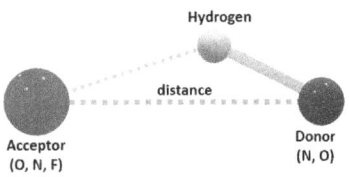

Fig. 3. Formation of H-Bond between two molecules. The highly electro-negative atom (blue) and the attached H atom (white) belong to one molecule. The red acceptor atom comes from another molecule. (Color figure online)

RDP Method: RDP algorithm (also known as Douglas-Peucker (DP)) algorithm is a classical method for trajectory compression [7] which compresses a polyline given a user defined error-tolerance threshold, ε. It works with offline data - requiring the complete trajectory before the algorithm can be applied, and outputs a simplified version of a given initial trajectory or curve. It begins by defining a base line (p_0, p_L) from the first point, p_0, of the trajectory to the last point, p_L, of the trajectory and finding the point, p_f, that is farthest from the base line formed. To find this p_f, RDP calculates the Euclidean distance of each point p_i between the line (p_0, p_L). If distance $d(p_f) < \varepsilon$, then all p_i between the line (p_0, p_L) are removed from the trajectory. If there exists a point p_f with distance $d(p_f, \overline{p_0 p_L}) > \varepsilon$ then the one with maximum distance value is kept and the base line is divided into two lines (p_0, p_f) and (p_f, p_l) and the process is repeated for both of the new lines recursively, eventually resulting in a compressed trajectory with fewer points in the original trajectory.

SPM Method: Introduced in [19], this algorithm uses same settings as RDP. The idea is to create a baseline from the first point, p_0, to the last point, p_L and then "scan" the trajectory by iterating over each point p_i from first to last and calculating its distance to the base line. If $d(p_i) > \varepsilon$, then that point is kept and the baseline is moved from (p_0, p_L) to (p_i, p_L); otherwise p_i is removed from the trajectory. Figure 4 shows an example of SPM, with points p_1, p_2, p_3 and their distances from the baseline. Given a threshold ε, the algorithm (in this instance) calculates that $d(p_1) < \varepsilon$ and $d(p_2) < \varepsilon$ and so it rejects those points (colored red) in Fig. 5 and moves the baseline to $\overline{p_3 p_L}$ because $d(p_3, \overline{p_0 p_L}) > \varepsilon$. The method is repeated from p_3 towards the last point p_L and, eventually, would yield a final version of the compressed polyline (cf. Fig. 6).

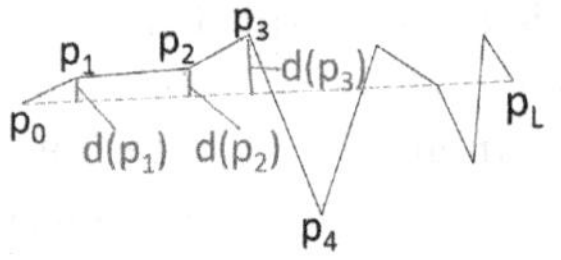

We note here that the main difference between the SPM and RDP algorithms is that RDP would first scan the entire original polyline and find the vertex that is at the furthest distance $\geq \varepsilon$ from the $\overline{p_0 p_L}$, instead of finding

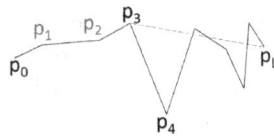

Fig. 4. Step 1

Fig. 5. Step 2 (Color figure online)

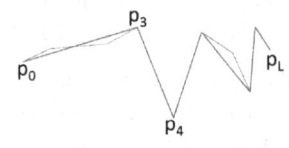

Fig. 6. Final step

the very first one e.g., p_4 in Fig. 4. Then, the original polyline is subdivided into two parts $\overline{p_0 p_4}$ and $\overline{p_4 p_L}$ which are solved recursively. The stopping criterion is that all the points in a given subdivision are within distance $\leq \varepsilon$ from the line segment between the first and the last point of that subdivision.

Related Work: We observe that when the polyline corresponds to a motion of an object (i.e., there is a time-value associated with the vertices, and the constant speed motion along edges is assumed) RDP and SPM need to be slightly adjusted. Namely, instead of time-oblivious Euclidean distance between the vertices of the polyline and the attempted compressed representation, the distances are evaluated at the corresponding temporal instants. The time-aware variant of the RDP was detailed in [4] (along with the impact of the uncertainty on popular spatio-temporal queries). As mentioned in Sect. 1, the problem of trajectory compression has been studied extensively, and a recent experimental comparison is presented in [26]. Incorporating the constraint of motion along road network in trajectory compression has also been considered (cf. [16]). What separates our work is that while the motion is constrained, it is due to the properties of molecules (and atoms within), not a road network. Different from the recent work on compressing MD simulation data in [9], we introduce the notion of generalization (i.e., domain-aware semantic pre-processing) and, while our compression is lossy, we focus on its impact on the within-distance query as an indicator for a potential interaction.

Problem Description

Existing trajectory compression techniques are usually used for GPS generated location-in-time data from land vehicles (the motion of which is sometimes constrained by a road network), as well as other free-space motion such as waterborne vessels [14,19]. Unlike such trajectory data where the positional information of single (uniquely identifiable) object is considered, the internal structural behavior of molecules makes the molecular trajectory rather different from the traditional trajectories. Firstly, each molecule contains a number of atoms with their own trajectories – and their intra-molecular relative positions are not rigid. Secondly, inter-molecular interactions may occur, resulting in both attraction as well as repulsion – and, in extreme cases, may yield creation of bonds between atoms from different molecules.

To tackle the challenges of storing, processing, and analyzing the molecular trajectory data, one can readily apply any of the existing compression methods (e.g., RDP or SPM) to separately recorded trajectories of individual atoms, regardless of their adherence to specific molecules. These compression methods reduce the amount of data. However, such "blind application"

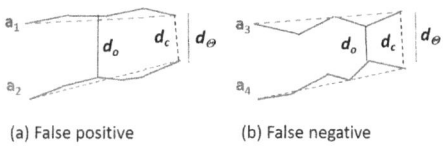

(a) False positive   (b) False negative

Fig. 7. Original vs. compressed trajectory distance.

of trajectory compression to each individual atom may ignore certain contexts/properties of the chemical interactions. In turn, without properly considering molecular properties, one may not only lose relevant information during compression but, as we will demonstrate, also lose benefits in the processing efficiency. A simple example is illustrated in Fig. 7, part (a) of which shows the original trajectories of two atoms a_1 and a_2 before the compression (solid polyline) and after compression (dashed segments). The minimal distance between the original trajectories (d_o) is greater than the minimal distance between the compressed versions (d_c), and we have $d_o > d_\theta > d_c$ – which yields a false positive. This, however, can be resolved by double-checking the original data at the respective time instant (i.e., apply refinement).

In contrast, Fig. 7(b) shows a scenario where for another pair of atoms, a_3 and a_4, we have a situation that $d_o < d_\theta$ but $d_c > d_\theta$. When processing *within-*

distance queries with d_θ over the compressed versions this situation yields a false negative. To avoid such an issue, one can incorporate the impact of the compression/uncertainty (cf. [4]) – i.e., by adjusting the value of d_θ in the query syntax by ε. In other words, instead of finding pairs points *within-distance* (d_θ), we query for *within-distance* $(d_\theta + 2\varepsilon)$ – and then refine to verify for potential proximity in the actual data. In this work, we introduce a preprocessing method based on the knowledge of domain experts, which is applied to the MD simulation data before proceeding with the compression – and, subsequently, processing of the *within-distance* query of interest over the compressed data.

In other words, the proposed generalization based preprocessing retains only the relevant data, in terms of chemical significance, to be compressed and analyzed. We can then judiciously apply the compression to the generalized data, yielding further gains in processing time and storage requirements.

3   Methodology

We now present the details of the proposed preprocessing, introduce a naïve compression algorithm over the preprocessed data, and discuss the application of existing trajectory compression algorithms. The settings are illustrated in Fig. 8 showing 3 drug molecules in the top, and a large polymer molecule in the bottom part, corresponding to an instantaneous time frame of the simulation.

Molecular Features
To provide a compact representation of MD simulation data, we rely on two basic contexts, described next.

`Molecular Center of Mass (CoM):` The CoM is obtained by calculating a weighted average of the coordinates by the respective atomic masses. For a given molecule with n atoms, let (x_i, y_i, z_i) denotes the coordinates of the i–th atom present in the molecule, and m_i denotes its mass. The x-coordinate of the CoM of the molecule is then calculated as: $x_{CoM} = \dfrac{1}{\sum\limits_{i=1}^{n} m_i} \sum\limits_{i=1}^{n} m_i \times x_i$ (similarly for the y_{CoM} and z_{CoM}).

The top portion in Fig. 8 illustrates the CoM for one of the drug molecules.

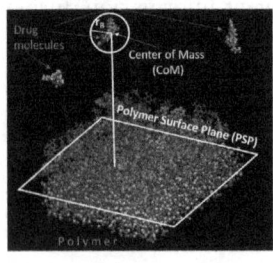

Fig. 8. Polymer and drug molecules.

`Polymer Surface Plane (PSP):` In MD simulations, the drug molecules move faster while the polymer substrates remain relatively stationary as the polymer backbones themselves are attached to a solid base. Hence, for practical purposes, chemistry experts consider the polymer backbone as static which, in turn, enables the use of plane along xy axis as a representative surface for the polymer at a (relatively) constant height. Thus, the equation of the PSP can be approximated by $z = Constant$. In our case study (flavanone drug separations on the polymer surface), the constant value of z is set to 44 Å and it corresponds to the top (horizontal) plane of the polymer surface. The concept of PSP is illustrated by the white rectangle in the bottom of Fig. 8. We note that, in general, drug molecules can be positioned both above and below the PSP.

Generalization and Compression

We now introduce the generalization approaches that will enable obtaining a compact representation of the molecular trajectories. Subsequently, we discuss the adaptation of the compression algorithms used in this work.

Generalization-Based Compact Trajectory Representation (GCTR): The two features specific to the MD domain that we described can be used as a preprocessing (i.e., generalization) steps to generate a compact representation of molecular trajectories, which we describe next.

Generalization Based on Drug CoM and PSP: Based on the properties of the flavanone drug, we can generate a simplified/compact representation of the MD simulation data as follows:

1. The drug molecule is approximated as a bounding sphere, centered at CoM and with a radius r_B equal to the distance between CoM and the farthest atom of the drug.
2. The polymer is approximated with its PSP.

Figure 8 illustrates the approximated representation for one of the drug molecules and the PSP of the polymer. We obtain this compact representation for every time frame of the MD simulation data and, for different values of r_B across different time frames we keep the maximum among them, denoted r_B^{max}.

As mentioned, certain molecular interactions are only possible when atoms from a drug molecule and polymer molecule are within certain distance d_θ. Once the individual atoms from the drug molecules are substituted by the CoM and the sphere with radius r_B^{max}, one cannot simply use d_θ from the CoM as a threshold, as it may yield false negatives. As a specific example, if the distance between the CoM and the PSP is $> d_\theta$ in Fig. 8 and, based on that, we rule out the possibility of the interaction – we will lose the potential of an atom being in the lower-part of the sphere being considered for an interaction with some of the polymer atoms. Thus, instead of looking for pairs of atoms from drug molecules and the polymer potentially satisfying *within-distance* (d_θ), we process *within-distance* (d_θ'), where $d_\theta' = d_\theta + r_B^{max}$, between the CoM and the PSP. In effect, we are ensuring to consider the top of the bounding sphere as a limiting distance threshold, instead of the CoM point only.

Generalization Based on Representative Atom (RA) and PSP: Due to the structure of the flavanone drug molecule, in this particular case study, the CoM in each time frame is always close to the coordinates of the ether oxygen atom, shown as the red colored atom labeled 'O' in Fig. 9. This yields an opportunity for another generalization, in the sense that instead of calculating the CoM, we can simply substitute it with an RA corresponding to the ether oxygen atom. To obtain the radius of the bounding sphere, we calculate the maximum distance between RA and the rest of the atoms in the drug molecule, and still use the PSP to represent the polymer surface. We reiterate that, this "alternative" compact representation is possible due to the specific properties of the flavanone drug and may not be generalizable to other MD simulation data – and we have used it in our experiments for comparison.

Generalization Aware Trajectory Compression (GATCo): We now describe how we combine the proposed generalization GCTR representation of the atomic trajectories with compression techniques.

Naïve GATCo (N-GATCo): The perpendicular distance between the CoM and the PSP can be calculated using the standard analytic geometry approaches – i.e., for a plane specified by the equation $Ax + By + Cz + D = 0$ and a point (x_1, y_1, z_1), the perpendicular distance is $\frac{|Ax_1 + By_1 + Cz_1 + D|}{\sqrt{A^2 + B^2 + C^2}}$. The naïve compression is based on a simple threshold-based criterion:

1. If the perpendicular distance between CoM (respectively, RA) and the PSP at a time frame t is smaller than a threshold d_δ, then the particular CoM (respectively, RA) is retained as a representative.
2. Otherwise – i.e., the perpendicular distance between CoM (respectively, RA) and the PSP at a time frame t is greater than d_δ – that CoM (respectively, RA) is excluded from the compressed representation.

The molecular motion is then approximated by the retained CoM (respectively, RA) time frames, with a linear interpolation in-between them. We provide quantitative observations regarding the values of d_δ in Sect. 4.

GATCo with Classical Trajectory Compression: After the preprocessing in which the molecules are compactly represented by a sphere (weighted CoM or RA) for each time frame, we apply modified versions of the RDP and SPM to the trajectories of CoM (respectively, RA).

Specifically, the adaptation of the original versions cater to the fact that the distances between a given point and the corresponding line segment are actually calculated in 3D space *and* with time-awareness [4]. The formula needed to calculate the distance is $d_P = \frac{|\overrightarrow{AP} \times \overrightarrow{AB}|}{|\overrightarrow{AB}|}$ where: A and B denote the starting and ending points of a segment considered as a candidate for compressed representation, and P denotes the point on the original trajectory which is a candidate to be eliminated from the compressed representation (i.e., "absorbed" by the segment AB).

We refer to the modified versions as SPM-3D-GATCo and RDP-3D-GATCo. While the complexity in terms of $O()$ bound as a function of the number of time frames is not changed for any of them, the main speed-ups in the execution of compression algorithms (and the storage savings), result from the fact that due to the preprocessing. SPM-3D and RDP-3D are applied only to the generalized (CoM or RA, with PSP) trajectories, not to the ones of every single atom.

To cater to the fact that the trajectory compression of CoM (respectively, RA) with a given tolerance ε may introduce false negatives, when processing the *within-distance* query over the compressed representations, we again use a modified parameter $d'_\theta = d_\theta + 2\varepsilon$ to identify potential candidates for interaction. Subsequently, we refine each such candidate using the actual values of the atoms coordinates instead of the compact representation. We note that, as part of our experimental evaluation, we also investigated the combined effect of applying SPM-3D and RDP-3D to the output of N-GATCo (denoted as SPM-3D-N-GATCo and RDP-3D-N-GATCo, respectively).

4 Experiments

We now present the details of our experimental observations, including datasets description, experimental setup and the discussion of the results. For reproducibility, the source code of our implementations is publicly available at https://github.com/abdullahshamail/GATCo. Datasets can also be obtained upon request, by contacting the authors – and we describe them next, along with the rest of the setup (queries, simulation platform and evaluation criteria).

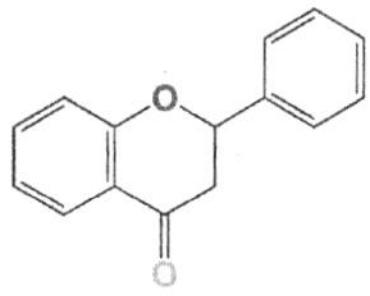

Fig. 9. Flavanone drug (Color figure online)

Trajectories Data and Platform

The flavanone drug molecule in the MD trajectory data [22,23] consists of 29 atoms. The structure of the drug molecule is shown in Fig. 9. The polymer substrate is Amylose Tris (3,5-dimethylphenyl carbamate) commonly known as ADMPC and its structure is shown in Fig. 10. Each polymer model in the simulation contains 1459 atoms. The contents of the drug and polymer molecules are given in Table 1. The MD simulation produces 200,000 time frames and each time frame captures the instantaneous 3D coordinates of all the atoms of the four polymers and drug molecules. For the purpose of analysis, we select only

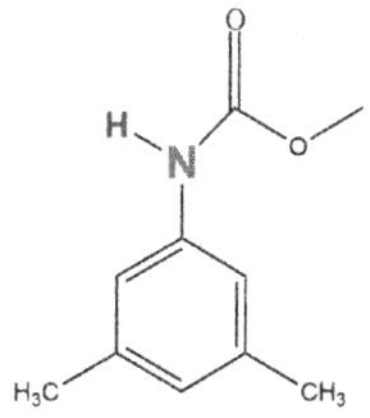

Fig. 10. One part of the repeating unit of ADMPC Polymer

216 N atoms (referred to as donors) and one ether O atom (referred to as the acceptor) to construct the uncompressed trajectories used in our experiments.

The experiments are conducted on a PC with Intel(R) Xeon(R) CPU E3-1240 v5 @3.50 GHz, 16 GB RAM, 512 GB disk storage and Windows 10 Enterprise 64-bit OS. The algorithms are implemented by Python 3.9.7.

Table 1. Contents of the drug and polymer

Atoms	Atomic mass	No. of atoms in drug	No. of atoms in polymer
Carbon (C)	12.0107	15	594
Hydrogen (H)	1.00794	12	667
Oxygen (O)	15.9994	2	144
Nitrogen (N)	14.0067	0	54

Within-Distance Threshold Query for H-Bond

The formation of an H-Bond can only occur when the donor and the acceptor atoms are within a certain proximity. Thus, we can translate the detection of a possible formation of H-Bond into processing of a *within-distance* query for a given threshold, for which the domain expertise (cf. [22]) suggests a value of $d_\theta = 3.5$ Å. We note that the *within-distance* threshold query, after applying the periodic boundary adjustments, outputs the first requirement of H-Bond formation; preprocessing based on angle threshold is left for a future work.

Evaluation Criteria

The following performance evaluation criteria are used to evaluate the efficiency of traditional and proposed compression approaches.

`Compression Ratio`: This is the relative reduction of the original data after applying a data compression algorithm. If the size required to store the trajectory data before compression is p and after compression is q, then the compression ratio R is defined as: $R = \frac{p}{q}$.

`Speedup Ratio of Query Processing Time`: Let T_1 denote the time for processing the *within-distance* query over the original uncompressed trajectory data and T_2 denote the respective time over the compressed trajectory data. The speedup ratio is defined as: $S_T = \frac{T_1}{T_2}$.

Results

We now report the quantitative observations from the comparative study of our proposed approaches using the trajectory data described in Sect. 4 when evaluating the *within-distance (3.5 Å)* query for the purpose of detecting a formation of H-Bond instances. We note that, the H-Bond query processing time for uncompressed trajectories is 38.83 s.

`Generalization`: For the dataset used in this study: (a) When CoM is used as a spatial representative of the drug molecule, the average value of r_B is 5.89 Å with a standard deviation of 0.057 Å, and the r_B^{max} value is 6.12 Å; and (b) When the drug molecule is represented via RA, the average value of r_B is 6.04 Å with a standard deviation of 0.106 Å, and the r_B^{max} value is 6.49 Å.

Table 2. N-GATCo using CoM and RA

Value of d_δ	Compress. ratio	
	CoM	RA
6 Å	2.15	2.12
5 Å	2.74	2.69
4 Å	3.16	3.13
3 Å	4.42	4.4
2 Å	9.52	8.03
1 Å	26.67	24.09

Table 3. SPM-3D vs. RDP-3D

Compress. technique	Epsilon (ε)	Compress. ratio	Compress. time (mins)	Query proc. time (s)
SPM-3D	0.5	1.47	508.37	22.67
	0.75	2.50	241.07	18.06
	1.0	4.21	145.95	14.56
RDP-3D	0.5	1.39	1391.73	23.52
	0.75	2.39	994.81	16.85
	1.0	4.08	842.43	15.09

Compression with Traditional Algorithms: We apply both SPM-3D and RDP-3D with varying epsilons (ε) to compress the original trajectories of all the atoms – i.e., without any preprocessing/generalization. The values of epsilon (ε) are selected to be 0.5 Å, 0.75 Å, and 1.0 Å. For the individually compressed trajectories (donor and acceptor atoms), we need linear interpolation in time for proper distance comparisons, to avoid losing a relevant instance (cf. Fig. 7).

The respective times required to compress and interpolate (which we jointly consider as compression time) for the original trajectories are shown in Table 3, for different values of ε. We see that, for $\varepsilon = 0.5$ Å, SPM-3D and RDP-3D can yield a compression ratio of 1.47 and 1.39 times, however, for $\varepsilon = 1$ Å the compression ratio becomes > 4. We also note that, in general:

– SPM-3D is completed rather faster than RDP-3D. This is because SPM-3D iterates over the trajectory once, as opposed to RDP-3D which keeps iterating (recursively) over the trajectory based on the split-point at the furthest distance from a candidate-segment.
– In this particular case study, SPM-3D even yields slightly higher compression ratio than RDP-3D. We note, however, that this need not be the case in general, as they are both heuristics (i.e., none of them is an optimal algorithm).

However, the values shown in Table 3 correspond to the settings in which no generalization has been applied. In the sequel, we provide observations regarding the GATCo-based compression approaches.

N-GATCo: We now report our experimental observations regarding the naïve compression introduced in Sect. 3. Recall that we either use the equations to compute the molecular CoM of the flavanone drug molecule at every time frame, or we use the ether O atom as RA at every time frame. We note that when it comes to the RA, the ether O atom is actually the one that acts as an acceptor for H-Bond. Also, recall that as the polymer is almost static compared to the drug molecules, the PSP is set to a fixed value. The compression ratios for the respective values of d_δ are shown in Table 2. As the distance from PSP declines (for both CoM and RA), we observe increase in the compression ratios, with quite higher values when the distance is ≤ 3 Å – going to 26.67 for CoM and 24.09 for RA. However, these come at a cost. Namely, when one ignores CoMs and RAs at distance 4 Å and above "blindly" (i.e., without considering r_B values), there is a risk of introducing false negatives. Double-checking for them, in effect, amounts to processing almost the entire dataset. Conversely, if one incorporates the value of r_B (which, as mentioned, on the average was 5.89 Å for CoM and 6.02 Å for RA), then it amounts to keeping all the instances which are within 6 Å from the PSP. This yields an effective storage savings of ~50% – i.e., a compression ratio of 2. In turn, the speedup ratios for N-GATCo become 2.95 for CoM and 3.03 for RA, as a consequence of the fact that we can prune some of the donor atoms from the polymer from consideration.

Table 4. SPM-3D and RDP-3D for CoM and RA GCTR

Compress. type	ε	Compress. ratio		Compress. time (s)		Query proc. time (s)	
		CoM	RA	CoM	RA	CoM	RA
SPM-3D-GATCo	0.5	1.49	1.61	230.68	101.38	21	21.9
	0.75	2.32	2.64	182.09	40.47	11.84	14.16
	1.0	3.40	4.47	162.10	20.85	6.82	8.34
RDP-3D-GATCo	0.5	1.48	1.58	517.28	381.26	20.68	22.8
	0.75	2.34	2.61	479.35	354.43	12.08	14.6
	1.0	3.46	4.45	471.61	333.49	6.72	8.53

Table 5. Compression efficiency

Compress. techniques	Compress. ratio	Speedup ratio	Query proc. time (s)
SPM-3D	1.47	1.71	22.67
RDP-3D	1.39	1.66	23.52
N-GATCo (CoM)	2.15	2.95	13.15
N-GATCo (RA)	2.12	3.03	12.82
SPM-3D-GATCo (CoM)	1.49	1.86	21.01
RDP-3D-GATCo (CoM)	1.48	1.89	20.40
SPM-3D-GATCo (RA)	1.61	1.78	21.90
RDP-3D-GATCo (RA)	1.58	1.71	22.80
SPM-3D-N-GATCo (RA)	**4.98**	**3.42**	**11.40**
RDP-3D-N-GATCo (RA)	4.58	**3.47**	**11.30**

GATCo with 3D Adapted Trajectory Compression: Here, we show the results when using GATCo in the settings in which SPM-3D and RDP-3D are applied for compression over GCTR. Table 4 presents the result, and we see that the compression time for each of SPM-3D-GATCo and RDP-3D-GATCo is significantly improved in comparison to the SPM-3D and RDP-3D compression over raw trajectories data without generalization. For example, the compression time for SPM-3D-GATCo (CoM) with $\varepsilon = 0.5$ is about 132 times less than SPM-3D with same value of ε (cf. Table 3).

Lastly, in Table 5 we show an aggregated summary of the compression ratio and speedup ratio of all the approaches used in this work, for a fixed value of $\varepsilon = 0.5$ Å. As can be seen, combining SPM-3D and RDP-3D with N-GATCo yield the largest compression ratios (about 5 times less space compared to uncompressed trajectories) for RA. The speedup ratio for both cases is ~3.4. Although the compression ratios are largest for the last two rows in Table 5, the gains in speedup ratios for processing the *within-distance* query are approximately 24% smaller, relative to the gains in the compression ratios. Most likely, the reason for this is that, additional false positives may occur as we expand d_θ range by $2 \cdot 0.5$ Å after applying SPM-3D (respectively, RDP-3D) on the outcome of N-GATCo.

5 Conclusions and Future Work

We presented an approach for using domain-specific knowledge to preprocess trajectory data of atoms belonging to molecules from MD simulation domain and generate compact representations of evolving drug molecules. The semantic generalization (preprocessing) consists of approximating the polymer substrates with PSP and the drug molecules with spheres centered at CoM or RA, and with radii equal to the distance between the center and the furthest atom in the respective drug molecule. We provided experimental observations of the improvements in terms of storage space savings for different combinations of compression and generalization, along with the speedup in terms of processing the *within-distance* query (essential for detecting a potential chemical interaction). Running ∼30 queries over compressed data – which may be useful for different types of interaction will amortize the cost of compression. There are multiple extensions that we plan to address in the future. The immediate next steps include incorporating the angle requirement for H-Bonds in the generalization and devising another volume-boundary of the drug molecule, with less of a dead space than a sphere (e.g., a bounding ellipsoid in a general position). We will also exploit adoption of other compression algorithms from the existing works in time series and trajectory [6,26], including the optimal 3D polyline compression algorithm [2] (an extension of the optimal polyline compression [5]). We also plan to evaluate the velocity autocorrelation between the uncompressed and (generalized) compressed trajectories and investigate the impact of compression on the quality of prediction of the diffusion in MD. This preliminary study was confined to the case of flavanone drug – and we plan to include other MD data sources for the purpose of developing more robust compression approaches. In the long run, we plan to extend the approaches to real-time compression of the partial simulation outputs.

References

1. Aminpour, M., Montemagno, C., Tuszynski, J.A.: An overview of molecular modeling for drug discovery with specific illustrative e.g's of apps. Molecules **24**, 1693 (2019)
2. Barequet, G., Chen, D.Z., Daescu, O., Goodrich, M.T., Snoeyink, J.: Efficiently approx. polygonal paths in 3+ dimensions. Algorithmica **33**, 150–167 (2002)
3. Bibelayi, D.D., Lundemba, A.S., Tsalu, P.V., Kilunga, P.I., Tshishimbi, J.M., Yav, Z.G.: Hydrogen bonds of C=S, C=Se and C=Te with C-H in small-organic molecule compounds derived from the Cambridge structural database (CSD) (2021)
4. Cao, H., Wolfson, O., Trajcevski, G.: Spatio-temporal data reduction with deterministic error bounds. VLDB J. **15**(3), 211–228 (2006)
5. Chan, W.S., Chin, F.: Approximation of polygonal curves with minimum number of line segments or minimum error. Int. J. Comput. Geom. Appl. **6**, 59–77 (1996)
6. Chiarot, G., Silvestri, C.: Time series compression: a survey (2021). https://doi.org/10.48550/ARXIV.2101.08784
7. Douglas, D.H., Peucker, T.K.: Algos for the reduction of the no. of points required to represent a digitized line or its caricature. Cartographica **10**, 112–122 (1973)

8. Guerrero-Corella, A., Fraile, A., Alemán, J.: Intramolecular HB activation: strategies, benefits, and influence in catalysis. ACS Organic & Inorganic Au (2022)

9. Hagita, K., et al.: Efficient compressed database of equilibrated configurations of ring-linear polymer blends for md simulations. Sci. Data **9**, 1–9 (2022)

10. Jeffrey, G.: An Introduction to Hydrogen Bonding. Oxford University Press, Oxford (1997)

11. Knight, K.J.: Pharma chemistry. Pharm. J. **282**, 105–128 (2021)

12. Kostal, J.: Computational chemistry in predictive toxicology: status quo et quo vadis? In: Advances in Molecular Toxicology, vol. 10 (2016)

13. Mcree, D.E.: Comp techniques. Practical Protein Crystallography (1999)

14. Muckell, J., Olsen, P.W., Hwang, J.H., Lawson, C.T., Ravi, S.S.: Compression of trajectory data: a comprehensive evaluation and new approach. GeoInformatica **18**, 435–460 (2013)

15. Pauling, L.: The Nature of the Chemical Bond, an Introduction to Modern Structural Chemistry, 3 edn. Cornell University Press, Ithaca (1960)

16. Sandu Popa, I., Zeitouni, K., Oria, V., Kharrat, A.: Spatio-temporal compression of trajectories in road networks. GeoInformatica **19**(1), 117–145 (2014). https://doi.org/10.1007/s10707-014-0208-4

17. Saalfeld, A.: Topologically consistent line simplification with the Douglas-Peucker algorithm. Cartogr. Geogr. Inf. Sci. **26**(1), 7–18 (1999)

18. Sayood, K.: Intro to Data Compression. Morgan Kaufmann Publisher, Burlington (2017)

19. Singh, A.K., Aggarwal, V., Saxena, P., Prakash, O.: Performance analysis of trajectory compression algorithms on marine surveillance data. In: ICACCI 2017 (2017)

20. Steiniger, S.: Enabling pattern-aware automated map generalization (2007)

21. Trajcevski, G.: Compression of spatio-temporal data (tutorial). In: IEEE International Conference on Mobile Data Management (MDM) (2016)

22. Wang, X., et al.: Md sims of the chiral recognition mechanism for a polysaccharide chiral stationary phase in enantiomeric chromatographic separations. Mol. Phys. **117**(23–24), 3569–3588 (2019)

23. Wang, X., Jameson, C.J., Murad, S.: Modeling enantiomeric separations as an interfacial process using amylose tris (3, 5-dimethylphenyl carbamate) (ADMPC) polymers coated on amorphous silica. Langmuir **36**, 1113–1124 (2020)

24. Weibel, R.: Generalization of spatial data: principles and selected algorithms. In: van Kreveld, M., Nievergelt, J., Roos, T., Widmayer, P. (eds.) CISM School 1996. LNCS, vol. 1340, pp. 99–152. Springer, Heidelberg (1997). https://doi.org/10.1007/3-540-63818-0_5

25. Wibowo, E.S., Park, B.D.: Two-dimensional nuclear magnetic resonance analysis of hydrogen-bond formation in thermosetting crystalline urea-formaldehyde resins at a low molar ratio. ACS Appl. Polym. Mater. **4**(2), 1084–1094 (2022)

26. Zhang, D., Ding, M., Yang, D., Liu, Y., Fan, J., Shen, H.T.: Trajectory simplification. Proc. VLDB Endow. **11**, 934–946 (2018)

Analysing Workload Trends for Boosting Triple Stores Performance

Ahmed Al-Ghezi$^{(\boxtimes)}$ and Lena Wiese(iD)

Goethe University Frankfurt, Robert-Mayer-Str. 10,
60629 Frankfurt am Main, Germany
{alghezi,lwiese}@cs.uni-frankfurt.de

Abstract. The Resource Description Framework (RDF) is widely used to model web data. The scale and complexity of the modeled data emphasized performance challenges on the RDF-triple stores. Workload adaption is one important strategy to deal with those challenges on the storage level. In all the current adaptation approaches, the workload statistics are built collectively, and the analysis process is not aware of old or recent items in the workloads. However, that does not simulate the timely trends that exist naturally in user queries and causes the analysis process to lag behind the rapid workload development. In this work, we model the workload statistics as time series and apply well-known smoothing techniques allowing the importance of the workload to decay over time. We apply the proposed approach on UniAdapt [1] which follows a unified and comprehensive storage adaption process.

Keywords: RDF · Triple-stores · Workload adaption

1 Introduction

The resource description framework (RDF) is increasingly used to model web-scale data in our digital universe. Despite its simple triple-based structure, it has shown high efficiency in modeling the resources and their complex relationships. Such RDF data are often cached from their sources into a triple store, where queries are processed. However, due to the big size of the data sets and their complex relationships, they need to be properly structured into multiple types of indexes, caches as well as replications. Those requirements of huge data structures are often faced by the high storage space consumption. Workload adaption has emerged as a vital approach to deal with that problem. The first works considered workload to enhance the RDF partitioning in a distributed environment. In this context, Partout [6] converted a collection of queries into global queries graph which then was used to estimate which data fragments are more probable to be targeted by the same queries. The basic aim was to reduce the communication cost during query execution. The same approach is extended by WARP [9], AdPart [7], Peng [14] and UniAdapt [1]. However, UniAdapt presented a comprehensive approach to adapt the indexes, cache, and replication

© Springer Nature Switzerland AG 2022
S. Chiusano et al. (Eds.): ADBIS 2022, LNCS 13389, pp. 285–298, 2022.
https://doi.org/10.1007/978-3-031-15740-0_21

in one optimization problem. In all the given systems, including UniAdapt, the workload is cumulatively collected from the time the system starts, and up to the current time. The collected workload is then analysed to detect frequent patterns and assign them numerical values that represent their relative importance. However, since the statistics are collected accumulatively and with no awareness of their collection's times, any trend changes would have to be compared to the whole history of workload. That makes the workload analysis inefficient in detecting changes in workload trends. Those timely trends are already recognized in real-world queries [3,18]. They can be processed by analyzing their seasonal factors [17] or their decay factors [8] over time. The seasonal effect means that some parts of the data could receive heavy access within some time window that is repeated after some duration. On the other hand, the decay effect means that the volume of access would decay over time. In this work, we propose transferring the workload collection into a set of time series by adding timestamps to each collected query. That allows applying well-known smoothing techniques to simulate the trend changes occurring in the real-world users' queries and thus boosting the adaption process speed in catching up with those trends.

The rest of this paper is structured as follows: In the next section, we provide essential background about RDF, query processing, and the unified adaption process. Section 3 describes our proposed method to change the workload into time series and apply the smoothing methods. We then provide a comprehensive experimental evaluation in Sect. 4 where we show the impact of the proposed approach. Finally, we provide a conclusion and future works in Sect. 5.

2   Background and Related Work

In this section, we review the essential methods used to implement a storage layer adaption based on the workload.

2.1   RDF Graph and Queries Processing

RDF is a data model that has been widely used to represent web-data, by making statements about resources using a triple-based format. Each triple is composed of Subject, Predicate, and Object. Those triple elements are abbreviated by (S, P, O). The subject represents a certain resource that has a relationship to the object, which is either another resource or a literal. The predicate is describing this relationship. A triple is often modeled as a graph G where both its subject and object are vertices connected by one directed edge. Since the object can be a subject in other triples, the data set or the data graph grows large and more triples can be defined to describe each resource and its relationships.

The big data graph G can be queried for certain subgraphs using a SPARQL query. Such a query q is an RDF graph on its own with some parts of it are left as variables. Since an RDF triple set can be mapped to an RDF graph (and vice versa), a query graph can be mapped to a set of triple patterns. The answer of a query q is all the subgraphs in data graph G that match q and substitute its variables. Since the subgraph isomorphism problem is known to be NP-complete [4], the query execution is only feasible with the assistance of suitable indexes.

Given the multiple evaluations and joining paths, each with different costs, the query optimizer tries to select an approximate optimal plan using dynamic programming [13]. However, an execution plan is only possible if the necessary indexes exist, which highlights the importance of optimizing those structures in RDF triple stores.

2.2 Indexing and Cache

A typical RDF index is a hash table where the RDF set of triples are hashed either on the subject, predicate, or object. The chosen triple's element serves as the key of the index, while the value is all the triples that have the key in the correct position. Based on that, we have three basic types of indexes: S, P, or O. Each index is also sorted on the remaining two triple elements. Since that there are two possible sorting orders to each of the three basic indexes, we end up with six possible indexes that are: SPO, SOP, OPS, OSP, POS, and PSO. Although the query execution engine can sometimes live with two indexes, a more efficient query execution requires all the given indexes. RDF-3X [13] and Hexastore [19] decided to fully implement the given six indexes. However, given that each index contains a duplicate of all the data triples, such a system may cause a lot of storage space consumption. Thus, other systems selected only a subset of indexes and restricted their query engine to exploit them. The selection was mainly based on observations of certain workloads. UniAdapt [1] allowed the system to dynamically build its indexes structure and adapt it, upon the workload's status and the status of other storage needs aiming for the best performance. Besides the normal indexes, UniAdaptbuilds cache indexes which are used to cache join results and save the most expensive cost. However, it requires much more space compared to normal indexes. Nevertheless, UniAdapt integrates the cost and benefit of such a cache in its storage adaption engine.

2.3 Replication

The RDF data set can grow very large which increases the problem of scalable data management. One important approach to deal with this problem is to exploit the capabilities of distribution systems. Among the multiple methods to distribute RDF management, is the federation of multiple centralized RDF stores. In such a method, each working node receives a partition of the global RDF graph and is responsible for managing and querying it [1,6,7,9,10]. However, the problem of partitioning has become the point of attention for such systems. Some of them preferred to use hash-based partitioning [7] aiming for fast completion. Unfortunately, it might cause a lot of communication costs. TUNABLE-LSH [2] used locality-sensitive hashing (LSH) to assign records that are accessed across related queries to close physical pages in the storage system. Other approaches depended on using more sophisticated graph partitioning methods by relying on METIS [12]. It aims to produce balanced (in size) partitions, that have minimum number of edges between them. That could increase the chance of a query being locally executed in one working node. However, the problem is now moved to the border regions: the regions where edges connecting vertices belong to different partitions. H-RDF-3X [10] made each node replicate from its border up to a given depth. Unfortunately, the required storage space for such a replication increases exponentially with that given depth. WARP [9] proposed to use H-RDF-3X method for a given small depth, then use the workload to select only the most important triples for further replications. The main problem in such an approach is related to finding suitable values for the assumed to be given parameters like the depth. UniAdapt [1] proposed to allow each node to replicate as much as it can allocate from its storage. It orders the triples (that are proposed for replication) by their importance, and greedily fills the replication containers. The expected benefit is derived by rules that are based on the workload and the distance from the border. Those replication rules are also competing with other rules about indexes and join cache. In such a way, the system always tries to optimize its storage with the best assignment of data to structures.

2.4 Workload Adaption

UniAdapt [1] implements a universal adaption approach by putting the indexes, cache, replication into a single optimization problem. That maps the problem into the knapsack problem [5]. The knapsack models the storage space which is to be filled with best assignments of indexes, cache, or replication aiming to achieve the best total value within the limited storage. The value of each assignment is the product of the previous access rate and its derived benefit. The access rate is calculated from the workload using a workload analysis engine that is based on the heat queries (Subsect. 2.5). The system collects and builds its workload statistic during *execution phases.* At some point in time, the system has enough resources to perform an adaption round. The access rules for each index are evaluated and sorted by their importance, and the highest important data replaces the lowest importance. Running this adaption process whenever possible makes the system choose the best employment to its storage resources.

2.5 Heat Query

The heat query is used by UniAdapt [1] to collect and store the queries that have been executed by the system. It keeps the information about the queries structures, frequencies, used indexes, as well as how the queries relate to each other. The basic structure of a heat query is a graph, since that the SPARQL queries are modeled as a graph. Figure 1 shows an example for a heat query evolving from four queries: $\langle Q_1, Q_2, Q_3, Q_4 \rangle$.

First, the system receives Q_1 and creates a new heat query h for it (otherwise it tries first to combine it with an existing heat query). Next, Q_2 is received, and it is combined with h such that the matching part between Q_2 and h becomes hotter – this is shown as darker color in the right side of Fig. 1. The above applies also in the same manner for Q_3 and Q_4. The variables within any query (here $?x$, $?y$ and $?z$) are substituted with a single variable $?x$. This normalization method unifies the variables and allows different queries variables to be directly combined. When Q_4 is combined, the heat value of C_1 is increased in Fig. 1 but a

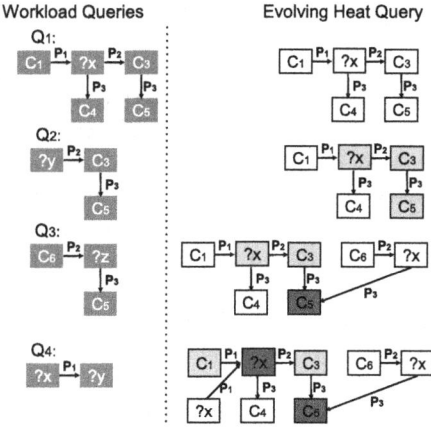

Fig. 1. Heat query evolving from four queries

new node is created that contains variable $?x$ and has a heat value equal to 1. With more workload queries received and executed by the system, related heat queries would be combined and thus be bigger in size regardless of their order of arrival. The heat values stored within the heat query represent the access frequencies. However, it stores also other execution statistics needed by the optimizer to calculate the effective benefit of each item.

Table 1. List of abbreviations

$h(t)$	Accumulative heat values up to time t
H	Heat time series
H_r	Resulted heat time series after smoothing and seasonality analysis
H_T	Heat time series after exponential smoothing
y	A time series
T	The index of the current value of a time series
SSE	Sum of squared error
e	The calculated error
α	First smoothing parameter
β	Second smoothing parameter
γ	Third smoothing parameter
s	Seasonality component
b	Trend component
ℓ	Level component
τ	Time span between two consecutive adaption operations
M	Uniform time span to sample the accumulative heat query function $h(t)$
m	Seasonality frequency

3   Heat Item Gets Colder

The problem of the heat query is that it collectively builds the heat values over time. Assume for example that a sudden high workload queries certain data parts only for a very short and limited time; we call this a spark query. The effect of this spark query would continue to influence the collective calculation of heat query for a long period of time until it eventually collects enough workload to overcome that spark. That makes it very difficult for the workload analysis process to detect the timely changes in query trends. To overcome this problem, we allow the importance of the collected workload to decay over time, or the hot items to get colder with time by decaying heat. To simulate this, we add a timestamp to the heat value for each heat query, then aggregate the heat values for each item over uniform sampling time spans M. This transfers the heat values into time series $y(t)$. All notation used in this paper is summarized in Table 1.

Definition 1. *If $h(t)$ is the accumulative heat value up to time t, and M is a given sampling time-span, then we define H as heat time series such that: $H_i = h(M \cdot i) - h(M \cdot (i - 1))$.*

A time-series is referred to a set of successive data points where traditionally the data points are used to represent many types of quantifiable items like sale transactions or temperature. Nevertheless, time series have been also used to study web search queries [20]. Thus, many well-studied techniques are already available to process time series. The most relevant concerning our application is the smoothing functions.

3.1   Smoothing Methods

There are many smoothing functions that can be applied to a time series to simulate decaying. We select two well-known smoothing methods, adapt them to our problem, and analyze their behaviour both theoretically and practically.

3.2   Exponential Smoothing

The collective heat query assumes that all parts of the collected workload are of equal importance independent of its time. This is modeled in a time series by the average method; that is for T collected observations, the expected value of the next r'th observation $\hat{y}_{T+r}$ is given by the following formula:

$$\hat{y}_{T+r} = \frac{1}{T} \sum_{t=1}^{T} y_t \tag{1}$$

To simulate the timely decay factor, the uniform weights of the average method are changed into exponential weights that favour the most recent values over the newest ones. That is reflected in well-known exponential smoothing method [11]:

$$\hat{y}_{T+1|T} = \alpha y_T + \alpha(1 - \alpha)y_{T-1} + \alpha(1 - \alpha)^2 y_{T-2} + \dots \tag{2}$$

where $0 < \alpha < 1$ is the smoothing parameter. The value of the equation at time $T + 1$ is the one-step-ahead forecast, and equal to a weighted average of all of the observations in the series $y_1, \ldots, y_T$. The parameter α controls the decrease rate of the weights. A very high value of α (close to one) would highly filter non recent values. A practical approach for choosing a proper value of α is to make the system choose it. One way to accomplish that is by minimizing the sum of squared errors on a training set. The error is then defined as follows:

$$SSE = \sum_{t=1}^{T} e_t^2 = \sum_{t=1}^{T} (y_t - \hat{y}_{t|t-1})^2 \tag{3}$$

We aim to use this exponential smoothing approach for modeling the decay of items' heat with respect to time within each heat query. For that purpose, the current heat is then calculated from the previous instead of the future values. Thus Eq. 2 is then rewritten as:

$$\hat{H}_T = \alpha H_t + \alpha(1 - \alpha)H_{t-1} + \alpha(1 - \alpha)^2 H_{t-2} + \ldots \tag{4}$$

However, the heat expressed by the above equation does not consider the periods in which some series are repeating themselves. Those periods are more analysed in the next approach.

3.3   Holt-Winters' Additive Method

Holt-winter method [11] captures seasonality by decomposing the forecasting formula into three components: level ℓ, trend b, and seasonal component s. The forecasting formula according to the additive method will be:

$$\hat{y}_{t+h} = \ell_t + h b_t + s_{t+h+m(k-1)} \tag{5}$$

where m is the number of measurement seasons in a term. The components are given by the following:

$$\ell_t = \alpha(y_t - s_{t-m}) + (1 - \alpha)(\ell_{t-1} + b_{t-1})$$
$$b_t = \beta(\ell_t - \ell_{t-1}) + (1 - \beta)b_{t-1}$$
$$s_t = \gamma(y_t - \ell_{t-1} - b_{t-1}) + (1 - \gamma)s_{t-m}$$

where k is the integer part of $\frac{(h-1)}{m}$, which keeps the values of the seasonal indices within the final term of considered timely data. The seasonal equation shows a weighted average between $(y_t - \ell_{t-1} - b_{t-1})$ (representing the current seasonal index), and the seasonal index of the same season of the last term (m time periods ago) [11].

We recall that Eq. 5 models the heat of each item of the heat query. UniAdapt runs its adaption operation on time spans – when the system has free processing resources. Thus, the system should use the heat values to predict the data access rate for the next τ time units. That is the time span until the system starts the next round of its adaption operation.

To predict the next expected value of τ, we present the values of τ as a time series on its own. We may simply consider the average as in Eq. 1 as a proper value for expectation. However, that can perform poorly when the system has to rapidly change its adaption span periods. That is highly related to the workload arrival rates. Thus, the exponential smoothing allows the system to fast adapt its predication of the next τ with the possible timely changes. That is given in the following:

$$\hat{\tau}_{T+1} = \sum_{t=0}^{T} \alpha(1 - \alpha)^t \tau_{T-t} \tag{6}$$

A suitable value of α is again found by minimizing the sum of square error as in Eq. 3. Having the time series of Eq. 6, we can estimate the time period until the next round of adaptation begins. The effective heat value of the heat query results from the intersection of this adaption series with the seasonality series s_t of Holt-Winter's method that is given in Eq. 5. That is the summation of the seasonality component values of Eq. 5 in the period from now until τ:

$$\sum_{i=0}^{i=\frac{\tau}{M}} s_{t+i+m(k-1)}, \tag{7}$$

where M is the time spans of the heat values' time series H (see Definition 1). For each of the heat query items, the heat H_r will be the summation of the exponential smoothed value of Eq. 4 and the seasonality component given in the above equation:

$$H_r = \hat{H}_T + \sum_{i=0}^{i=\frac{\tau}{M}} s_{t+i+m(k-1)} \tag{8}$$

4   Experimental Evaluation

In this section, we evaluate the effect of the time smoothing approach on the performance of UniAdapt. We measure the enhancement in the workload changes detection and how fast the system reacts to those changes. For this purpose, we apply the exponential smoothing method (given in Sect. 3.2) and our modified Holt-Winter-Winters' additive method (given in Sect. 3.3), then compare them to the original collective method of the heat queries. We use the LUBM [15] data set, which is a generated RDF data set that contains data about universities. The data set can be easily increased in size by increasing the number of universities and/or their members (student, teacher, publications...). However, the number of properties (the unique predicates) is kept small and limited. Two important workload properties are referenced in our evaluation: query length, and workload quality. The query graph's length is the value of the maximum shortest distance that can be found between any two vertices in the graph. The workload quality is the standard deviation of the workload's frequencies distribution.

4.1   Exponential Smoothing

In the first part of the evaluation, we test the effect of the exponential smoothing on the heat queries' ability to detect workload change, or more specifically, how fast and accurate it detects those changes. For that purpose, we generate the workloads given in Table 2. First, the system receives one of the workloads given in the table for some initial adaption rounds between three and six rounds, then it rapidly changes to another workload. Figure 2 depicts the behavior of the adaption system when moving from one workload to the next. The first part is the transition from W_{g1} to W_{g2}. Both of the workloads contain general average access to given indexes in Table 2. The specific accesses to the data parts are uniform. In such a case, the specific rule of UniAdapt is not in action but rather the general rules. The transition form Workload W_{g1} to W_{g2} in adaption period numbered 4 has caused a big increase in queries execution time. This is because the system could not provide the OPS index which was mainly not built in the first adaption periods that mainly required the SPO index. The accumulative adaption of the original heat query slowly adapts to the required change by building more of the OPS index in the next adaption round. That is seen mainly as linear behavior. On the other hand, the smoothed adaption showed an exponential decrease in the queries execution time. That resulted in a much faster adaption that has saved a lot of latency. We consider then the workloads W_{g1} and W_{g3} in Fig. 2. The transition from Workload W_{g1} to W_{g3} caused a direct decrease in execution time due to that W_{g3} is better utilizing the join cache and still using the same indexes of W_{g1}. However, that is rapidly enhanced in the next rounds within the exponential smoothing adaption compared to a slower linear accumulative adaption. For the two remaining parts of Fig. 2, a big increase in execution time is again recorded when transitioning from Workload W_{g2} towards both of workloads W_{g3} and W_{g4} in adaption round numbered 4. The afterward exponential enhancement is clear when moving towards Workload W_{g3}, when the system quickly learned to build index SPO that is highly required by Workload W_{g3} and replaced OPS index, then it further optimized the cache-usage to boost the performance. That cache optimization was less effective in Workload W_{g4}. Nevertheless, the exponential smoothing was highly effective for speeding up the adaption of the indexes structures towards the needs of workloads W_{g3} and W_{g4} with respect to the former Workload W_{g2}.

4.2   Hot Data Parts

In this section of the experimental evaluation, we test the behavior of the adaption system with respect to the workload that targets specific areas of the data graph. Those parts are referred to as the hot parts which are widely found in real-world SPARQL queries [14,16]. To simulate such a workload, we generate it with a frequencies distribution that follows a normal distribution to the data graph. Such a normal distribution of edges density in the RDF graph is also observed by [21]. We aim to test how fast the system responds to the changes in the hot regions. Those changes are either shifts in the locality of the hot

Table 2. Generated workloads properties

Workl.	Properties
W_{g1}	General workload requires mainly the indexes: SPO and POS
W_{g2}	General workload requires mainly the indexes: OPS and POS
W_{g3}	general workload requires mainly the indexes: OPS, POS and cache index
W_{g4}	General workload requires mainly the indexes: SPO, POS and cache index
W_{s1}	Data-specific workload with quality (the standard deviation frequency distribution) of 0.1
W_{s2}	Data-specific workload with quality (the standard deviation of frequencies distribution) of 0.01, and average query length of 3
W_{s3}	Data-specific workload with quality (the standard deviation of frequencies distribution) of 0.1, and average query length of 4
W_h	Data-specific workload with quality (the standard deviation of frequencies distribution) of 0.01, and average query length. of 4
W_l	Data-specific workload with quality (the standard deviation of frequencies distribution) of 0.1, and the average query length of 4

region (new hot regions replace the existing) or changes in the ratio of distribution which is mapped to variations in the standard deviation of the given normal distribution. The smaller is the standard deviation, the more is the ratio of the workload targeting a small region of the data. That case recognized by UniAdapt [1] as high workload *quality*. This is because the optimizer can easily detect those small hot regions of the data and efficiently structure them into multiple indexes and caches for better queries execution performance, and without paying the high cost of storage space. We first test the behavior of the adaption system assuming a workload that first targets a specific region of the data graph, then rapidly changes to another region that does not overlap with the first region. However, the distribution quality is kept the same during that change. The system performance with respect to Workload W_{S1} of Table 2, is recorded in Fig. 3. It starts with five rounds called $W_{S1\text{-}a}$ targeting a random region of the workload, then targets another region in the next five rounds with Workload $W_{S1\text{-}b}$. However, both of $W_{S1\text{-}a}$ and $W_{S1\text{-}b}$ are identical to the properties of W_{S1}, which has high workload quality; that means that only a small portion of data is the highly accessed part. Thus, despite the target-change of that portion, the system was able to efficiently adapt to it even with the collective heat query. However, there is still some recognizable performance difference recorded for the smoothed method. Moving to Workload W_{S2} and W_{S3} of Fig. 3, we tested data-specific workloads with less quality. The effect of the low workload quality is more significant on the collective adaption. That is because the system required a lot of storage space to handle the big size of targeted data. When that targeted data are changed, the old ones would continue to affect the statistic calculation for a longer time in the accumulative adaption with respect

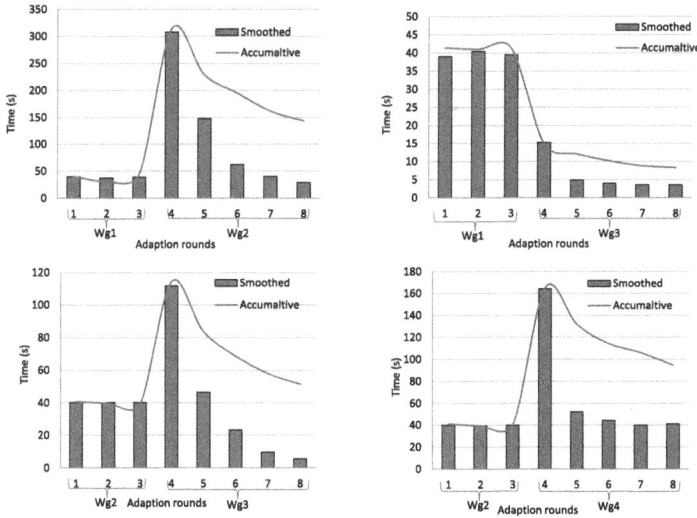

Fig. 2. Adaption behaviour when transiting between the given workloads

to the smoothed adaption. That is clearly seen in the W_{S2} and W_{S3} parts of Fig. 3, where the accumulative adaption showed linear behavior as compared to a faster-smoothed adaption. The next experiment deals with a change in the quality of the workload. That case is shown in Fig. 3 where the system starts with Workload W_h that has high quality, then changes to W_l after five rounds of adaption. The initial high performance is directly related to the ability of the adaption process to optimize the storage resources especially the cache and indexes making use of the high quality of the workload. The adaption process is totally affected when the quality suddenly changes.

4.3 Seasonality Factors and Capacity

In this part, we show the effect of changing system capacity on the adaption process and as a result on the smoothing process. UniAdapt is adaptive with space, which means that the system can adapt its storage structures to suit the current status of storage availability enlightened by the workload, and aiming always for best performance. According to [1], the capacity is defined as the number of full indexes that the memory can hold. That value is relative to the data set size and the available storage size.

It is clear from [1] that whenever the relative capacity is too tight, the adaption role is more effective. This is because a limited resource is more precious, and its optimization is more beneficial. The same capacity context is generally reflected on the smoothing effects in Fig. 4a. We have six runs, each with a different level of capacity. For each run, we sketch the total seconds saved by the smoothed adaption Starting with capacity levels of 6 and 4, the high capacity

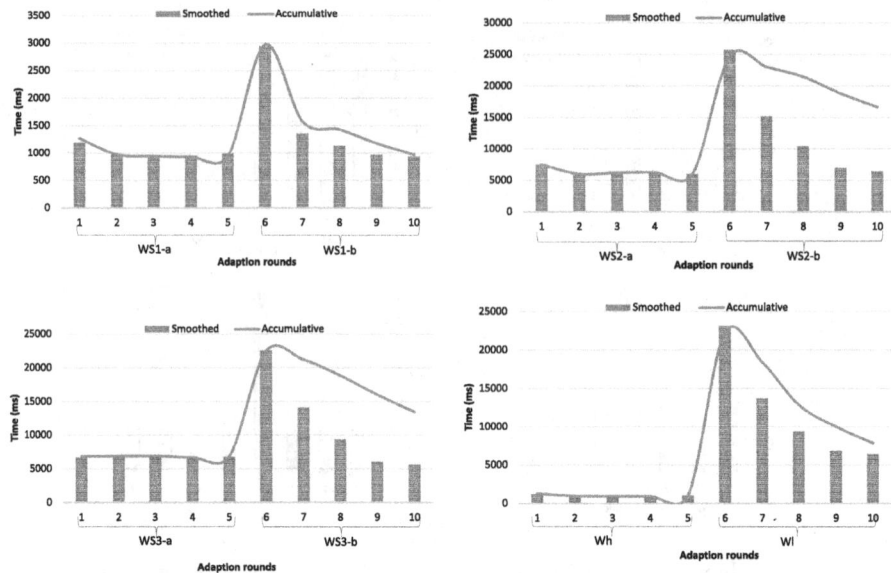

Fig. 3. Adaption behaviour with respect to the given workloads

relatively decreases the role of adaption and as a result, the system is less vulnerable to the effects of the workload changes. Nevertheless, the saved times are still avoiding considerable delays that are reflected in better throughput and more performance. However, going to lower levels of capacity, the adaption role becomes very vital in optimizing the use of the limited storage. That is shown in capacity levels of 2.7 and 2.2. The time until the system detects any new changes in the workload in such a limited capacity, and then adapts its indexes and storage structure, is reflected in a high cost on the query execution performance. The smoothed adaption clearly helped with the quick adaption to the new workload changes, and thus saved a considerable amount of delays.

In the last part of our evaluation, we consider the behavior of the workload that has repeated patterns. In this context, we allowed a repeated workload every four adaption rounds and sketched the results in Fig. 4b. The accumulative adaption line showed decreasing trend after Round 4. It slowly and steadily absorbed the shock of sudden workload change, making use of the steady accumulated workload stats. On the other hand, the exponentially-smoothed adaption enhanced the response between the periods. However, it has a very slow reaction to the periodic component and showed a periodic increase in response time. The response was much better when it was supported by the Holt-Winters' additive method. That was clear after adaption round 12, where the detection of the periodic workload trend was reflected into smoothing the periodic increase in the response time.

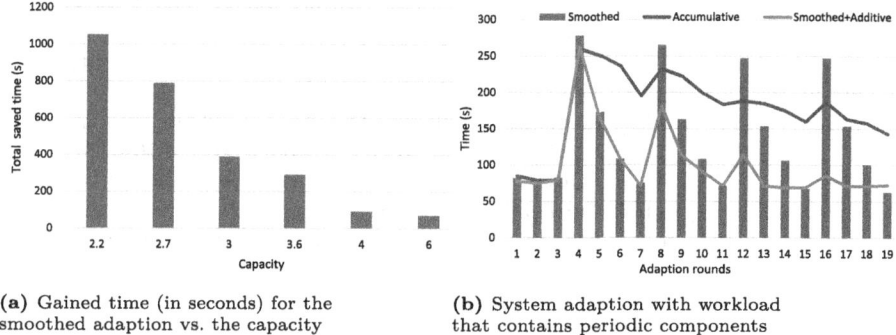

(a) Gained time (in seconds) for the smoothed adaption vs. the capacity

(b) System adaption with workload that contains periodic components

Fig. 4. Runtime measurement for smoothed adaption and periodic components

5 Conclusion and Future Work

A fully adaptable RDF triple-store adapts its storage structures (indexes, cache, and replications) with the current status of the workload. In this paper, we present a powerful approach to boost the performance of such a triple store. We convert the simple collective workload analysis module into time-aware sets of time series. By applying well-known smoothing methods, the system becomes much faster to adapt to changes in queries trends. That results in a high boost to the query execution performance as we have validated expediently. In the next steps, more methods could be developed or integrated to enhance the detection of periodicity in the workload. Moreover, other temporal behavior of the user queries could be detected and analyzed like the sequencing in query generations from single users.

References

1. Al-Ghezi, A., Wiese, L.: Universal storage adaption for distributed RDF-triple stores. In: Golfarelli, M., Wrembel, R., Kotsis, G., Tjoa, A.M., Khalil, I. (eds.) DaWaK 2021. LNCS, vol. 12925, pp. 97–108. Springer, Cham (2021). https://doi.org/10.1007/978-3-030-86534-4_8

2. Aluç, G., Özsu, M.T., Daudjee, K.: Building self-clustering RDF databases using tunable-LSH. VLDB J. **28**(2), 173–195 (2019). https://doi.org/10.1007/s00778-018-0530-9

3. Bonifati, A., Martens, W., Timm, T.: An analytical study of large SPARQL query logs. Proc. VLDB Endow. **11**(2), 149–161 (2017)

4. Cook, S.A.: The complexity of theorem-proving procedures. In: Harrison, M.A., Banerji, R.B., Ullman, J.D. (eds.) Proceedings of the 3rd Annual ACM Symposium on Theory of Computing, Shaker Heights, Ohio, USA, 3–5 May 1971, pp. 151–158. ACM (1971)

5. Dasgupta, S., Papadimitriou, C.H., Vazirani, U.V.: Algorithms. McGraw-Hill (2008)

6. Galárraga, L., Hose, K., Schenkel, R.: Partout: a distributed engine for efficient RDF processing. In: Proceedings of the 23rd International Conference on World Wide Web, WWW 2014 Companion, pp. 267–268. ACM, New York (2014)

7. Harbi, R., Abdelaziz, I., Kalnis, P., Mamoulis, N., Ebrahim, Y., Sahli, M.: Accelerating SPARQL queries by exploiting hash-based locality and adaptive partitioning. VLDB J. 25(3), 355–380 (2016). https://doi.org/10.1007/s00778-016-0420-y

8. Hashavit, A., Levin, R., Guy, I., Kutiel, G.: Effective trend detection within a dynamic search context. In: Perego, R., Sebastiani, F., Aslam, J.A., Ruthven, I., Zobel, J. (eds.) Proceedings of the 39th International ACM SIGIR Conference on Research and Development in Information Retrieval. ACM (2016)

9. Hose, K., Schenkel, R.: WARP: workload-aware replication and partitioning for RDF. In: ICDE Workshops, pp. 1–6. IEEE Computer Society (2013)

10. Huang, J., Abadi, D.J., Ren, K.: Scalable SPARQL querying of large RDF graphs. Proc. VLDB Endow. 4(11), 1123–1134 (2011)

11. Hyndman, R., Athanasopoulos, G.: Forecasting: Principles and Practice, 2nd edn. OTexts, Australia (2018)

12. Karypis, G., Kumar, V.: A fast and high quality multilevel scheme for partitioning irregular graphs. SIAM J. Sci. Comput. 20(1), 359–392 (1998)

13. Neumann, T., Weikum, G.: The RDF-3X engine for scalable management of RDF data. VLDB J. 19(1), 91–113 (2010). https://doi.org/10.1007/s00778-009-0165-y

14. Peng, P., Zou, L., Chen, L., Zhao, D.: Query workload-based RDF graph fragmentation and allocation. In: EDBT, pp. 377–388. OpenProceedings.org (2016)

15. SWAT Projects: The Lehigh University Benchmark (LUBM). http://swat.cse.lehigh.edu/projects/lubm/

16. Rietveld, L., Hoekstra, R., Schlobach, S., Guéret, C.: Structural properties as proxy for semantic relevance in RDF graph sampling. In: Mika, P., et al. (eds.) ISWC 2014. LNCS, vol. 8797, pp. 81–96. Springer, Cham (2014). https://doi.org/10.1007/978-3-319-11915-1_6

17. Shokouhi, M.: Detecting seasonal queries by time-series analysis. In: Proceeding of the 34th International ACM SIGIR Conference on Research and Development in Information Retrieval, SIGIR 2011, Beijing, China, 25–29 July 2011, pp. 1171–1172. ACM (2011)

18. van Kleef, P.: DBpedia usage report (2018). https://medium.com/virtuoso-blog/dbpedia-usage-report-as-of-2018-01-01-8cae1b81ca71

19. Weiss, C., Karras, P., Bernstein, A.: Hexastore: sextuple indexing for semantic web data management. Proc. VLDB Endow. 1(1), 1008–1019 (2008)

20. Zhang, R., Konda, Y., Dong, A., Kolari, P., Chang, Y., Zheng, Z.: Learning recurrent event queries for web search. In: Proceedings of the 2010 Conference on Empirical Methods in Natural Language Processing, EMNLP 2010, MIT Stata Center, USA, 9–11 October 2010, pp. 1129–1139. ACL (2010)

21. Zloch, M., Acosta, M., Hienert, D., Dietze, S., Conrad, S.: A software framework and datasets for the analysis of graph measures on RDF graphs. In: Hitzler, P., et al. (eds.) ESWC 2019. LNCS, vol. 11503, pp. 523–539. Springer, Cham (2019). https://doi.org/10.1007/978-3-030-21348-0_34

Machine Learning

Comparision of Models Built Using AutoML and Data Fusion

Anam Haq[1]([✉])(iD), Szymon Wilk[1]([✉])(iD), and Alberto Abelló[2](iD)

[1] Poznan University of Technology, Street Piotrowo 3, 60-965 Poznan, Poland
anam.haq@doctorate.put.poznan.pl, szymon.wilk@cs.put.poznan.pl
[2] Universitat Politècnica de Catalunya, Carrer de Jordi Girona,
08034 Barcelona, Spain
aabello@essi.upc.edu

Abstract. Automated machine learning (AutoML) has made life easier for data analysts or scientists by providing quick insights into data by building machine learning (ML) models. AutoML techniques are applied to vast areas from image processing, speech recognition, natural language processing reinforcement learning, and more. However, there is still room for many improvements. AutoML techniques focus only on problems related to predictive modeling, and most of them are designed to work with structured data. AutoML techniques are also time-consuming as they require time to select the appropriate ML pipeline. This paper presents an alternative time-efficient approach for mixed data (both categorical and numerical features obtained from UCI and Kaggle repository) using a data fusion process, which provides high macro average accuracy in less time as compared to AutoML. The AutoML tool considered here is autoscikit-learn (auto-sklearn). This specific library is built in Python using scikit-learn. The implementation of data fusion is also done in Python using scikit-learn. We conclude from the experimental analysis that the pipeline constructed provides better results than the auto-sklearn. This obtained conclusion is supported by a statistical test (Wilcoxon signed ranks test) based on macro average accuracy obtained for both approaches.

Keywords: Automated machine learning · AutoML tools · Auto-sklearn · Hyperparameter optimization · Data fusion · Combination of interpretation · Prediction models

1 Introduction

Machine learning (ML) has revolutionized many aspects of computer applications such as computer vision, speech recognition, and gaming during the past decade. Much progress has been made in developing deep learning models and automated ML models. However, there is still need for expert knowledge that is required to implement the "shallow" decision models successfully.

© Springer Nature Switzerland AG 2022
S. Chiusano et al. (Eds.): ADBIS 2022, LNCS 13389, pp. 301–314, 2022.
https://doi.org/10.1007/978-3-031-15740-0_22

To understand the benefits of AutoML, the best way is to start by considering the fundamental limitations and issues of ML models. It is a well-known fact that ML models are often "hand-crafted" - as they are built on an ad-hoc basis to resolve a specific problem [5]. It takes much consideration to decide whether certain steps such as uncertainty estimation and missing data imputation are necessary or not. Also, what type of feature selection and classifier should be used depending on the data characteristics. ML engineers or data scientists must carefully select the accurate ML model and optimization processes (applying different parameters and hyperparameters to a model and choosing the one with the best outcome) for all these components to achieve the desired results using the designed ML model. The Regularization process requires episodes of trial and error until a good choice of parameters for a particular problem is obtained [22].

Automated ML (AutoML) is designed to build suitable ML models in an automated way based on the data characteristics. The user only provides data for a specific problem to the process. The AutoML determines the most suitable pre-processing, feature selection and classification algorithm with its parameters and hyperparameters that gives the best outcomes for the problem at hand. There is no doubt that AutoML has freed the data scientist from the tedious task of determining the best configuration for the ML pipeline and has made their work easier and more reliable [11].

Auto-Weka tool considered the problem of simultaneously selecting an ML algorithm and optimizing its hyperparameters. They dubbed this problem into the combined algorithm selection and hyperparameter optimization (CASH) problem [22]. In this particular tool, typically, four decisions need to be made to construct the whole AutoML pipeline for the CASH problem [11]:

1. Pre-processing algorithm to be used,
2. Feature engineering approach,
3. Learning algorithm to be used,
4. Appropriate values for parameters and hyperparameters.

A detailed description of parameters and hyperparameters is as follows:

1. Parameters: These parameters are optimized during the training of the model and,
2. Hyperparameters: Hyperparameters are the parameters that the user defines to control the behavior of the learning algorithm. The values of these hyperparameters are set before the learning processing begins.

One of the main challenges in AutoML is that configuration space is huge for hyperparameter and parameter selection. We can explore certain configuration space areas, but findings these specific combinations (hyperparameters and parameters) is challenging. Also, it is often unclear which hyperparameter of an algorithm needs to be optimized and in which range. Moreover, in current AutoML techniques, the objective kept in mind while building such AutoML pipelines is to optimize the prediction performance. However, this results in neglecting the computation cost that it carries along.

To address the issue of computational cost, Auto-Scikit was introduced. Auto-sklearn relates closely with Auto-Weka as it uses the same bayesian optimization scheme. However, to speed up the optimization performance, the AutoML tool employs the concept of meta-learning. It collects information from similar datasets, which helps construct the AutoML pipeline relatively quickly compared to Auto-Weka.

AutoML provides a way to democratize the use of ML so it can be within reach of the non-ML expert [8]. Major tech companies are now implementing their AutoML tools. After taking a look into the challenge of the computational cost associated with AutoML tools, it can be said that there is still much need for improvements.

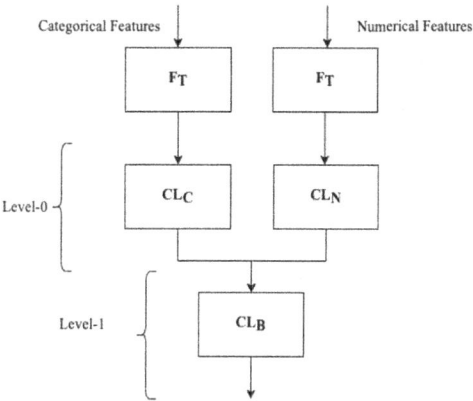

Fig. 1. Data fusion S-COI model

This paper presents an alternative approach to an AutoML tool (auto-sklearn), which helped us build more effective ML pipelines by consuming less time and producing higher macro accuracy. In our approach, we have used the concept of data fusion (DF) [7] which handles diversified data sources (data present in different formats). The DF approaches have been tested in [7] in a specific clinical problem where different sources corresponded to results of lab tests, physical examinations, etc. From these approaches we have selected a simple combination of interpretation model (S-COI) to construct the ML pipeline for our current research. For this specific DF process, we have a limitation to work with mixed structure datasets (feature set with numerical and categorical values). Minimal pre-processing is done for each numerical (missing value imputation and normalization) and categorical features (missing value imputation and one-hot encoding). This pre-processing layer in the DF process is labeled as feature transformation layer. The idea behind the DF process is that we build a separate ML pipeline for numerical and categorical features, and combine the outcome by using a base classifier. A limited number of classifiers are used to construct a DF pipeline with a fixed set of hyperparameters and parameters.

The study aims to provide an alternative to the present AutoML tool by considering DF as a simple form of AutoML, which provides an optimized ML pipeline with less computational cost. We have compared the ML pipeline built using auto-sklearn (which uses the same Python library, i.e., scikit learn) with the DF pipeline. Our DF approach gives better macro average accuracy for almost all datasets in a shorter time. The parameters and hyperparameters for the DF pipelines used in our approach are set to specific values. They remain consistent throughout the DF process for the datasets mentioned in Sect. 1 (i.e., we do not perform hyperparameter optimization).

The paper is structured as follows, in the next section we discuss the related work done with respect to AutoML process and tools. After that we present the problem statement from Sect. 3. In Sect. 4 we explained the proposed approach in detail which includes the description of datasets in Sect. 4.1, DF process in Sect. 4.2, and experimental design in Sect. 4.3. After discussing the proposed approach in detail, we performed performance evaluation using statistical analysis in Sect. 4.4. In the last Section, we conclude the paper and provide a glimpse of the future work.

2   Related Work

The most challenging task that appears within AutoML is the optimization of hyperparameters, and various approaches are used to configure these parameters for learning algorithms.

The very first techniques for hyperparameter optimization were greedy depth analysis [15] and pattern analysis [17], both of which were developed to improve the standard configuration of hyperparameters.

Later, grid analysis was introduced, which was also used to optimize hyperparameters since the 1990s [12] and was implemented in ML tools [16]. In grid search, the user defines a finite set of values for each hyperparameter under consideration - then the Cartesian product of these values is established. The major drawback of the grid search approach was that it suffers from the curse of dimensionality, exponentially growing the search space as the number of hyperparameters were increased. An alternative to grid search is the random search approach that was described by James Bergstra et al. in [1]. In this alternative approach the hyperparameters values are searched at random until the search budget is ended [5,9,10].

In 2009, Escalante et al. [3] extended the hyperparameter optimization (HPO) work to the full model selection problem. This method includes the selection of pre-processing, feature selection, and learning algorithms and all their hyperparameters. The authors constructed an ML pipeline by utilizing multiple ML algorithms using hyperparameter optimization and empirically found that the created models were comparable to those created using expert knowledge; hence no domain knowledge was required to apply their method to any dataset. The first AutoML tool that we have considered initially was Auto-WEKA [19]. Auto-WEKA is an automated tool built using WEKA libraries and uses sequential model-based algorithm configuration (SMAC) to resolve the CASH problem. It identifies the automated ML pipeline along with parameters and hyperparameters of algorithms. The automated ML pipeline constructed using Auto-WEKA consists of pre-processing data and selection of ML algorithm (parameters and hyperparameter optimization). However, there was no feature engineering present there. In [18], Silva et al. detected the presence of breast cancer among patients with time-series information and used Auto-WEKA to construct a classifier.

In 2015, Feurer et al. [6] presented an alternative approach to Auto-WEKA. They developed an automated ML pipeline called auto-sklearn using the scikit-learn (sklearn) library. It contains some feature engineering as it calculates meta-features from the datasets; these meta-features provide add-on information while selecting the different parts of the ML pipeline (pre-processing, ML algorithm selection, hyperparameter optimization). In the end, the automated ML tool creates ensemble classifiers using the selected best ML models.

Kietz et al. [13] presented the first approach for the configuration of Rapid Miner modules using hierarchical task networks (HTN) planning. The algorithm is driven by a ranking obtained from the frequency of users' usage of the Rapid-Miner tool. ML Plan is another automated ML tool that uses HTN planning. In ML plan, AutoML is reduced to a graph search problem. More specifically, ML Plan uses a best-fit search algorithm on the graph stimulated by a forward decomposition procedure of the HTN planning problem. ML Plan splits the AutoML problem into algorithm selection and algorithm configuration issues. The ML Plan process starts with a fixed set of pre-processing algorithms, classification methods, and the associated parameters. In the first phase, a pre-processing algorithm and classification algorithm are selected and in the second phase they are configured.

However, the approach we will be presenting in this paper is more straightforward and is based on our previous research work completed in [7], based on DF process (a process of "fusing" data coming from diversified sources). During experimental analysis we found out that DF is a competitor for AutoML in case when we have diversified data sources. DF models can be of two types which are combination of data (COD), and combination of interpretation (COI) approaches. These approaches are defined as follows:

1. COD assumes that all the extracted features from multiple data sources are initially aggregated into a uniform data space (we refer to this phase as to aggregation). A single classifier (base classifier) is constructed from this space.
2. COI assumes that a separate classifier is constructed for every data source. All individual outcomes are then subject to the aggregation carried out by a "combiner." The latter can be regarded as a base classifier that generates a final decision or outcome. COI resembles an ensemble of classifiers (in particular, the stacking scheme). However, the difference is that there is no feature transformation layer in the ensemble.

We concluded that it would be best to use the COI model if we have datasets coming from different modalities (different sources, i.e., images, laboratory reports, and results from physical examination). Also, these DF models perform well with an imbalanced class distribution.

3   Problem Statement

As discussed before, the current AutoML tools available consume a lot of time in creating an optimized ML pipeline. Even after using hours of allocated time to build an optimized pipeline, there is no guarantee that the presented ML pipeline would be the best. This restricts the ML engineers and data scientists to use AutoML tools in routine work. In our paper, we focuse on specific AutoML tool (auto-sklearn). Auto-sklearn model should be executed for at least 90 min to produce an optimal ML pipeline. However, the presented pipeline is not always the best possible solution. As auto-sklearn is based on bayesian optimization and meta learning approach it takes time to find the best pre-processing, ML algorithm selection, hyperparameter for classifiers used, which results in current time consumption.

The time constraint is not the only downside; prolonged run time means high resource utilization and significant power consumption, which results in high ecological cost.

To these above mentioned challenges in AutoML, we proposed an alternative methodological solution, i.e., using a DF model to create a ML pipeline, which acts as a simplified AutoML process. We have implemented the DF solution using sklearn library and for consistency we have performed the comparison to an AutoML (auto-sklearn) tool which is also built using sklearn.

4   Proposed Approach

In this section, we will explain the proposed approach in detail. The idea behind constructing a DF ML pipeline comes from our previous paper [7] and this paper aims to compare the performance of the decision models (based on macro average accuracy) and DF process (based on execution time) with auto-sklearn.

The experimental flow consists of a DF process and an auto-sklearn process. Only one process can be activated at a time as shown in the Fig. 2. Initial feature pre-processing (feature transformation) steps are similar for both processes. As stated earlier datasets consist of mixed structure feature set (numerical and categorical), we apply different feature transformation schemes based on feature characteristics, i.e.,

1. For categorical features, we perform missing value imputation (replacing categorical attribute values with the most frequently occurring) and later apply one-hot encoding (as ML algorithm present in sklearn do not deal with categorical attributes so the transformation was necessary).

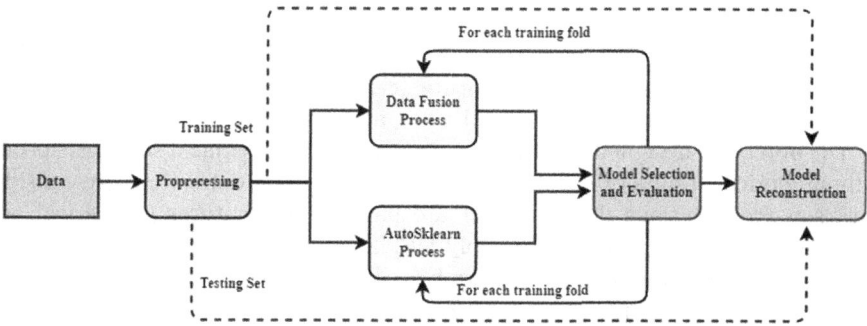

Fig. 2. Experimental flow

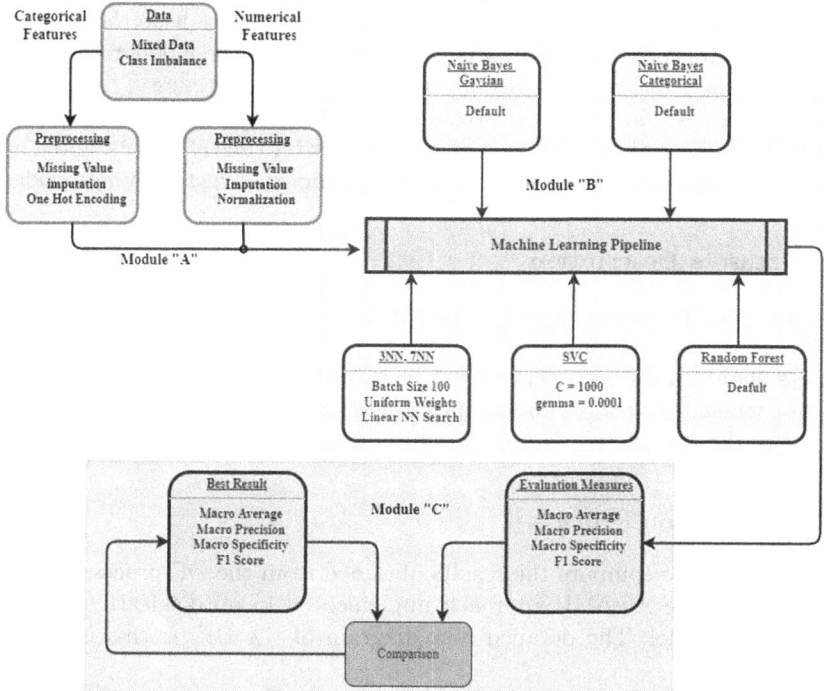

Fig. 3. Data fusion pipeline

2. For numerical features, we use median value imputation for missing values and later apply normalization (normalization is applied so all numerical features have similar range and cannot effect the model performance).

Once the transformation is done, separate pipelines are constructed for autosklearn process and DF process.

For auto-sklearn, pipeline is constructed after the model parameter settings are configured, more details are provided in Sect. 4.3. For DF separate pipelines are constructed for each type of feature (categorical and numerical), and the outcome of each pipeline is combined using a base classifier.

DF in [7] was designed for diversified data sources in clinical settings, specifically numerical results of lab tests and other categorical data. Here, we consider data sets from various domains that are have mixed features (numerical and categorical) – in this way we can apply DF. The limitation is that our approach was developed for a clinical problem where data format was associated with its "meaning" and we do not have it here.

When evaluating the performance of DF and auto-sklearn process, we considered the macro-averages of the following unique measures: accuracy equal to mean sensitivity, precision, specificity, and F1 score. The macro-averaged measure is often used in the context of imbalanced data sets as a better alternative to overall accuracy [4]. The comparison and selection of the model rely on macro average accuracy; other measures are reported for completeness.

The paper also aims to demonstrate the benefits of the DF process over auto-sklearn. These benefits are better performance, better time efficiency, and a much simpler pipeline for DF than the complex pipeline constructed by auto-sklearn.

4.1   Datasets Description

We have worked on ten datasets collected from UCI and Kaggle repository, which are publicly available[1,2]. The limitation of the data selection was that the datasets should always contain mixed features, i.e., numerical and categorical features, which restricts us in the data selection process. The detailed characteristics of the considered datasets are provided in Table 1.

4.2   Data Fusion Process

In this section, we compare the results obtained from the DF process presented in Fig. 1 with the AutoML approach implemented in auto-sklearn. Figure 1 is a basic COI model. The detailed flow diagram of the DF process is shown in Fig. 3.

We apply 5-fold cross validation (cv) externally. By following COI model, we build stacking ensemble classifiers and in order to choose the best ensemble we employ 10-fold cv scheme internally.

The DF process shown in Fig. 3 is divided into three major parts. Each one of them is explained in detail below:

---

[1] UCI repository https://archive.ics.uci.edu/ml/index.php.

[2] Kaggle repository https://www.kaggle.com/datasets.

Table 1. Characteristics of datasets considered in the study

Dataset	Instances	Features	Categorical features	Numerical features	No. of classes	Data set source
Auto-University (AU)	2500	100	42	58	3	UCI
Student Performance (SP)	480	16	12	4	3	UCI
Breast Cancer(BC)	286	10	8	2	2	UCI
Cryotherapy (CRYO)	90	6	3	3	2	UCI
Census Income (CENSUS)	45211	16	9	7	2	UCI
Insurance (INS)	1338	6	4	2	2	Kaggle
Heart Disease (HEART)	270	13	5	8	2	UCI
Diabetes (DIA)	8973	48	32	16	3	UCI
Contraceptive Method Choice (CMC)	1473	9	4	5	3	UCI
Hepatitis (HEP)	155	19	13	6	2	UCI

1. Module A: Consists of features transformation part which is applied on categorical and numerical attributes present in data. For each type of attribute a separate set of pre-processing schemes is used as mentioned in Sect. 4.2.
2. Module B: Five different classifiers are present in the pool of available classifiers with a set of hyperparameters defined (see Table 2). When pre-processed input is passed to this module, it builds a DF model Fig. 1 using the different combinations of classifiers that are used at level-0 (CL_C and CL_N) and level-1 (CL_B) as shown in Fig. 1. In the next step, the evaluation measures are calculated.
3. Module C: The results for each run are compared with the previously stored maximum result (macro average accuracy) obtained using a different combination of classifiers. If the newly acquired result is more than the previous result, the pre-stored maximum values are overwritten by the new outcomes, and the respective pipeline is selected.

4.3   Experimental Design

In this section, we will discuss the experimental setup for DF and auto-sklearn models. The auto-sklearn model is configured using the following settings:

1. The per runtime limit is set to 5400 s (90 min).
2. As the datasets are already pre-processed, so the feature pre-processing option is set to no pre-processing.
3. The Resampling strategy is set to cv with 10 folds internal cv.

For each outer folds (5-folds cv) the auto-sklearn is executed for 90 min uses 10-fold cv on each inner fold. Once the best ensemble combination has been identified, the model is reconstructed from the entire learning set and evaluated using the testing datasets.

The DF models are constructed by combing different types of classifiers into stacking ensembles. Classifiers employed in our study are described in Table 2; the selection of classifiers are based on our experience with analysis of clinical data [21], and on the results of other studies related to data fusion [20]. However, in our current study we have also considered datasets from different domains and applied the same set of classifiers on them. We used default values for most of the classifiers parameters and hyperparameters (NB-G, NB-C and RF), as they performed well on default parameter settings. For remaining (KNN and SVM) we have used the outcome of the grid search performed in [7], to find the best possible values of C (cost) and γ (hyperparameters) for SVM. We have considered the same values which were used on transformed features. For the KNN classifier, we checked values of k ranging from 3 to 11 and observed the best performance for k 3 and 7 neighbors. This range of the parameter k was defined as arbitrary based on the literature, and our previous experience with other data sets [21]. The respective parameter and hyperparameters for each classifier used in DF process are mentioned below:

Table 2. Classifiers considered in the paper and their parameters (scikit-learn)

Symbol	Description	Parameters
KNN	A k-nearest neighbor classifier with Euclidean distance	$k = 3, 7$
NB-G, NB-C	A naive Bayes classifier where distribution is Gaussian and Categorical provided in sklearn library with default parameters	default parameters in sklearn
RF	A random forest classifier	default parameters in sklearn
SVM	An SVM (called SVC in Python) classifier with a radial basis kernel function	$C = 1000, \gamma = 0.0001$

For evaluation of the ML pipelines, aside from macro accuracy, we have also considered computational time and size of the stored model to measure model complexity. The computational time is the total time consumed (during external 5-fold cv) by a given AutoML or DF pipeline to provide the final output.

4.4   Performance Evaluation

By comparing the output obtained using the suggested DF approach shown in Table 3 and using auto-sklearn shown in Table 4, we can indicate that in

Table 3. Results obtained for DF (all values of evaluation measures are macro averages)

Data set	CL_C	CL_N	CL_B	Evaluation measures [%]			
				Accuracy	Precision	Specificity	F1 measure
AU	NB-C	RF	RF	41.0	37.0	72.0	39.0
SP	7NN	SVM	RF	83.0	82.4	68.0	78.0
BC	3NN	RF	RF	52.3	69.0	92.3	59.5
CRYO	NB-G	SVM	1NN	92.0	92.0	93.0	92.0
CENSUS	3NN	RF	NB-G	73.0	69.7	90.0	71.2
INS	NB-G	RF	1NN	68.2	68.3	67.7	68.5
HEART	3NN	NB-G	RF	85.0	86.2	91.0	85.5
DIA	NB-G	SVM	NB-G	42.0	37	72.5	38.2
CMC	NB-C	7NN	RF	58.6	45.7	60.8	51.6
HEP	7NN	NB-G	RF	89.1	86.8	44.4	87.9

Table 4. Results obtained for AutoML for the learning time of 90 min (all values of evaluation measures are macro averages)

Data set	Evaluation measures [%]			
	Accuracy	Precision	Specificity	F1 measure
AU	40.1	38.7	76.0	39.3
SP	78.6	80.4	87.6	80.3
BC	43.0	40.0	60.3	41.3
CRYO	92.0	85.7	85.7	88.9
CENSUS	77.4	36.0	84.2	49.1
INS	50.2	65.2	72.0	57.0
HEART	81.2	79.8	81.4	80.0
DIA	43.0	36.2	72.3	36.5
CMC	56.4	55.9	77.9	55.5
HEP	81.0	93.7	77.7	87.2

most cases, the DF process performs better, as we create separate pipelines for categorical and numerical features. We use different pre-processing schemes and different classifier based on the type of attributes. The running time is kept consistent, i.e., 5400 s for all the datasets when running auto-sklearn.

The comparison between both approaches concerning time and model size is presented in the Table 5.

We considered calculating computational complexity based on the time required for a given scheme (DF or auto-sklearn) and on the size of the models generated by both schemes. Model size depends on the complexity of type of pre-processing, feature selection and classification algorithms used. Conclusions from results presented in Table 5 are summarized below:

1. Out of 10 datasets for 7 datasets (AU, SP, BC, INSURANCE, HEART, CMC, and HEP) DF process performed better with higher macro average accuracy as compared to auto-sklearn.

Table 5. Computational complexity obtained for auto-sklearn and Data Fusion (DF) process

Data set	Auto ML		DF	
	Running time (sec)	Size (MB)	Running time (sec)	Size (MB)
AU	54000	777	3.31	0.12
SP	54000	305	1.00	0.10
BC	54000	162	4.21	0.10
CRYO	54000	300	2.89	0.04
CENSUS	54000	870	168.55	0.51
INSURANCE	54000	248	3.33	0.10
HEART	54000	236	1.00	0.07
DIA	54000	300	60.20	0.98
CMC	54000	427	4.35	0.06
HEP	54000	109	1.00	0.041

2. For one dataset, i.e., CRYOTHERAPY, the obtained macro average accuracy was the same using both processes.
3. For 2 datasets (DIABETES and CENSUS), the macro average accuracy obtained using auto-sklearn was better than for the DF. approach.

From the results shown in Table 3, 4 and 5 we can conclude that DF performance exceeds for majority of datasets. The only datasets where the performance in terms of macro-accuracy is less than auto-sklearn are DIABETES and CENSUS. However, the difference in macro-accuracy is not large (only 4.4% for CMC and only 1% more for DIABETES).

To check if there is a statistically significant difference (in terms of macro avg) between DF and auto-sklearn approaches, we used a two-tailed Wilcoxon test as recommended by [2]. We were able to reject the null hypothesis at $\alpha = 0.05$ ($W_{crit} = 5.9$, $W_{stat} = 8$), thus confirming the difference between auto-sklearn and data fusion.

5   Conclusion and Future Work

This paper has compared two AutoML processes – a simple form of AutoML (DF process) and a more complex one (auto-sklearn). For majority of datasets the DF process produces better results which high macro average accuracy as compared to AutoML. However, for only a few (two datasets), the performance of DF models was lower.

Following are the lesson learned from our experimentation.

– From the performance comparison between the two approaches, we can conclude that the DF process works better than auto-sklearn for seven datasets.

- From the results obtained in Sect. 4.4, we have observed that the performance of auto-sklearn is slightly better on two datasets (DIABETES and CMC) as compared to DF models. However, we should keep in mind that the results obtained through auto-sklearn consumed 90 min of computational time before producing the outcome, which is higher than the time consumed while using DF models (less than 3 min).
- From the comparison of model complexity, we can conclude that DF models produce results quickly and consume way less memory and time as compared to auto-sklearn models, as shown in Sect. 4.4.
- Statistical results show compelling proof that the two approaches do not produce similar outcomes, and the difference between these two approaches is statistically significant.

For future work, we would like to extend the experiment to more datasets; we want to implement the concept of a mixture of experts (MoE) [14], which will help us automate classifier selection. It would be also an interesting to identify guidlines on when it is better to use DF than full AutoML (whether each are suited for a particular class problems).

Acknowledgement. This research has been funded by the European Commission through the Erasmus Mundus Joint Doctorate - "Information Technologies for Business Intelligence - Doctoral College" (IT4BI-DC).

References

1. Bergstra, J., Bengio, Y.: Random search for hyper-parameter optimization. J. Mach. Learn. Res. **13**(10), 281–305 (2012). http://jmlr.org/papers/v13/bergstra12a.html
2. Demšar, J.: Statistical comparisons of classifiers over multiple data sets. J. Mach. Learn. Res. **7**, 1–30 (2006)
3. Escalante, H.J., Montes, M., Sucar, L.E.: Particle swarm model selection. J. Mach. Learn. Res. **10**(15), 405–440 (2009). http://jmlr.org/papers/v10/escalante09a.html
4. Ferri, C., Hernández-Orallo, J., Modroiu, R.: An experimental comparison of performance measures for classification. Pattern Recogn. Lett. **30**(1), 27–38 (2009). https://doi.org/10.1016/j.patrec.2008.08.010
5. Feurer, M., Hutter, F.: Hyperparameter optimization. In: Hutter, F., Kotthoff, L., Vanschoren, J. (eds.) Automated Machine Learning. TSSCML, pp. 3–33. Springer, Cham (2019). https://doi.org/10.1007/978-3-030-05318-5_1
6. Feurer, M., Klein, A., Eggensperger, K., Springenberg, J., Blum, M., Hutter, F.: Efficient and robust automated machine learning. In: Advances in Neural Information Processing Systems, vol. 28, pp. 2962–2970 (2015)
7. Haq, A., Wilk, S., Abelló, A.: Fusion of clinical data: a case study to predict the type of treatment of bone fractures. Int. J. Appl. Math. Comput. Sci. **29**(1), 51–67 (2019). https://doi.org/10.2478/amcs-2019-0004
8. He, X., Zhao, K., Chu, X.: AutoMl: a survey of the state-of-the-art. Knowl. Based Syst. **212**, 106622 (2021)

9. Hutter, F., Hoos, H., Leyton-Brown, K.: An efficient approach for assessing hyper-parameter importance. In: Xing, E.P., Jebara, T. (eds.) Proceedings of the 31st International Conference on Machine Learning. Proceedings of Machine Learning Research, Bejing, China, 22–24 June 2014, vol. 32, pp. 754–762. PMLR (2014). https://proceedings.mlr.press/v32/hutter14.html

10. Hutter, F., Hoos, H.H., Leyton-Brown, K.: Sequential model-based optimization for general algorithm configuration. In: Coello, C.A.C. (ed.) LION 2011. LNCS, vol. 6683, pp. 507–523. Springer, Heidelberg (2011). https://doi.org/10.1007/978-3-642-25566-3_40

11. Hutter, F., Kotthoff, L., Vanschoren, J. (eds.): Automated Machine Learning: Methods, Systems, Challenges. TSSCML, Springer, Cham (2019). https://doi.org/10.1007/978-3-030-05318-5

12. John, G.H.: Cross-validated c4.5: using error estimation for automatic parameter selection. Technical report, Stanford University (1994)

13. Kietz, J.U., Serban, F., Bernstein, A., Fischer, S.: Towards cooperative planning of data mining workows. In: Proceedings of the Third Generation Data Mining Workshop at the 2009 European Conference on Machine Learning (ECML 2009) (2009). https://doi.org/10.5167/uzh-25868

14. Kim, Y.J., et al.: Scalable and efficient MoE training for multitask multilingual models. CoRR abs/2109.10465 (2021). https://arxiv.org/abs/2109.10465

15. Lee, H.K.H., Gramacy, R.B., Linkletter, C., Gray, G.A.: Optimization subject to hidden constraints via statistical emulation. Pacific J. Optim. 7(3), 467–478 (2011). https://scholar.google.com/scholar?cluster=6376843206533956665

16. Madrid, J.G., et al.: Towards automl in the presence of drift: first results. CoRR abs/1907.10772 (2019). http://arxiv.org/abs/1907.10772

17. Momma, M., Bennett, K.P.: A pattern search method for model selection of support vector regression. In: SDM (2002)

18. Silva, L.F., Santos, A.A.S., Bravo, R.S., Silva, A.C., Muchaluat-Saade, D.C., Conci, A.: Hybrid analysis for indicating patients with breast cancer using temperature time series. Comput. Methods Programs Biomed. 130, 142–153 (2016). https://doi.org/10.1016/j.cmpb.2016.03.002. https://www.sciencedirect.com/science/article/pii/S0169260715300249

19. Thornton, C., Hutter, F., Hoos, H.H., Leyton-Brown, K.: Auto-weka: combined selection and hyperparameter optimization of classification algorithms. In: Proceedings of the 19th ACM SIGKDD International Conference on Knowledge Discovery and Data Mining, pp. 847–855. ACM (2013). https://arxiv.org/pdf/1208.3719

20. Tiwari, P., Viswanath, S.E., Lee, G., Madabhushi, A.: Multi-modal data fusion schemes for integrated classification of imaging and non-imaging biomedical data. In: IEEE International Symposium on Biomedical Imaging (ISBI), pp. 165–168 (2011)

21. Wilk, S., Stefanowski, J., Wojciechowski, S., Farion, K.J., Michalowski, W.: Application of preprocessing methods to imbalanced clinical data: an experimental study. In: ITIB (2016)

22. Yao, Q., et al.: Taking human out of learning applications: a survey on automated machine learning (2019)

Dimensional Data KNN-Based Imputation

Yuzhao Yang[1]([⊠]) , Jérôme Darmont[2] , Franck Ravat[1] ,
and Olivier Teste[1]

[1] IRIT-CNRS (UMR 5505), Université de Toulouse, Toulouse, France
{Yuzhao.Yang,Franck.Ravat,Olivier.Teste}@irit.fr
[2] Université de Lyon, Lyon 2, UR ERIC, Lyon, France
jerome.darmont@univ-lyon2.fr

Abstract. Data Warehouses (DWs) are core components of Business Intelligence (BI). Missing data in DWs have a great impact on data analyses. Therefore, missing data need to be completed. Unlike other existing data imputation methods mainly adapted for facts, we propose a new imputation method for dimensions. This method contains two steps: 1) a hierarchical imputation and 2) a k-nearest neighbors (KNN) based imputation. Our solution has the advantage of taking into account the DW structure and dependency constraints. Experimental assessments validate our method in terms of effectiveness and efficiency.

Keywords: Data imputation · Data warehouses · Dimensions · KNN

1   Introduction

Data warehouses (DWs) are widely used in companies and organizations as a significant Business Intelligence (BI) tool to help them building their decision support systems. Data in DWs are usually modelled in a multidimensional way, which allows the user to analyse data through On Line Analytical Processing (OLAP). An OLAP model organizes data according to analysis subjects (facts) associated to analysis axis (dimensions). Each fact is composed of measures. Each dimension contains one or several analysis viewpoints (hierarchies).

Missing data may exist in a DW. There are 2 types of DW missing data: **dimensional missing data** which are missing data in the dimensions and **factual missing data** which are in the facts. These missing data have impact on OLAP analyses. It is important to complete the missing data for the sake of a better data analysis.

Data imputation is the process of replacing the missing values by some plausible values based on information available in the data [12]. The current DW data imputation research mainly focuses on factual data [4,21,25]. Yet the dimensional missing data make aggregated data incomplete and make it hard

This work is supported by the French National Research Agency (ANR), Project ANR-19-CE23-0005 BI4people (Business intelligence for the people).

S. Chiusano et al. (Eds.): ADBIS 2022, LNCS 13389, pp. 315–329, 2022.
https://doi.org/10.1007/978-3-031-15740-0_23

to analyse them with respect to hierarchy levels. Therefore the imputation for DW dimensions is also necessary. However the DW dimension has a complex structure containing different hierarchies with different granularity levels having their dependency relationships. When we complete the dimensional missing data, we have to take the DW structure and the dependency constraints into account. We proposed a hierarchical imputation based on the inter- and intra-dimensional hierarchical dependency relationships [27] for the imputation of dimensional missing data. To the best of our knowledge, there is no other specific data imputation method for DW dimensions. The hierarchical imputation is convincible because we use accurate data based on real functional dependency relationships. However, this method is limited owing to the sparsity problem which means that for an instance to be completed, there may not be an instance sharing the same value on a lower-granularity level of the hierarchy.

In order to complete as many values as possible, in this paper, we propose H-OLAPKNN, an imputation method for DW dimensions by extending the hierarchical imputation with a novel dimension imputation method called OLAPKNN. OLAPKNN is based on K-nearest neighbours (KNN) algorithm. KNN imputation finds the K nearest neighbors of an instance with missing data then fills in the missing data based on the mean or mode of the neighbors' value [23]. We choose KNN because it is a non-parametric and instance-based algorithm, which is widely applied for data imputation [3] and has been proved to have relatively high accuracy [2,23]. Compared to the basic KNN imputation, OLAPKNN considers the structure complexity and the dependency constraints of the dimension hierarchies. Moreover, the dimensional data are usually qualitative on which we focus in this paper.

The remainder of this paper is organized as follows. In Sect. 2, we review the related work about data imputation algorithms. In Sect. 3, we formalize the DW dimension model. In Sect. 4, we propose a distance calculation method for dimension instances. In Sect. 5, we explain in detail our proposed dimension imputation algorithm. In Sect. 9, we validate our proposal by some experiments. In Sect. 7, we conclude this paper and hint at future research.

2   Related Work

There are various data imputation methods [16]: statistic based imputation, machine-learning based imputation, rule based imputation, external source based imputation and hybrid methods etc. The statistic based imputation completes the missing values by applying the statistical methods like filling average, the most frequent value or with the value of the most similar record; there are also methods using the regression to predict the missing values [19]. The machine learning based imputation methods use algorithms like k-nearest neighbor (KNN) [2,10,17,23], regression models [13], Naive Bayes [9] to predict the missing values. The rule based imputation methods [5,8,22] complete the missing values by some business rules, similarity rules or dependency rules. Concerning the external source based methods, the crowdsourcing [14] can be applied for the data imputation by putting forward the queries in the crowdsourcing frameworks and collecting answers to complete the missing data. There are also methods which realize the imputation through web information [26,29] like web pages, web lists and web tables. What's more, there are hybrid methods which mix different imputation methods to provide a higher performance.

The statistic and machine learning based methods mainly focus on the numerical data, which fit for the imputation of facts where the data are mostly numerical. However, in the dimensions, there are mainly qualitative data which make it difficult to process the data imputation by such imputation methods. The rule based and external source based imputation methods may be suitable for the imputation of dimensions, but they need time and efforts to create rules or find the appropriate sources. Hence we propose H-OLAPKNN which combines the hierarchical imputation with a KNN-based imputation method.

3   DW Dimension

As a DW is composed of dimensions and facts and we focus on the dimension imputation, we introduce the DW dimension concepts used in this paper [20].

Definition 1 (Dimension). *In a data warehouse, a **dimension**, denoted by D, is defined as (A^D, H^D, I^D). $A^D = \{a_1, ..., a_u\} \cup \{id\}$ is a set of attributes, where id represents the dimension's identifier; $H^D = \{H_1, ..., H_v\}$ is a set of hierarchies; I^D is a matrix of dimension instances, for a given row r, the row instance vector is denoted as i_r; for a given attribute a_u, their joint instance value is denoted as i_{r,a_u}.*

Definition 2 (Hierarchy). *A **hierarchy** of dimension D, denoted by $H \in H^D$, is defined as $(Param^H, Weak^H)$. $Param^H =< id^D, p_2^H, ..., p_v^H >$ is an ordered set of dimension attributes, called **parameters**, which set granularity levels along the dimensions, $\forall k \in [1...v], p_k^H \in A^D$. Parameter p_1^H rolls up to p_2^H in H is denoted as $p_1^H \preceq_H p_2^H$; $Weak^H = Param^H \rightarrow 2^{(A^D - Param^H)}$ is a mapping possibly associating each parameter with one or several **weak attributes**, which are also dimension attributes providing additional information; All parameters and weak attributes of H constitute the hierarchy attributes of H, denoted by $A^H = Param^H \cup (\bigcup_{p_v^H \in Param^H} Weak^H[p_v^H])$*

There exists different types of hierarchy, but the most basic and common one is the strict hierarchy [15] where a value at a hierarchy's lower-granularity belongs to only one higher-granularity value [24]. Thus in this paper, we only consider the case of the strict hierarchy.

4   Distance Between Dimension Instances

Since the KNN imputation select the k-nearest neighbors of the missing data instance for the imputation, we should calculate the distance between dimension instances containing missing data to be completed and other instances. In a dimension D, for an instance $i_1 \in I^D$ containing missing data on a hierarchy $H_1 \in H^D$, and another instance $i_2 \in I^D$, we propose to calculate their distance by 4 levels:

- The **dimension instance distance** is the final distance between two instances i_1 and i_2, denoted by $\Delta(i_1, i_2)$. Since the attributes on the same hierarchy have their dependency relationships, we consider the attributes of a hierarchy as an entirety. $\Delta(i_1, i_2)$ is thus calculated by the weighted sum of the **hierarchy instance distances**.
- The **hierarchy instance distance** is the distance of the attributes of a hierarchy $H_2 \in H^D$ i.e. distance between $\{i_{1,a_1} \in i_1 : a_1 \in A^{H_2}\}$ and $\{i_{2,a_1} \in i_2 : a_1 \in A^{H_2}\}$, denoted by $\Delta_{H_2}(i_1, i_2)$. It is calculated by the weighted sum of the **hierarchy level instance distances**. The lowest-granularity level of each hierarchy is the same i.e. id with its weak attributes, so we consider the hierarchy instance distance from the second level of the hierarchy and we regard each weak attribute of id as a hierarchy containing only one parameter.
- The **hierarchy level instance distance** is the instance distance between the attributes of a level l on a hierarchy H_2 i.e. distance between $\{i_{1,a_2} \in i_1 : a_2 \in p_l^{H_2} \cup Weak^{H_2}[p_l^{H_2}]\}$ and $\{i_{2,a_2} \in i_2 : a_2 \in p_l^{H_2} \cup Weak^{H_2}[p_l^{H_2}]\}$, denoted by $\Delta_{p_l^{H_2}}(i_1, i_2)$. It is calculated by the average of the instance distances of the level's parameter and weak attributes (**attribute distances**).
- The **attribute distance** is the instance distance of an individual attribute $a_u \in A^D$ i.e. distance between i_{1,a_u} and i_{2,a_u}, denoted by $\Delta(i_{1,a_u}, i_{2,a_u})$.

Based on the explanation of the distances, we then give the formulas and some examples to illustrate them in detail.

Example 1. Given a dimension *Product* containing two hierarchies H_1 and H_2 whose schema and instances are shown in Fig. 1. Instance i_1 contains missing values on H_1, Fig. 2 shows the calculation of the distance $\Delta(i_1, i_2)$ between i_1 and another instance i_2.

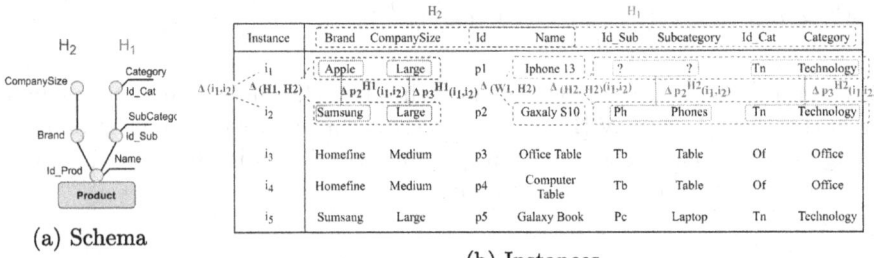

(a) Schema

(b) Instances

Fig. 1. Schema and instances of dimension *Product*

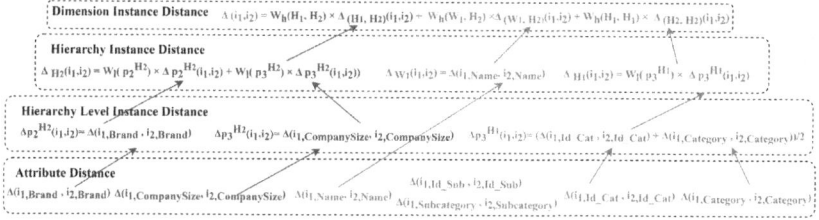

Fig. 2. Distance between i_1 and i_2

4.1 Attribute Distance

There are different attribute data types which can be mainly classified into numerical and textual. For numerical data, we use normalized distance of numerical data [1]. For textual data, we first apply semantic distance e.g. cosine distance based on word2vec [11]. If the attribute value cannot be found in the model, we can then use the syntactic distance e.g. normalized Levenshtein Distance [28].

For an attribute a_{u1}, if $i_{1,a_{u1}}$ is missing, then $\Delta(i_{1,a_{u1}}, i_{2,a_{u1}})$ cannot be calculated and is not taken into count for the distance calculation. For an attribute a_{u2}, if $i_{2,a_{u2}}$ is missing, then $\Delta(i_{1,a_{u2}}, i_{2,a_{u2}})$ is obtained by the average distance between $i_{1,a_{u2}}$ and other instances whose value of a_{u2} is not missing.

Example 2. Following the calculation rules of the attribute distance, we obtain $\Delta(i_{1,Brand}, i_{2,Brand}) = 0.71$, $\Delta(i_{1,CompanySize}, i_{2,CompanySize}) = 0$, $\Delta(i_{1,Name}, i_{2,Name}) = 0.8$, $\Delta(i_{1,Id\_Cat}, i_{2,Id\_Cat}) = 0$, $\Delta(i_{1,Category}, i_{2,Category}) = 0$. Since i_{1,Id_Sub} and $i_{1,Subcategory}$ are missing, $\Delta(i_{1,Id\_Sub}, i_{2,Id\_Sub})$ and $\Delta(i_{1,Subcategory}, i_{2,Subcategory})$ cannot be calculated and are not taken into count for the calculation of $\Delta(i_1, i_2)$.

4.2 Hierarchy Level Instance Distance

The hierarchy level instance distance $\Delta_{p_l^{H_2}}(i_1, i_2)$ is calculated as (1).

$$\Delta_{p_l^{H_2}}(i_1, i_2) = \frac{\Delta(i_{1,p_l^{H_2}}, i_{2,p_l^{H_2}}) + \sum_{w \in Weak[p_l^{H_2}]} \Delta(i_{1,w}, i_{2,w})}{1 + |Weak[p_l^{H_2}]|} \tag{1}$$

As we mentioned that we only consider the levels from the second level of each hierarchy, we do not calculate the distance for the first level of hierarchies. The weak attributes of the first hierarchy levels are regarded as hierarchies containing only one parameter, so their level distance is not needed to be calculated neither.

Example 3. According to (1), for the levels in H_1, we have $\Delta_{p_3}^{H_1}(i_1, i_2) = (0 + 0)/2 = 0$. As the parameter and weak attribute value of the second level i_{1,Id_Sub} and $i_{1,Subcategory}$ are missing, the distance of this level is not taken into account. For H_2, since the two levels contain only one parameter without weak attribute, their hierarchy level is equal to the attribute distance of the parameter, so we have $\Delta_{p_2}^{H_2}(i_1, i_2) = 0.71$, $\Delta_{p_3}^{H_2}(i_1, i_2) = 0$.

4.3 Hierarchy Instance Distance

The hierarchy instance distance is calculated as (2), where $W_l(p_l^{H_2})$ is the hierarchy level weight.

$$\Delta_{H_2}(i_1, i_2) = \sum_{p_l^{H_2} \in H_2 \setminus \{id\}} W_l(p_l^{H_2}) \Delta_{p_l^{H_2}}(i_1, i_2) \tag{2}$$

For a weak attribute $w \in Weak^{H_2}[id]$ of the first hierarchy level, $\Delta_w(i1, i_2) = \Delta(i_{1,w}, i_{2,w})$.

Hierarchy Level Weight. Since the parameters on the lower levels have thinner granularity, their weight for measuring the hierarchy instance distance should be higher. Here, we propose two hierarchy level weights: one is based on the cardinalities of the parameters and another is an incremental weight.

- For the cardianlity-based weight, we consider the number of the distinct values of the level as the portion of the weight. Thus for the cardianlity-based hierarchy level weight of the lth level at H_2 is calculated as (3), where $dv(n)$ denotes the number of distinct values of the nth level.

$$W_l^c(p_l^{H_2}) = \frac{dv(l)}{\sum_{j=2}^{|Param^{H_2}|} dv(j)} \tag{3}$$

- However, when the cardinality ratio between certain parameters is very large, the cardinality-based weight may be biased. So we also propose another type of incremental hierarchy level distance weight. For the incremental weight, we consider the weight of the highest-granularity as one portion and it increases by one portion for each neighboring lower-granularity level. The total weight should be equal to 1, thus the incremental hierarchy level weight of the lth level at H_2 is calculated as (4).

$$W_l^i(p_l^{H_2}) = \frac{2(|Param^{H_2}| - l + 1)}{|Param^{H_2}|^2 - |Param^{H_2}|} \tag{4}$$

Example 4. Our example has only 5 instances, so we can use cardinality-based weight to get hierachy level weight. We thus have for H_1: $W_l(p_2^{H_1}) = 3/(3 + 2) = 0.6$ and $W_l(p_3^{H_1}) = 2/(3 + 2) = 0.4$. For H_2: $W_l(p_2^{H_2}) = 3/(3 + 2) = 0.6$ and $W_l(p_3^{H_2}) = 2/(3 + 2) = 0.4$. We can then calculate the hierarchy instance distances: $\Delta_{H_1}(i_1, i_2) = 0.4 \times 0 = 0$, $\Delta_{H_2}(i_1, i_2) = 0.6 \times 0.71 + 0.4 \times 0 = 0.426$, $\Delta_{w_1}(i_1, i_2) = 0.8$.

4.4 Dimension Instance Weight

The dimension instance weight $\Delta(i_1, i_2)$ is calculated as (5), where $W_h(H_1, H_2)$ and $W_h(H_1, w)$ are hierarchy weights of H_2 and w with respect to H_1.

$$\Delta(i_1, i_2) = \sum_{H_2 \in H^D} W_h(H_1, H_2)\Delta_{H_2}(i_1, i_2) + \sum_{w \in Weak^{H_2}[id]} W_h(H_1, w)\Delta_w(i_1, i_2)$$

$$(5)$$

Hierarchy Weight. The dependency degree in the rough set theory [18] measures the degree of the dependency between attributes, so it is applied for the hierarchy weight.

When calculating the hierarchy distance weight, we can consider a decision system $S = (I^D, A_n^{H_2}, A_n^{H_1})$, since we do not take the first level of a hierarchy into account, $A_n^{H_1} = A^{H_1} \setminus (\{id\} \cup Weak^{H_1}[id])$, $A_n^{H_2} = A^{H_2} \setminus (\{id\} \cup Weak^{H_2}[id])$. The second hierarchy level parameters $p_2^{H_1}, p_2^{H_2}$ determine all the other hierarchy attributes in $A_n^{H_1}$ and $A_n^{H_2}$, we can reduce the attribute sets of $A_n^{H_1}$ and $A_n^{H_2}$ to the sets containing only the values of the second hierarchy level parameter $p_2^{H_1}, p_2^{H_2}$. According to [18], the degree k to which H_1 depends on H_2, denoted $H_2 \Rightarrow_k H_1$ is thus defined as:

$$k = \gamma(A_n^{H_2}, A_n^{H_1}) = \gamma(p_2^{H_2}, p_2^{H_1}) = \frac{card(POS_{p_2^{H_2}}(p_2^{H_1}))}{card(I^D)}$$

$$(6)$$

where $POS_{p_2^{H_2}}(p_2^{H_1}) = \bigcup_{X \in I^D/p_2^{H_1}} p_{2*}^{H_2}(X)$ and $card(X)$ is the cardinality of an non-empty set X, the missing second level parameter values are not taken into account. For H_1 itself, we have $\gamma(A_n^{H_1}, A_n^{H_1}) = 1$.

The hierarchy distance weight of H_2 with respect to H_1 is the ratio of their dependency degree with respect to the sum of the dependency degrees of the all hierarchies and first level weak attributes in D with respect to H_1 as (7).

$$W_h(H_1, H_2) = \frac{\gamma(A_n^{H_2}, A_n^{H_1})}{\sum_{H_3 \in H^D} \gamma(A_n^{H_3}, A_n^{H_1}) + \sum_{w \in Weak^{H_1}[id]} \gamma(w, A_n^{H_1})}$$

$$(7)$$

Example 5. In our example, we have $card(I^D) = 5$, $card(POS_{p_2^{H_2}}(p_2^{H_1})) = 2$, so $\gamma(A_n^{H_2}, A_n^{H_1}) = 2/5 = 0.4$. In the same way, we can get $\gamma(w_1, A_n^{H_1}) = 2/5 = 0.4$, we also have $\gamma(A_n^{H_1}, A_n^{H_1}) = 1$. We can thus get the hierarchy weights: $W_h(H_1, H_2) = 0.4/(0.4 + 0.4 + 1) = 0.22$, $W_h(H_1, H_1) = 1/(0.4 + 0.4 + 1) = 0.56$ and $W_h(w_1, H_2) = 0.4/(0.4 + 0.4 + 1) = 0.22$. We can finally obtain the dimension instance distance $\Delta(i_1, i_2) = 0.22 \times 0.46 + 0.22 \times 0.8 + 0.56 \times 0 = 0.28$

5   H-OLAPKNN Imputation

5.1   H-OLAPKNN Overview

The H-OLAPKNN imputation is shown in Algorithm 1. It is composed of three steps where the first is the hierarchical imputation and the next two steps concern the OLAPKNN imputation.

1. The hierarchical imputation is based on the functional dependencies of the hierarchy attributes. It searches for an instance having the same value on a lower-granularity level parameter of the missing value and whose attribute of the missing value is not empty, we can then replace the missing value with this non-empty value ($line_1$).
2. The weak attributes' values are determined by their parameters' values, so we complete the parameters before completing their weak attributes. Thus then, for missing data of each hierarchy ($line_2$), we create candidate lists of the instances containing possible replaced values and select the k nearest neighbors in the candidate lists to complete the missing data ($line_3$).
3. There are weak attributes which can be completed together with their parameter. Finally for the remaining missing weak attribute data, they are completed in the similar way ($line_4$).

Next, we explain in detail the OLAPKNN imputation algorithm. A weak attribute of a hierarchy can be regarded as a "highest level parameter" of a part of the hierarchy whose imputation is similar to the parameter imputation. So we only explain the parameter imputation in this paper.

5.2   Imputation for Parameters by OLAPKNN

Parameter Imputation Order. We first introduce the continuous missing parameter group in order to explain the imputation order for parameters.

Definition 3 (Continuous missing parameter group). *For an instance $i_r \in I^D$ in the dimension D containing missing values on parameters of a hierarchy H, all these parameters are in a set $Pm_r^H = \{p_v^H \in Param^H : i_{r,p_v^H}$ is empty\}. For the parameters in Pm_r^H, they can be divided into one or several continuous missing parameter groups. A **continuous missing parameter group (CG)** contains one or several parameters which are neighbors on H and are maximal neighbors in Pm_r^H. By neighbors on H, we mean that for the parameter p_{lowest} having the lowest-granularity level in the CG on H and the one $p_{highest}$ having the highest-granularity level, if there exists any parameter $p_{middle} \in Param^H$, such that $p_{lowest} \preceq_H p_{middle} \preceq_H p_{highest}$, then $p_{middle} \in Pm_r^H$; By maximal neighbors in Pm_r^H, we mean that if there exists any parameter $p_{low_2} \in Param^H$, such that $p_{low_2} \preceq_H p_{lowest}$, then $p_{low_2} \notin Pm_r^H$, if there exists any parameter $p_{high_2} \in Param^H$, such that $p_{highest} \preceq_H p_{high_2}$, then $p_{high_2} \notin Pm_r^H$. We call all CGs of a hierarchy H containing a same number of parameters a n-CGs of H, where n denotes the number of parameters.*

Algorithm 2 shows the imputation of the parameters. For a given hierarchy H on a dimension D, we carry out the imputation for parameters in the n-CGs by the ascending order of n ($line_1$). We can thus make sure that all the $(n-1)$-CGs instances are completed so that we can carry out the imputation for the n-CGs based on the existing data. Then for each n-CGs, we look at all possible CG combinations ($line_{2-3}$). Next we verify if there are instances containing missing values for each possible CG ($line_{4-9}$). According to $Definition$ 3, the instances of a CG on H have missing values on all parameters of the group. If there is a neighboring lower-granularity or higher-granularity parameter of the group, the instances do not have missing value on them ($line_9$).

Algorithm 1: $H - OLAPKNN(D)$

1 $hierarchicalImputation(D)$;
2 **for** $H \in H^D$ **do**
3      $imputeParam(D, H)$;
4      $imputeWeak(D, H)$;

Algorithm 2: $imputeParam(D, H)$

1 **for** $ncontinuous \leftarrow 1$ to $|Param^H| - 1$ **do**
2   **for** $i \leftarrow 1$ to $|Param^H| - ncontinuous$ **do**
3         $P_{CG} \leftarrow Param^H[i : i + ncontinuous - 1]$;
4         $p_{low}, p_{high} \leftarrow \emptyset$;
5   **if** $i > 1$ **then**
6           $p_{low} \leftarrow Param[i - 1]$;
7   **if** $i < |Param^H| - ncontinuous$ **then**
8           $p_{high} \leftarrow Param[i + ncontinuous]$;
9         $I_{missing} = \{i_r \in I^D : (\forall p_{cg} \in P_{CG}, i_{r,p_{cg}} = null) \wedge (\exists p_{low} \implies i_{r,p_{low}} \neq null) \wedge (\exists p_{high} \implies i_{r,p_{high}} \neq null)\}$;
10        $lowMap \leftarrow Map$;
11   **for** $i_m \in I_{missing}$ **do**
12          $I_{candidate} \leftarrow getCandidateList(D, P_{CG}, p_{high}, i_m, 1)$;
13          $vWeightMap \leftarrow getVWeightMap(D, i_m, I_{candidate}, k, P_{CG})$;
14          $lowMap \leftarrow replaceNoPlow(D, H, lowMap, vWeightMap, i_m, P_{CG}, p_{low})$;
15   **if** $\exists p_{low}$ **then**
16          $replacePlow(lowMap, P_{CG}, H, D, p_{low})$;

Candidate List. Since some missing data are already completed by the hierarchical imputation, for the remaining missing data, they can no longer be completed with the aid of their lower-granularity parameters. For a value of one parent parameter, there may be several possible values on a child parameter of its. So for a missing data instance of a CG, we can find all possible replaced values based on their neighboring higher parameter and create a candidate list (Algorithm 2 $line_{11}$). The candidate list contains not only the candidate replaced values of CG attributes but also the values of all other attributes of the dimension because we need all attribute's value for the calculation of the distances.

Algorithm 3 shows the candidate list creation for an instance of a CG. If the neighboring higher-granularity parameter p_{high} of the CG exists, we search for all the instances having the same values on p_{high} as the CG instance, and containing non-missing values on the CG parameters. Then these instances can be added into the candidate list ($line_{1-3}$). If there does not exist a neighboring higher parameter for a CG, we add all the instances of the dimension which contain non-missing values on the CG parameters into the candidate list ($line_{4-5}$).

Creation of Replaced Value Weight Map. For the CG instance, we can get a map for each possible replaced values in the nearest neighbors with their distance-based weight for the selection of the final replaced value as described in Algorithm 4. We first create a map of each instance in the candidate list with its distance with respect to the missing instance ($line_{1-3}$). Then we can select the k nearest candidate instances to create a candidate list if the candidate list contains more than k instances, if not, we can keep all candidate instances

($line_{4-5}$). The selected candidate instances may contain same replaced values, so we create a map of each replaced values with their weight ($line_6$). According to [7], for an instance i_m of a CG, for a selected candidate list containing k instances, the distance weight of the n nearest instance i_{cn} can be calculated as (8), where i_{ck} denotes the kth nearest instance and i_{c1} denotes the nearest instance. It is to be noted that $W_d(i_m, i_c) = 1$ when $\Delta(i_m, i_{ck}) = \Delta(i_m, i_{c1})$.

$$W_d(i_m, i_c) = \frac{\Delta(i_m, i_{ck}) - \Delta(i_m, i_{cn})}{\Delta(i_m, i_{ck}) - \Delta(i_m, i_{c1})} \tag{8}$$

Thus the weight of a candidate of replaced values is the sum of the weight of the instances which contain them ($line_{4-5}$).

Algorithm 3: $getCanList(D, P_{CG}, p_{high}, i_m, parameter)$

1 **if** $parameter = 1$ **then**
2   **if** $\exists p_{high}$ **then**
3      $I_{candidate} \leftarrow \{i_r \in I^D : (\exists p_{cg} \in P_{CG}, i_{r,p_{cg}} \neq null) \wedge (i_{r,p_{high}} = i_{r_{missing},p_{high}})\}$;
4   **else**
5      $I_{candidate} \leftarrow \{i_r \in I^D : (\exists p_{cg} \in P_{CG}, i_{r,p_{cg}} \neq null)\}$;
6 **else**
7    $I_{candidate} \leftarrow \{i_r \in I^D : (i_{r,weak} \neq null)\}$;
8 **return** $I_{candidate}$

Algorithm 4: $getVWeightMap(D, i_m, I_{candidate}, k, P_{CG})$

1 $iDistanceMap \leftarrow Map$;
2 **for** $i_c \in I_{candidate}$ **do**
3    $iDistanceMap[i_{c,id}] \leftarrow \Delta(i_m, i_c)$;
4 **if** $|I_{candidate}| > k$ **then**
5    $iDistanceMap \leftarrow iDistanceMap.top(k)$;
6 $vWeightMap \leftarrow Map$;
7 **for** $i_{c,id} \in iDistanceMap.keys()$ **do**
   /* addMap(Map, key, value): Create the map if it does not exist. Add the value to the existing value if the key exists, assign the value to the key if not. */
8    $addMap(vWeightMap, \{i_{c,p_{cg}} : i_{c,p_{cg}} \in i_{c,P_{CG}}\}, Wv(i_m, i_c))$;
9 **return** $vWeightMap$

Replacement of Values. To fill in the missing values of CG, we have two cases: the first case (Algorithm 2 $line_{13}$) is that there is no lower non-id parameter of the missing parameter group, the second case (Algorithm 2 $line_{14-15}$) is that there is such parameter. The difference is that for the second case, we have to take the strictness of hierarchy into account by making sure that a lower parameter value of the CG has only one higher-granularity level parameter after the imputation.

The replacement of the values of the first case is described in Algorithm 5. We can take the values having the highest weight in the weight map ($line_1$) to fill in the missing values of the CG ($line_{2-3}$).

The replacement of the values for the second case is described in Algorithms 5 and 6. We create a map $lowMap$ for each neighboring lower-granularity parameter value which corresponds to another map of the each possible replaced value and its total weight (Algorithm 2 $line_{10}$). For each instance of the CG, we get the replaced values with the highest value weight (Algorithm 5 $line_1$). For the value of its neighboring lower-granularity parameter, we update the replaced values and the weight (Algorithm 5 $line_{8-10}$). When all the missing instances of a CG are treated, we get a final $lowMap$. For each value of the neighboring lower-granularity level parameter in $lowMap$, we can take the replaced values with the highest weight to fill in the missing values (Algorithm 6 $line_{1-5}$).

Algorithm 5: $replaceNoPlow(D, H, lowMap, vWeightMap, i_m, P_{CG}, p_{low})$

1 $i_{replace,P_{CG}} \leftarrow vWeightMap.top(1).key()$;
2 **if** $\not\exists p_{low}$ **then**
3       $i_{m,P_{CG}} \leftarrow i_{replace,P_{CG}}$;
4   **for** $p_{cg} \in P_{CG}$ **do**
5   **for** $w^{p_{cg}} \in Weak^H[p_{cg}]$ **do**
6   **if** $i_{m,w^{p_{cg}}} = \emptyset$ **then**
7                   $i_{m,w^{p_{cg}}} \leftarrow \{i_{r,w^{p_{cg}}} \in I^D : i_{r,p_{cg}} = i_{m,p_{cg}}\}.getOne()$;

8 **else**
9       $addMap(lowMap[i_{m,p_{low}}], i_{replace,P_{CG}}, vWeightMap[i_{replace,P_{CG}}])$;
10      $addMap(lowMap, i_{m,p_{low}}, lowMap[i_{m,p_{low}}])$;
11 **return** $lowMap$

Algorithm 6: $replacePlow(lowMap, P_{CG}, H, D, p_{low})$

1 **for** $i_{m,p_{low}} \in lowMap.keys()$ **do**
2       $vWeightMap \leftarrow lowMap[i_{m,p_{low}}].top(1)$;
3       $i_{replace,p_{low}} \leftarrow vWeightMap.key()$;
4   **for** $i_m \in \{i_r \in I^D : i_{r,p_{low}} = i_{m,p_{low}}\}$ **do**
5           $i_{m,P_{CG}} \leftarrow i_{replace,P_{CG}}$;
6   **for** $p_{cg} \in P_{CG}$ **do**
7   **for** $w^{p_{cg}} \in Weak^H[p_{cg}]$ **do**
8   **if** $i_{m,w^{p_{cg}}} = \emptyset$ **then**
9                       $i_{m,w^{p_{cg}}} \leftarrow \{i_{r,w^{p_{cg}}} \in I^D : i_{r,p_{cg}} = i_{m,p_{cg}}\}.getOne()$;

6 Experimental Assessment

6.1 Technical Environment and Datasets

To evaluate the effectiveness and efficiency of H-OLAPKNN, we implement our algorithms and conduct experiments with different datasets and compare it to other imputation methods. Our code is developed in Python 3.7 and is executed on a Intel(R) Core(TM) i5-10210U 1.60 GHz CPU with a 16 GB RAM. Data are integrated in R-OLAP format with Oracle 11g. The distance metrics that we use are like described in Sect. 4.1, for the word embedding based distance, we use Google's pre-trained word2vec model[1].

We employ 3 real world datasets. The first dataset is a regional sale dataset (**RegionalSales**) storing sales data for a company across US regions . The second (**IBRD**) and the third (**MIGA**) ones are data of world bank which are respectively the International Bank for Reconstruction and Development balance sheet data and the Multilateral Investment Guarantee Agency issued projects data. For each one of the datasets, we create a DW for our experiment. The link of the dataset source and more information can be found in our github[2].

6.2 Experimental Methodology

We apply different missing rates (1%, 5%, 10%, 20%, 30%, 40%) for each attribute except for the primary keys. To generate missing data for an attribute, we sort randomly all the instances and remove attribute values of the first certain percentage of instances. For the effectiveness, since we focus on the qualitative data, we apply the accuracy (number of correctly replaced values divided by number of missing values) as the metric instead of the root mean square error (RMSE) [16] which is widely used but is only suitable for quantitative data

---

[1] https://drive.google.com/file/d/0B7XkCwpI5KDYNlNUTTlSS21pQmM/edit?resourcekey=0-wjGZdNAUop6WykTtMip30g.
[2] https://github.com/Implementation111/H-OLAPKNN/.

imputation. For the efficiency, we get the run time of each algorithm. We carry out 20 tests for each dataset and get the average accuracy and run time.

We compare our proposed method with some other methods in the literature as baseline. **H-OLAPKNN(MI)**: Since the mutual information is widely used in the KNN based data imputation [10,17]. In this baseline, we apply our proposed OLAPKNN algorithm by using the mutual information instead of the dependency degree as the hierarchy distance weight. **KNN** [6]: This method use the basic KNN algorithm to generate the replaced values for missing data **Mode**: The Mode method simply replace the missing data with the most frequent non-empty value of the attribute in the table.

6.3   Results and Analysis

For each dataset, the optimal k of KNN is different. So we test with different k values between 1 and 20 and choose the best one for the experiment of each data set which are respectively 5, 4 and 8. For the W_l, we choose the weight with a better result as the weight for each dataset which are respectively W_l^i, W_l^c and W_l^i. Then we compare the accuracy and the run time of each algorithm.

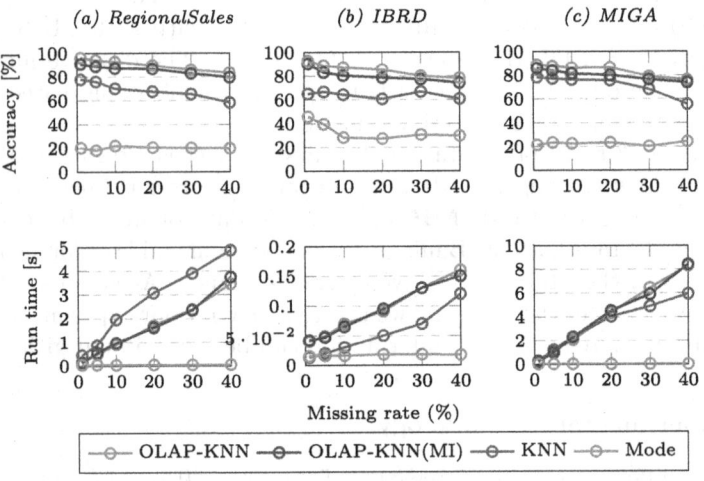

Fig. 3. Results of experiments

Accuracy. The accuracy result is shown as Fig. 3. We can see that the proposed H-OLAPKNN algorithm outperforms all the other baseline algorithms for each dataset. The Mode method has the worst result since it is too simple and it takes nothing into account. Compared to the mutual information, we observe that using the dependency degree as the hierarchy distance weight can help us get a more accurate result as it considers the ordered dependency instead of the inter-dependency. Compared to the basic KNN method, the H-OLAPKNN returns a better accuracy results since it considers the structure of the DW and take the dependencies among the attributes into account. The accuracy of H-OLAPKNN, H-OLAPKNN(MI) and KNN decreases with the increase of the missing rate because when there are more missing data, 1) we have less complete data for getting more precise distance scores and 2) it is more likely that the proper replaced data do not exist in the table.

Run Time. The run time result is shown as Fig 3. The Mode algorithm costs less time since it is the simplest method. The run time of the other three algorithms changes linearly with respect to the missing rate. There is no big difference between H-OLAPKNN and H-OLAPKNN(MI) since they are only different in terms of the calculation of hierarchy distance weight. The OLAPKNN costs less time than KNN for dataset **RegionalSales**, but more time for the other two datasets. This is because the hierarchical imputation complete most of the data of **RegionalSales** so that it takes less time for OLAPKNN to create the candidate list and compare the similarities.

7   Conclusion and Future Work

In this paper, we propose a DW dimensional data imputation method by combining hierarchical imputation with a novel KNN-based algorithm. We first define a 4-level distance calculation method for dimension instances by taking advantage of the DW dimension structure. Then by applying the proposed distances, we define the KNN-based algorithm. The advantage of the proposed algorithm is that it takes the dimension structure complexity into account and is able to make replaced values conform to the dependency constraints of the hierarchies. Our proposal is validated by a series of experiments and is proved to outperform other baselines in the literature. It increases the dimension data imputation accuracy by up to 25.2% with respect to the basic KNN imputation.

In the future, we will extend our method for the imputation of numerical data in the dimensions and facts. We also intend to generalize the method for non-DW data by proposing an approach automatically modelling them in OLAP.

References

1. Preface. In: Han, J., Kamber, M., Pei, J. (eds.) Data Mining. 3rd edn. (2012)
2. Li, Y.Y., Parker, L.E.: Nearest neighbor imputation using spatial-temporal correlations in wireless sensor networks. Inf. Fusion **15**, 64–79 (2014)
3. Beretta, L., Santaniello, A.: Nearest neighbor imputation algorithms: a critical evaluation. BMC Med. Inform. Decis. Making **16**(3), 197–208 (2016)
4. Bimonte, S., Ren, L., Koueya, N.: A linear programming-based framework for handling missing data in multi-granular data warehouses. Data Knowl. Eng. **128**, 101832 (2020)

5. Breve, B., Caruccio, L., Deufemia, V., Polese, G.: RENUVER: a missing value imputation algorithm based on relaxed functional dependencies. In: EDBT, pp. 1–52 (2022)
6. Domeniconi, C., Yan, B.: Nearest neighbor ensemble. In: ICPR, vol. 1 (2004)
7. Dudani, S.A.: The distance-weighted k-nearest-neighbor rule. IEEE Trans. Syst. Man Cybern. **4**, 325–327 (1976)
8. Fan, W., Jianzhong, L., Shuai, M., Nan, T., Wenyuan, Y.: Towards certain fixes with editing rules and master data. VLDB J. **21**, 173–184 (2010)
9. Farhangfar, A., Kurgan, L.A., Pedrycz, W.: A novel framework for imputation of missing values in databases. IEEE SMC **37**(5), 692–709 (2007)
10. García-Laencina, P.J., Sancho-Gómez, J.L., Figueiras-Vidal, A.R., Verleysen, M.: K nearest neighbours with mutual information for simultaneous classification and missing data imputation. Neurocomputing **72**(7), 1483–1493 (2009)
11. Jatnika, D., Bijaksana, M.A., Suryani, A.A.: Word2vec model analysis for semantic similarities in English words. Proc. Comput. Sci. **157**, 160–167 (2019)
12. Li, D., Deogun, J., Spaulding, W., Shuart, B.: Towards missing data imputation: a study of fuzzy k-means clustering method. In: IJCRS, pp. 573–579 (2004)
13. Little, R., Rubin, D.: Statistical Analysis with Missing Data. Wiley, New York (2002)
14. Lofi, C., El Maarry, K., Balke, W.-T.: Skyline queries over incomplete data - error models for focused crowd-sourcing. In: Ng, W., Storey, V.C., Trujillo, J.C. (eds.) ER 2013. LNCS, vol. 8217, pp. 298–312. Springer, Heidelberg (2013). https://doi.org/10.1007/978-3-642-41924-9_25
15. Malinowski, E., Zimányi, E.: OLAP hierarchies: a conceptual perspective. In: Persson, A., Stirna, J. (eds.) CAiSE 2004. LNCS, vol. 3084, pp. 477–491. Springer, Heidelberg (2004). https://doi.org/10.1007/978-3-540-25975-6_34
16. Miao, X., Gao, Y., Guo, S., Liu, W.: Incomplete data management: a survey. Front. Comput. Sci. **12**(1), 4–25 (2018). https://doi.org/10.1007/s11704-016-6195-x
17. Pan, R., Yang, T., Cao, J., Lu, K., Zhang, Z.: Missing data imputation by K nearest neighbours based on grey relational structure and mutual information. Appl. Intell. **43**(3), 614–632 (2015). https://doi.org/10.1007/s10489-015-0666-x
18. Pawlak, Z., Skowron, A.: Rudiments of rough sets. Inf. Sci. **177**(1), 3–27 (2007)
19. Garcia-Laencina, P.J., Sancho-Gómez, J.L., Figueiras-Vidal, A.R.: Pattern classification with missing data: a review. Neural. Comput. App. **19**(2), 263–282 (2010)
20. Ravat, F., Teste, O., Tournier, R., Zurfluh, G.: Algebraic and graphic languages for OLAP manipulations. Int. J. Data Warehousing Mining **4**, 17–46 (2008)
21. de S. Ribeiro, L., Goldschmidt, R.R., Cavalcanti, M.C.: Complementing data in the ETL process. In: Cuzzocrea, A., Dayal, U. (eds.) DaWaK 2011. LNCS, vol. 6862, pp. 112–123. Springer, Heidelberg (2011). https://doi.org/10.1007/978-3-642-23544-3_9
22. Song, S., Zhang, A., Chen, L., Wang, J.: Enriching data imputation with extensive similarity neighbors. Proc. VLDB Endow. **8**(11), 1286–1297 (2015)
23. Troyanskaya, O., et al.: Missing value estimation methods for DNA microarrays. Bioinformatics **17**(6), 520–525 (2001)
24. Trujillo, J., Palomar, M., Gomez, J., Song, I.Y.: Designing data warehouses with OO conceptual models. Computer **34**(12), 66–75 (2001)
25. Wu, X., Barbará, D.: Modeling and imputation of large incomplete multidimensional datasets. In: DaWak, pp. 286–295 (2002)
26. Yakout, M., Ganjam, K., Chakrabarti, K., Chaudhuri, S.: Infogather: entity augmentation and attribute discovery by holistic matching with web tables. In: ACM SIGMOD, pp. 97–108 (2012)

27. Yang, Y., Abdelhédi, F., Darmont, J., Ravat, F., Teste, O.: Internal data imputation in data warehouse dimensions. In: DEXA, pp. 237–244 (2021)

28. Yujian, L., Bo, L.: A normalized Levenshtein distance metric. IEEE Trans. Pattern Anal. Mach. Intell. **29**(6), 1091–1095 (2007)

29. Zhixu, L., Sharaf, M.A., Sitbon, L., Sadiq, S., Indulska, M., Zhou, X.: A web based approach to data imputation. World Wide Web **17**(5), 873–897 (2014)

Feature Ranking from Random Forest Through Complex Network's Centrality Measures
A Robust Ranking Method Without Using Out-of-Bag Examples

Adriano Henrique Cantão$^{(\boxtimes)}$ ⓘ, Alessandra Alaniz Macedo ⓘ, Liang Zhao ⓘ, and José Augusto Baranauskas$^{(\boxtimes)}$ ⓘ

Department of Computer Science and Mathematics, Faculty of Philosophy, Sciences and Letters at Ribeirao Preto, University of Sao Paulo, Bandeirantes Avenue, 3900, Ribeirao Preto, SP 14040-901, Brazil
{cantao,ale.alaniz,zhao,augusto}@usp.br
https://www.usp.br/

Abstract. The volume of available data in recent years has rapidly increased. In consequence, datasets commonly end up with many irrelevant features. That increase may disturb human understanding and even lead to poor machine learning models. This research proposes a novel feature ranking method that employs trees from a Random Forest to transform a dataset into a complex network to which centrality measures are applied to rank the features. That process takes place by representing each tree as a graph where all the tree features are vertices on this graph, and the links within the nodes (father → child) of the tree are represented by a weighted edge between the two respective vertices. The union of all graphs from individual trees leads to the complex network. Then, three centrality measures are applied to rank the features in the complex network. Experiments were performed in eighty-five supervised classification datasets, with a variation in the feature noise level, to evaluate our novel method. Results show that centrality measures in non-oriented complex networks are comparable and may be correlated to the Random Forest's variable importance ranking algorithm. Vertex strength and eigenvector outperformed the Random Forest in 40% noise datasets, with a not statistically different result at a 95% confidence level.

Keywords: Feature ranking · Random Forest · Complex networks · Centrality measures

1 Introduction

The constant increase in technology and applications has rapidly generated massive amount of data in domains, such as social media, healthcare, bioin-

This study was financed in part by the Coordenação de Aperfeiçoamento de Pessoal de Nível Superior - Brasil (CAPES) - Finance Code 001.

formatics, and online education. These data are mostly numbers, videos, image, text, and eventually involve many irrelevant and redundant features. This high-dimensional data poses a significant challenge for the machine learning methods. The model-building process can lead to poor performance when the dataset contains redundant, unnecessary, and noisy features [16,20,25,34]. In scenarios like this, it is common to use either a ranking to better understand the quality level of each of the features or a feature selector to find a subset of relevant features.

Considering that we propose a ranker using a complex network, obtained from a Random Forest, from which features are extracted for analysis of their relevance. On one hand, Random Forests can provide measures of feature importance [6]. However, few works have studied the theoretical properties and statistical mechanisms of these measures [21]. On the other hand, the topological structure of a complex network allows the use of centrality measures [4] that may contribute to a better understanding of the features' importance. Once we analyze a complex network using any centrality measure's perspective, it's possible to rank these features from the most central to the last. The ranking generated reflects the relevance of each feature to that complex network.

This work is organized as follows: we present related studies in Sect. 2, including basic concepts about Random Forest and Complex Networks. In Sect. 3 we describe our methodological approach to generating a complex network from trees in a Random Forest. Section 4 shows the empirical setup used to evaluate the proposed algorithm. Section 5 shows the experiments and discusses the results; finally, Sect. 6 presents the main conclusions of this study.

2   Related Work

Studies have been representing data into complex networks to solve several problems in the Machine Learning field. For this reason, new methodologies have been proposed for converting traditional sets of examples into graphs, such as KNN-Graph and K-Associated graphs [3]. Some methods use hybrid networks whose vertices can be either examples or features of the dataset [27]. However, those studies focus on representing the instances of the dataset as vertices of the network. In contrast, some studies [26,38] use networks of features and apply distance-based measures to find the links between features when performing the representation of a dataset into a complex network. From an already created network, [26] apply the Laplace centrality measure between the vertices of the same cluster, then the vertices are sorted by their centrality value, and vertices with values above the defined threshold are removed. In [38], similarity measure is employed, which initially calculates the distance of the features by applying kernel measure in the network adjacency matrix, then a threshold is applied to remove all vertices considered not relevant, with a value above the established threshold. In [16] the PageRank centrality was applied for feature selection in 7 multi-label datasets. Firstly, a complete, non-oriented and weighted graph is created, in which each vertex of the graph represents an attribute of the dataset and is adjacent to all other vertices. The weight of the edges is obtained by the

Euclidean distance of the correlation matrix between the attributes × classes of the dataset. Then, the PageRank metric is applied to generate a score for each vertex and, finally, the vertices are ordered in descending order of score, generating a ranking of the attributes; the user defines the number of attributes of interest to be selected. There are other ways to perform feature selection in complex networks, such as a path-based method [23], spectral [37], which selects the features based on the graph structure, and dominant-set clustering associated with multidimensional interaction information [35]. Our work focuses on centrality measures. We have extensively tested the feature ranking efficiency of strength, eigenvector and Katz centralities on oriented and non-oriented feature networks generated through Random Forests.

2.1   Feature Selection and Ranking

Methods for selecting features can be classified into three groups: wrappers, filters, and embedded [9]. Wrappers choose and test the features in the same step and, in the end, select a subset of features that results in a better classification performance. Filters work in two stages, at first, the features are selected or ranked, and, in a second stage, the features are used by an independent machine learning algorithm. Embedded selects the features during the training process of the algorithm itself.

A ranker is a two-step method also used for feature selection. It firstly generates a ranking using a metric $\mathcal{M}$. We assume features are better scored with larger values of the relevance metric $\mathcal{M}$. Secondly, the best features in the generated ranking are selected, (i) using a threshold θ is applied to select the relevant features; (ii) alternatively, an integer k can be fixed, and only the top-k features of the ranking are selected [9]. The feature ranking generated by this method provides insight into the data by clearly presenting the relevance of each feature, and when the most relevant ones are selected, it can improve the performance of learning algorithms [34]. The ranker proposed in this research can be seen as a filter, since it orders features from the most relevant to the least ones. By performing two stages, filters act in the preprocessing stage of data mining tasks [2].

2.2   Random Forests

A Random Forest is formally defined as a classifier composed of a set of trees $\{h_{S_l}(x)\}$, $l = 1, 2, \ldots, L$, induced from $\{S_l\}$ independent random samples with identical distribution, generally a bootstrap sample. The tree is grown recursively repeating these steps: (i) selecting m' features at random from the m features ($m' \ll m$); (ii) picking the best feature split-point among the m' features; and (iii) splitting the node into children nodes [17].

Each tree predicts the class of the entry x and then the most popular class among the trees is chosen for this same entry [6,36]. Therefore, Random Forests apply the same method as bagging [5] to produce random samples (bootstrap samples) of training sets for each tree. Breiman [6] justifies the use of the bagging method in Random Forests for two reasons: the use of bagging seems to improve performance when random features are used; bagging can be used to provide continuous estimates of the generalization error of the combined set of trees, as well as estimates for strength and correlation, using the Out-Of-Bag estimator. The classification error of the forest depends on the strength of the individual trees in the forest, the correlation between any two trees in the forest and the features importance [6, 22].

The methodology proposed herein comprises the use of a Random Forest to determine the links between two features and the weight of these links. From this definition, the representation of the dataset into a complex network is made.

2.3   Complex Networks

Complex networks are represented through graph theory, which initially, in the 1950s, focused on regular graphs; still, in the same period, large-scale networks, which had no apparent structural models, initially studied by mathematicians Paul Erdős and Alfred Rényi were described as random graphs [13]. They then defined the Erdős-Rényi model with N nodes and a p probability of connection between pairs of nodes [10]; this model was a reference for a long time. With the growth of research in this area, it ended up being questioned. Intuitively, many systems showed to follow some principles of structural organization that would represent a certain topology and not a random structure [1]. In a simplified definition, a complex network can be understood as a graph with a nontrivial structure that is composed of a set of vertices connected by edges [24]. Trivial, in this case, means a graph with only one vertex and no edges.

Complex networks have been established as a powerful tool; its topological structure has been shown to be useful for the detection of classes and clusters, either by clustering or classification algorithms. Accordingly, there has been an increase in the use of Machine Learning methods based on networks [33], which has become an area of active research with a diversity of successful applications in the approach of global information such as: semi-supervised learning [27], data clustering [14], regression [28] and classification [11].

Common representations of complex networks are (i) weighted directed graphs (digraphs), and its derivations; (ii) unweighted digraphs; (iii) weighted graphs; and (iv) unweighted graphs. In a weighted (di)graph, the weight of an edge is represented by w_{ij} whenever there is an edge connecting a pair of vertices i and j [32].

Several discrete structures, such as lists and trees, can be represented by graphs. Therefore, there are several situations in the complex networks field where a structure of interest is firstly represented by a network, and then analyzed by the topological characteristics using informative measures of this network [10]. Some of those measures are: (i) centrality, in which the importance of a vertex for the network is calculated according to the number of paths that go through this vertex; (ii) agglomeration coefficient, which shows how connected is the neighborhood of a given vertex, this measure captures the local structure by counting the triangle connections formed by the vertex to be analyzed and

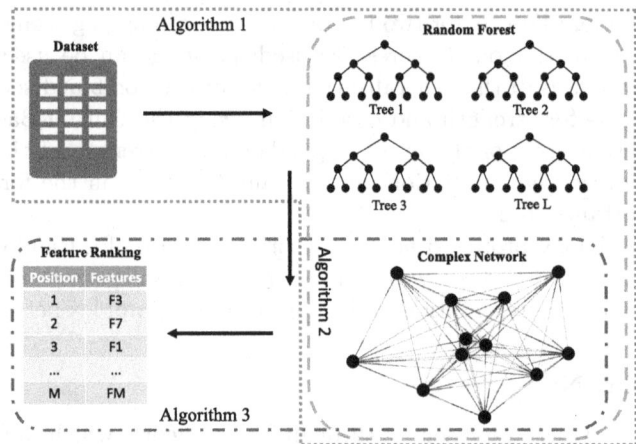

Fig. 1. General scheme of the ranker proposed. Initially, a Random Forest is trained on the input dataset (Algorithm 1). The features used and links generated by each tree are then represented as a complex network (Algorithm 2). Finally, centrality measures are applied to obtain the ranking of features (Algorithm 3).

two of its neighbors; and (iii) spectral, which is equivalent to the set of eigenvalues of the adjacency matrix of a network, these eigenvalues can also be used to determine the order momentum.

In this research we use centrality measures applied to weighted graphs and weighted digraphs as metric $\mathcal{M}$.

3   From Trees in a Random Forest to a Complex Network

Our method exploits the trees of a Random Forest as a strategy to identify the connections between the pairs of features, and the respective weights of these connections to compose a complex network. Afterward, centrality measures are applied to rank the features, as shown schematically in Fig. 1. This method is represented as a pseudocode, by Algorithms 1, 2, and 3.

According to Algorithm 1 (lines 4–5), a Random Forest is initially generated containing $L = 64$ trees [29]. Then, each tree in the forest is represented by an individual weighted graph (lines 6–7). Afterwards, the method convertTreeToNet (line 7) will convert each tree in the forest to its respective graph according to Algorithm 2, explained in detail below:

- Assume that a decision tree has E tree edges numbered $e = 1, 2, \ldots, E$. A key point to be emphasized is that in a decision tree the connection between two decision nodes (tree edge) that links features (X_i, X_j) can occur more than once at the same level of the tree or even at other (sub) levels. We denote the e-th tree edge (X_i, X_j) with weight $w_{ij}^{(e)}$ using the notation $(e, X_i, X_j, w_{ij}^{(e)})$

Algorithm 1. Dataset to complex network through Random Forest

---

Require: Instances: a set of n labeled instances $\{(\mathbf{x}_i, y_i), i = 1, 2, \ldots, n\}$ containing m features $\{X_1, X_2, \ldots, X_m\}$;
1: L: number of trees in the forest, where $L \geq 1$, *default* $L = 64$;
2: Directed: Boolean variable specifying if the trees are transformed into oriented graphs (`true`) or non-oriented graphs (`false`), *default* `false`.
Ensure: N: the complex network generated from Instances.
3: **function** convertData2Net(Instances, L, Directed)
4:      $T \leftarrow$ buildRandomForest(Instances, L)
5:      Let $\{T_1, T_2, \ldots, T_L\}$ be the individual trees of the forest T
6:      **for** $l \in \{1, 2, \ldots, L\}$ **do**
7:          $A^{(l)} \leftarrow$ convertTreeToNet(T_l, Directed)
8:      $\mathcal{N}_{ij} \leftarrow \sum_{l=1}^{L} A_{ij}^{(l)}$ for $1 \leq i, j \leq m$ $\triangleright$ $\mathcal{N}$ is the resulting complex network
9:      **return** $\mathcal{N}$

Algorithm 2. Decision Tree to graph

Require: Tree: a decision tree induced from a labeled dataset containing m features $\{X_1, X_2, \ldots, X_m\}$;
1: Directed: Boolean variable specifying if the trees are transformed into oriented graphs (`true`) or non-oriented graphs (`false`), *default* `false`.
Ensure: A: adjacency matrix representing the Tree transformed into a complex network.
2: **function** convertTreeToNet(Tree, Directed)
3:      Let $\mathcal{E} = \{(X_i, X_j, w_{ij}^{(1)}), \ldots, (X_p, X_q, w_{pq}^{(E)})\}$ be the list of tree edges in Tree, where $|\mathcal{E}| = E$
4:      $A_{ij} \leftarrow 0$ for $1 \leq i, j \leq m$
5:      **for** $e \in \{1, 2, \ldots, E\}$ **do**
6:          Let $(X_i, X_j, w_{ij}^{(e)})$ be the e-th tree edge in $\mathcal{E}$
7:          $A_{ij} \leftarrow A_{ij} + w_{ij}^{(e)}$
8:          **if** not Directed **then**
9:              $A_{ji} \leftarrow A_{ji} + w_{ji}^{(e)}$
10: **return** A

---

which can be simplified to $(X_i, X_j, w_{ij}^{(e)})$ where it is implicit the fact the reference is to the e-th tree edge. The set of tree edges will be denoted by $\mathcal{E} = \{(X_i, X_j, w_{ij}^{(1)}), (X_i, X_j, w_{ij}^{(2)}), \ldots, (X_p, X_q, w_{pq}^{(E)})\}$, where it is clear that $|\mathcal{E}| = E$.

- In the transformation of the tree into a graph performed by the Algorithm 2, initially, the adjacency matrix that will represent the tree converted to a graph contains only zeros, i.e., $A_{ij} \leftarrow 0$ for $1 \leq i, j \leq m$ (line 4). Then, for each edge e in the tree linking the features $(X_i, X_j, w_{ij}^{(e)})$ the weight $w_{ij}^{(e)}$ will be added to the current value of the edge (i, j) in the graph represented by the matrix A, $A_{ij} \leftarrow A_{ij} + w_{ij}^{(e)}$ (line 7). This process is repeated for all E tree edges (lines 5–9).

- To assign the weights of the tree edges various global metrics (global for each tree in the forest) can be used, such as (i) the value of the feature importance for the tree that the link occur [6]; (ii) assigning a unit weight whenever an edge exists; local metrics to the tree edge can also be analyzed, such as (a) Gini index [7] and (b) information gain ratio [31]. In this work we are using the unit weight metric.

Figure 2 shows the transformation of two trees of a given forest (left) into their respective weighted individual graphs (center), then into a directed and

---

Algorithm 3. Feature ranker through centrality measure

Require: $\mathcal{N}$: A complex network containing m vertices represented by its adjacency matrix

1: $C(\cdot)$: centrality measure, *default* $C(\cdot)$ = eigen$(\cdot)$. The measure considers the in-degree or out-degree (according to the value of the Directed and InDegree parameters)

2: Directed: Boolean indicating if the complex network is directed (true) or not directed (false), *default* false

3: InDegree: Boolean indicating if the centrality measure is calculated using the in-degree (true) or the out-degree (false), *default* true. Only applicable if Directed = true.

Ensure: Ranked: list of features with shape $< i, C(i) >$ ordered by the measure $C(\cdot)$

4: **function** featureRanker($\mathcal{N}$, $C(\cdot)$, Directed, InDegree)

5:     centralityList $\leftarrow \emptyset$ ▷ will hold tuples (feature, centrality measure value)

6:     **for** $j \in \{1, 2, \ldots, m\}$ **do**

7:         centralityList $\leftarrow$ centralityList $\cup \{ < j, C(j) > \}$

8:     Ranked $\leftarrow$ Order centralityList by measure $C(\cdot)$

9: **return** Ranked

---

non-directed complex networks (right) composed of the union of the two individual graphs, and lastly the strength centrality value for each node, considering in and out ways in the directed network. That union is performed by summing the individual graphs' edges weights (Algorithm 1, line 8). Finally, the Leaf nodes, labeled as 'L' in the trees, are not represented in the graphs. Nodes labeled with numbers indicate features.

The ranking method is applied to the complex network generated by Algorithm 1 using metrics $\mathcal{M}$ such as vertex strength (a weighted degree measure), eigenvector and Katz that calculate the centrality of the vertices (features) in the complex network. These metrics are denoted by the function $\mathcal{M} = C(\cdot)$ in Algorithm 3 (lines 6–7). The values resulting from the function $C(\cdot)$ are ordered (line 8) aiming to identify the most central features that tend to have greater importance than the other features according to these metrics. Two networks were generated (oriented and non-oriented) to analyze if the edges' orientations influence the centrality measures' final score.

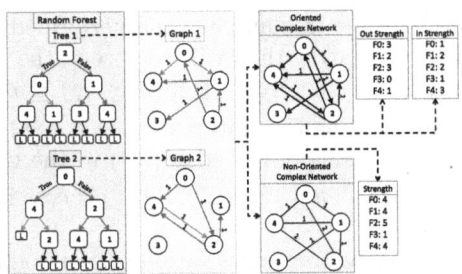

Fig. 2. Representation of two trees of a Random Forest (left) in their respective individual weighted graphs where edges represent the links between the features in the trees (center) and the oriented and non-oriented complex networks, generated from the union of the graphs (right). In trees the leaf nodes, labeled as 'L', are not represented in the graph. Nodes labeled with numbers indicate attributes.

4   Experimental Setup

In this work, 22 data generators were used (Table 1 in Appendix A) from which 85 distinct and noiseless classification datasets were generated, 10 times each; all of them for supervised learning. This number of datasets were generated to represent problems of different dimensions and complexities, as, datasets containing a large number of features, small number of examples, highly similar features, and overlapping classes. The number of instances was fixed to 300 and the number of relevant features ($\rho \in \{2, 3, 5, 7, 21\}$) and classes ($c \in \{2, 3, 4, 5, 7, 8\}$) was varied. After creating the datasets, additional features were inserted as a form of noise. All original features of each generated dataset are considered relevant features, and all additional features are noisy copies of the relevant features.

When working with artificially generated data, it is possible to know which features are the most relevant. The noise insertion in the datasets in a controlled manner makes it possible to evaluate the ranking's capability and the efficiency of the proposed ranker. The noise insertion procedure adopted in this research is based on the model used in [18] and was then adapted, according to the needs of our work.

The rates of 5%, 10%, 20%, and 40% of noise were used on the instances, and 11 levels of noise were used on features. The number of noisy features increases exponentially, where for each original dataset with ρ relevant features, $2^i \times \rho$ ($i = 0, 1, \cdots, 10$) new noisy features were generated and appended after the original (relevant) ones in the noisy dataset. That ensures the relevant features are always in the first ρ columns in the dataset. The total number of features in the noisy datasets is $m = \{\rho + 2^i \times \rho\}$. For example, considering an original dataset containing two relevant features ($\rho = 2$), for each instance noise rate, it was generated noisy datasets containing a total of $m = \{4, 6, 10, 18, 34, 66, 130, 258, 514, 1026, 2050\}$ features.

For each dataset, the noise insertion process was repeated 10 times and results averaged. In this way, each original dataset resulted in 4400 noisy datasets. Since there are 85 original datasets, a total of 374,000 noisy datasets were generated ($85 \times 10 \times 4 \times 11 \times 10$). The number of features in the datasets varies from $m = 4$, in the smaller ones, to $m = 21,525$, in the larger ones.

Every single dataset was then represented by a feature-feature complex network. The edge weight metric used here was the unit weight, which assigns a value of 1 whenever a new edge is added to the network or sums 1 to the edge weight in case it already exists. After that, three measures of centrality were applied to each network: i) strength, ii) eigenvector and iii) Katz. Each measure was then used to rank the features according to their centrality importance to the network.

After executing the proposed ranker, we evaluated the efficiency of each measure in finding the relevant features amidst the noisy features. For this purpose, the *ranking score* $\mathcal{R}$ ($0 \leq \mathcal{R} \leq 1$) is proposed (Eq. 1). The $\mathcal{R}$ value is based on the position (rank) of the relevant features in the ranking generated by the centrality measures and by the Random Forest. $\mathcal{S}$ (Eq. 2) represents the sum of the ρ relevant features' ranks (r_i is the rank found by Algorithm 3 for the relevant feature i); $\mathcal{B}$ (Eq. 3) is the sum of the best-case ranks, where all the relevant features are in the first positions of the ranking (top-ρ ranks), and $\mathcal{W}$ (Eq. 4) is the sum of the worst-case ranks, where all the relevant features are in the last positions (bottom-ρ ranks). The higher the relevant features are in the ranking the higher their $\mathcal{R}$ value will be. For instance, with $\rho = 2$ relevant in a total of $m = 10$ features, $\mathcal{B} = 1 + 2 = 3$, and $\mathcal{W} = 9 + 10 = 19$. $\mathcal{R} = 1$ if the two relevant are in the top-2 positions ($\mathcal{S} = \mathcal{B}$). $\mathcal{R} = 0$ if the two relevant are in the last two (bottom-2) ranks ($\mathcal{S} = \mathcal{W}$). If they are in any other ranks, $\mathcal{B} < \mathcal{S} < \mathcal{W}$, and $0 < \mathcal{R} < 1$.

$$\mathcal{R} = 1 - \left(\frac{\mathcal{S} - \mathcal{B}}{\mathcal{W} - \mathcal{B}} \right) \tag{1}$$

$$\mathcal{S} = \sum_{i=1}^{\rho} r_i \tag{2}$$

$$\mathcal{B} = \sum_{i=1}^{\rho} i = \frac{\rho + \rho^2}{2} \tag{3}$$

$$\mathcal{W} = \sum_{i=(m-\rho+1)}^{m} i = \frac{(2m - \rho + 1)\rho}{2} \tag{4}$$

5 Results and Discussion

In the following the notation A(noise,orientation,measure) is used, where A() means the results of running the Algorithms 1, 2 and 3 using one of the following: noise $\in \{5\%, 10\%, 20\%, 40\%\}$, orientation $\in \{g, in, out\}$, and measure $C(\cdot) \in \{eigen, katz, str\}$. Each value of the noise parameter means the percentage of instances which became noisy in those datasets; when it comes to the orientation parameter, 'g' indicates non-oriented networks, 'in' indicates oriented networks, generated as $i \leftarrow j$, and 'out' indicates oriented networks, generated as $i \rightarrow j$, as shown in Fig. 2; 'eig', 'katz' and 'str' indicate the centrality measure applied to the network for eigenvector, Katz, and vertex strength respectively.

To avoid text repetition within the same table or figure, the noise value has been replaced by a dot – A($\cdot$,orientation,measure). This dot can be read as the noise rate value described in the caption of its respective table or figure.

The Random Forest's feature importance ranker was used as a baseline to compare with the centrality measures obtained from the complex network.

The Friedman test [15] was used as the statistical test of hypotheses, which rejected the null hypothesis. Therefore, the Bonferroni-Dunn [12] post-hoc test was applied to detect any significant differences, represented by the Critical Difference Diagrams in Fig. 3 (a–d). Both tests were conducted with a confidence level of 95%.

- **40% noise**: The best average rank was obtained by A($\cdot$,out,str) followed by A($\cdot$,g,eigen), and then rf($\cdot$). Fig. 3a shows that all our proposal is not statistically different from the Random Forest feature ranker, except for A($\cdot$,out,katz), A($\cdot$,out,eigen), and A($\cdot$,in,str).

- **20% noise**: the best average rank is rf($\cdot$) followed by A($\cdot$,g,eigen), and then A($\cdot$,g,katz). In Fig. 3b we can notice that although rf($\cdot$) obtained the best average rank, it is not statistically different from A($\cdot$,g,eigen), A($\cdot$,g,katz), A($\cdot$,g,str), A($\cdot$,out,str), and A($\cdot$,in,eigen).
- **10% noise**: the best average rank was obtained by rf($\cdot$) followed by A($\cdot$,g, eigen), and after that, by A($\cdot$,g,katz). However, Fig. 3c shows that rf($\cdot$) is not statistically different from A($\cdot$,g,eigen), A($\cdot$,g,katz), and A($\cdot$,g,str).
- **5% noise**: the best average rank was obtained by rf($\cdot$) followed by A($\cdot$,g,eigen), and then A($\cdot$,g,katz). Again, Fig. 3d shows the A($\cdot$,g,eigen), A($\cdot$,g,katz), and A($\cdot$,g,str) are not statistically different from the Random Forest feature ranker.

In summary, considering all noise rates, A($\cdot$,g,eigen), A($\cdot$,g,Katz), A($\cdot$,g,str) presented results statistically not different from rf($\cdot$). That indicate that the centrality measures used in the experiments in non-oriented complex networks seem to capture the behavior of the Random Forest feature importance ranker, which is an interesting, and unexpected result.

First, let us remember how Random Forest computes feature importance. Following the induction of the forest, one of the features has its values permuted in the Out-Of-Bag (OOB) examples, which are then presented to the respective tree after the permutation, and, finally, it compares the rate of correct classification with and without the permutation of that specific feature. The permutation process is repeated for each feature in the dataset. The greater the increase in error rate generated by the permutation of the feature, the greater the importance of this feature for representing the class in this dataset [6].

In this sense, Random Forest calculates the importance of features in each tree using information beyond the forest itself (the OOB sample, i.e., using additional information in the form of future examples not seen during training, the tree induction process). On the other hand, our approach looks only for edges on trees without any additional information. Yet, the results obtained show a relation between these two distinct strategies.

If this correlation exists, in fact, then our results show that the Random Forest's feature importance is equivalent to ranking complex networks by centrality measures. In any way, our results show that it is possible to identify the importance of each feature without additional information beyond the forest itself, i.e., no Out-Of-Bag or future data is necessary.

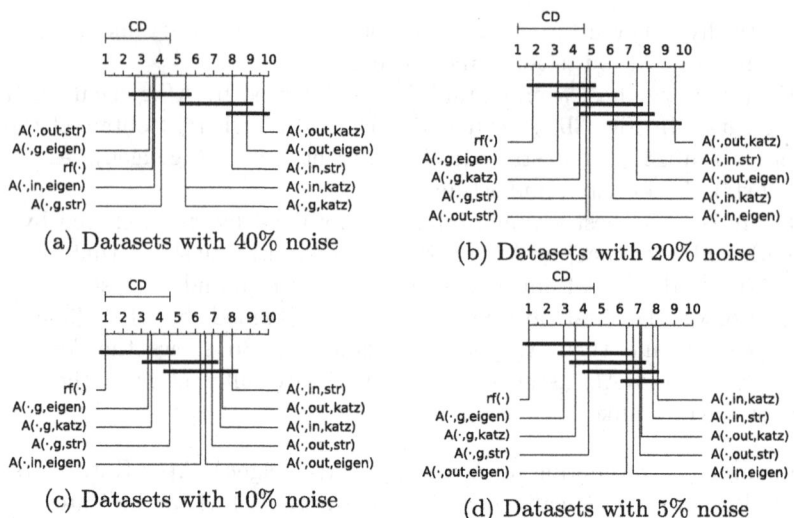

Fig. 3. Critical Difference Diagrams (CD) using Bonferroni-Dunn post-hoc test.

6   Conclusions

In this work, a feature ranking method was proposed using centrality measures in complex networks generated from Random Forests. Datasets with 40% noise in examples have a lower correlation between features once noisy features are copies of the relevant features with noise inserted. For this case, two measures, A(·,out,str) and A(·,g,eigen), outperformed, but not significantly, the Random Forest for ranking the relevant features. Our results allow us to conclude that eigenvector, Katz, and vertex strength centrality measures in non-oriented complex networks may capture the behavior of the Random Forest's feature importance ranker. Even if this is not the case, our results are strong enough to show that it is possible to identify the importance of each feature without additional information (unseen data) other than what is already represented in the forest itself. In future work, we are analyzing regression datasets as well as other edge weighting measures in our approach, such as the Gini index and the Random Forest's trees scores.

Supplementary Material

The source code, datasets, and other information to replicate the experiments carried out in this work can be found in the following the GitHub repository https://github.com/ahcantao/adbis2022_ranker.

A   Artificial Dataset Generators

Datasets were generated by the packages Scikit-learn [30], MLBench [19], and KODAMA [8].

Table 1. Description of the artificial dataset generators.

Package	Function name	#ρ Features	# Classes	Total
mlbench	spirals	2	2	1
mlbench	cassini	2	3	1
mlbench	shapes	2	4	1
mlbench	smiley	2	4	1
mlbench	2dnormals	2	2, 3, 5, 7	4
mlbench	cuboids	3	4	1
mlbench	waveform	21	3	1
mlbench	circle	2, 3, 5, 7	2	4
mlbench	ringnorm	2, 3, 5, 7	2	4
mlbench	threenorm	2, 3, 5, 7	2	4
mlbench	twonorm	2, 3, 5, 7	2	4
mlbench	xor	2,3	2, 4	2
mlbench	hypercube	2, 3	4, 8	2
KODAMA	spirals	2	2, 3, 5, 7	4
KODAMA	dinisurface	3	3	1
KODAMA	helicoid	3	3	1
KODAMA	swissroll	3	3	1
sklearn	make_circles	2	2	1
sklearn	make_moons	2	2	1
sklearn	make_classification	2, 3, 5, 7	2, 3, 5, 7	14
sklearn	make_gaussian_quantiles	2, 3, 5, 7	2, 3, 5, 7	16
sklearn	make_blobs	2, 3, 5, 7	2, 3, 5, 7	16
Total	22			85

References

1. Albert, R., Barabási, A.L.: Statistical mechanics of complex networks. Rev. Mod. Phys. **74**, 47–97 (2002). https://doi.org/10.1103/RevModPhys.74.47
2. Baranauskas, J.A., Netto, O.P., Nozawa, S.R., Macedo, A.A.: A tree-based algorithm for attribute selection. Appl. Intell. **48**(4), 821–833 (2017). https://doi.org/10.1007/s10489-017-1008-y

3. Bertini, J.R., Zhao, L., Lopes, A.A.: An incremental learning algorithm based on the k-associated graph for non-stationary data classification. Inf. Sci. **246**(Supplement C), 52–68 (2013). https://doi.org/10.1016/j.ins.2013.05.016

4. Bonacich, P.: Power and centrality: a family of measures. Am. J. Sociol. **92**(5), 1170–1182 (1987). https://doi.org/10.1086/228631

5. Breiman, L.: Bagging predictors. Mach. Learn. **24**(2), 123–140 (1996). https://doi.org/10.1007/BF00058655

6. Breiman, L.: Random forests. Mac. Learn. **45**(1), 5–32 (2001). https://doi.org/10.1023/A:1010933404324

7. Breiman, L., Friedman, J., Olshen, R., Stone, C.: Classification and Regression Trees. Wadsworth & Books, Pacific Grove (1984). https://doi.org/10.1201/9781315139470

8. Cacciatore, S., Tenori, L., Luchinat, C., Bennett, P.R., MacIntyre, D.A.: KODAMA: an R package for knowledge discovery and data mining. Bioinformatics **33**(4), 621–623 (2016). https://doi.org/10.1093/bioinformatics/btw705

9. Chandrashekar, G., Sahin, F.: A survey on feature selection methods. Comput. Electr. Eng. **40**(1), 16–28 (2014). https://doi.org/10.1016/j.compeleceng.2013.11.024

10. Costa, L.F., Rodrigues, F.A., Travieso, G., Boas, P.R.V.: Characterization of complex networks: a survey of measurements. Adv. Phys. **56**(1), 167–242 (2007). https://doi.org/10.1080/00018730601170527

11. Cupertino, T., Carneiro, M., Zheng, Q., Zhang, J., Zhao, L.: A scheme for high level data classification using random walk and network measures. Exp. Syst. Appl. **9** (2017)

12. Dunn, O.J.: Multiple comparisons among means. J. Am. Stat. Assoc. **56**(293), 52–64 (1961). https://doi.org/10.1080/01621459.1961.10482090

13. Erdős, P., Rényi, A.: On the evolution of random graphs. In: Publication of the Mathematical Institute of the Hungarian Academy of Sciences, pp. 17–61 (1960)

14. Ferreira, L.N., Zhao, L.: Time series clustering via community detection in networks. Inf. Sci. **326**, 227–242 (2016). https://doi.org/10.1016/j.ins.2015.07.046

15. Friedman, M.: A comparison of alternative tests of significance for the problem of m rankings. Ann. Math. Stat. **11**(1), 86–92 (1940). https://doi.org/10.1214/aoms/1177731944

16. Hashemi, A., Dowlatshahi, M.B., Nezamabadi-pour, H.: MGFS: a multi-label graph-based feature selection algorithm via pagerank centrality. Exp. Syst. Appl. **142** (2020). https://doi.org/10.1016/j.eswa.2019.113024

17. Hastie, T., Tibshirani, R., Friedman, J.: The Elements of Statistical Learning: Data Mining, Inference and Prediction, 2nd edn. Springer New York (2009). http://www-stat.stanford.edu/~tibs/ElemStatLearn/

18. Khoshgoftaar, T.M., Hulse, J.V.: Empirical case studies in attribute noise detection. IEEE Trans. Syst. Man Cybern. Part C (Applications and Reviews) **39**(4), 379–388 (2009). https://doi.org/10.1109/TSMCC.2009.2013815

19. Leisch, F., Dimitriadou, E.: mlbench: machine learning benchmark problems (2010), R package version 2.1-1(2010)

20. Li, J., et al.: Feature selection: a data perspective. ACM Comput. Surv. **50**(6) (2017). https://doi.org/10.1145/3136625

21. Louppe, G., Wehenkel, L., Sutera, A., Geurts, P.: Understanding variable importances in forests of randomized trees. In: Advances in Neural Information Processing Systems, vol. 26, pp. 431–439. Curran Associates, Inc. (2013). https://dl.acm.org,https://doi.org/10.5555/2999611.2999660

22. Ma, Y., Guo, L., Cukic, B., Lane: a statistical framework for the prediction of fault-proneness. In: AAdvances in Machine Learning Applications in Software Engineering. pp. 237–265. IGI Global (2006). https://doi.org/10.4018/978-1-59140-941-1.ch010

23. Mairal, J., Yu, B.: Supervised feature selection in graphs with path coding penalties and network flows. J. Mach. Learn. Res. **14**(39), 2449–2485 (2013). http://jmlr.org/papers/v14/mairal13a

24. METZ, J.e.a.: Redes complexas: conceitos e aplicações. Tech. Rep. 290, ICMC-USP January 2007. http://repositorio.icmc.usp.br//handle/RIICMC/6720

25. Miao, J., Niu, L.: A survey on feature selection. Procedia Comput. Sci. **91**, 919–926 (2016)

26. Moradi, P., Rostami, M.: A graph theoretic approach for unsupervised feature selection. Eng. Appl. Artif. Intell. **44**, 33–45 (2015). j.engappai.2015.05.005

27. Neto, F.A., Zhao, L.: Random Walk in Feature - Sample Networks for Semi-Supervised Classification, pp. 1–6 (2016). https://doi.org/10.1109/BRACIS.2016.41

28. Ni, B., Yan, S., Kassim, A.: Learning a propagable graph for semisupervised learning: classification and regression. IEEE Trans. Knowl. Data Eng., **24**(1), 114–126 (2012). https://doi.org/10.1109/TKDE.2010.209

29. Oshiro, T.M., Perez, P.S., Baranauskas, J.A.: How Many trees in a random forest? In: Perner, P. (ed.) MLDM 2012. LNCS (LNAI), vol. 7376, pp. 154–168. Springer, Heidelberg (2012). https://doi.org/10.1007/978-3-642-31537-4_13

30. Pedregosa, F., et al.: Scikit-learn: machine learning in Python. J. Mach. Learn. Res. **12**, 2825–2830 (2011)

31. Quinlan, J.R.: C4.5: Programs for Machine Learning. MA, San Francisco (1993). https://doi.org/10.1007/BF00993309

32. Rathkopf, C.: Network representation and complex systems. Synthese **195**(1), 55–78 (2015). https://doi.org/10.1007/s11229-015-0726-0

33. Silva, T.C., Zhao, L.: Machine Learning in Complex Networks. Springer, Cham (2016). https://doi.org/10.1007/978-3-319-17290-3

34. Venkatesh, B., Anuradha, J.: A review of feature selection and its methods. Cybern. Inf. Technol. **19**(1), 3–26 (2019). https://doi.org/10.2478/cait-2019-0001

35. Zhang, Z., Hancock, E.R.: A graph-based approach to feature selection. In: Graph-Based Representations in Pattern Recognition. pp. 205–214. Springer, Heidelberg (2011). https://doi.org/10.1007/978-3-642-20844-7_21

36. Zhao, Y., Zhang, Y.: Comparison of decision tree methods for finding active objects. Adv. Space Res. **41**(12), 1955–1959 (2008). https://doi.org/10.1016/j.asr.2007.07.020

37. Zheng, W., Zhu, X., Zhu, Y., Hu, R., Lei, C.: Dynamic graph learning for spectral feature selection. Multim. Tools Appl; **77**(22), 29739–29755 (2017). https://doi.org/10.1007/s11042-017-5272-y

38. Zhu, Y., Zhong, Z., Cao, W., Cheng, D.: Graph feature selection for dementia diagnosis. Neurocomputing **195**, 19–22 (2016). https://doi.org/10.1016/j.neucom.2015.09.126

Data Science Methods

Data Narrative Crafting
via a Comprehensive and Well-Founded Process

Faten El Outa[1]([envelope]) [ORCID], Patrick Marcel[1] [ORCID], Veronika Peralta[1] [ORCID],
Raphaël da Silva[2], Marie Chagnoux[3], and Panos Vassiliadis[4] [ORCID]

[1] University of Tours, Blois, France
{faten.elouta,patrick.marcel,veronika.peralta}@univ-tours.fr
[2] Rue89 Strastbourg, Strasbourg, France
[3] University Paris 8, Paris, France
marie.chagnoux@univ-paris8.fr
[4] University of Ioannina, Ioannina, Greece
pvassil@cs.uoi.gr

Abstract. Data narration is the activity of crafting narratives supported by facts extracted from data exploration and analysis, using interactive visualizations. While data narration has recently attracted much attention, the process of crafting data narratives is loosely documented and has not yet been formally described. In this article, we propose a comprehensive and well-founded process to fill this need. It aims at (i) supporting the complete cycle of data narration, from the exploration of data to the visual rendering of the narrative, (ii) being flexible enough to cover a wide range of crafting practices, and (iii) being well founded upon with a conceptual model of the domain.

Keywords: Data narrative crafting · Data journalism · Process

1 Introduction

Data narratives are receiving increasing interest from several research communities (e.g., visualization, data management, computer-human interfaces) [2] and many application domains (e.g. journalism, business, e-government, health). They are largely used by journalists, scientists, and other communicators, to convey striking messages to a given audience. In addition, the crafting of a data narrative includes a variety of activities, including the analysis of data, the drawing of relevant messages from data, the structuring of messages into a coherent story and its visual rendering. Despite this diversity of activities, sometimes even conducted by different people with varied professions and skills, there is no framework, workflow, or tool for supporting the crafting of data narratives.

In an effort to clarify the concepts of data narratives, we recently defined a data narrative *as a structured composition of messages that (a) convey findings over the data, and, (b) are typically delivered via visual means in order to*

© Springer Nature Switzerland AG 2022
S. Chiusano et al. (Eds.): ADBIS 2022, LNCS 13389, pp. 347–360, 2022.
https://doi.org/10.1007/978-3-031-15740-0_25

facilitate their reception by an intended audience, and we proposed a conceptual model describing and structuring the key concepts around data narratives [12]. This model (described in Sect. 2) is organized in 4 layers: factual, intentional, structural and presentational, which reflect the transition from raw data to the visual rendering of the story. With this definition and model in mind, our aim in this paper is to contribute with a study of the dynamic aspects of data narrative crafting. Like many works in the literature (e.g., [5,8,10]), we postulate that the different forms of data narration can be described by a comprehensive process encompassing the various activities ranging from data exploration to the rendering of the narrative. A formal description of this process will benefit novice data narrators, like e.g., non technical data journalists, and will be instrumental to the development of tools for supporting advanced data narrators.

Accordingly, we reviewed the literature around the process of crafting data narratives, and we conducted a survey with data journalists in order to understand how they craft a data narrative. As an outcome of the former, we found that globally the research communities agreed in the fact that the crafting process includes three main phases: (i) the *analyzing* phase that handles the activities of exploring data, retrieving findings and formulating messages learned from data, (ii) the *structuring* phase that includes the activities to organize the plot of the narrative in an understandable way and, (iii) the *presenting* phase that covers the activities to convey the structured messages visually. However, our bibliographical study revealed the absence of a comprehensive and well-founded process that covers the main activities of the crafting process, specially those dealing with user intentions and their tight relation to data analysis. From the survey, we observed the crafting workflows regularly followed by 18 data journalists, and we contrasted them to the literature. It turned out that journalists follow the same three phases, mostly in a linear way, attaching less attention to the structuring phase, while spending more time in the analyzing phase.

These considerations from the literature study and the survey with data journalists enabled us to identify the activities (and their chaining) for crafting data narratives. Based on those, we propose a comprehensive and well-founded process that (i) covers the whole cycle of data narrative crafting, from exploration of the data to the visual presentation of the narrative, (ii) accommodates a wide range of practices observed on the field, and, (iii) is founded on a conceptual model of the domain that clarifies the concepts involved in the process [12].

The scope of our method targets the population of data journalists or any other data enthusiast that constructs data narratives out of existing data. The reason for proposing the method is exactly the observed discrepancy between literature and practice, with omissions of important parts from both sides. *Thus one significant contribution of our work is the explicit treatment of all the steps that should be involved in the process.* Secondly, apart from providing a methodological guidance, our method can enable *the support of the process via tooling.* Indeed, there is a lack of integrated tools covering the whole crafting process and recommending actions to less-experienced narrators. In particular, an application that would automatically document the data exploration and narration crafting is desperately needed by data workers, who spend hours to document

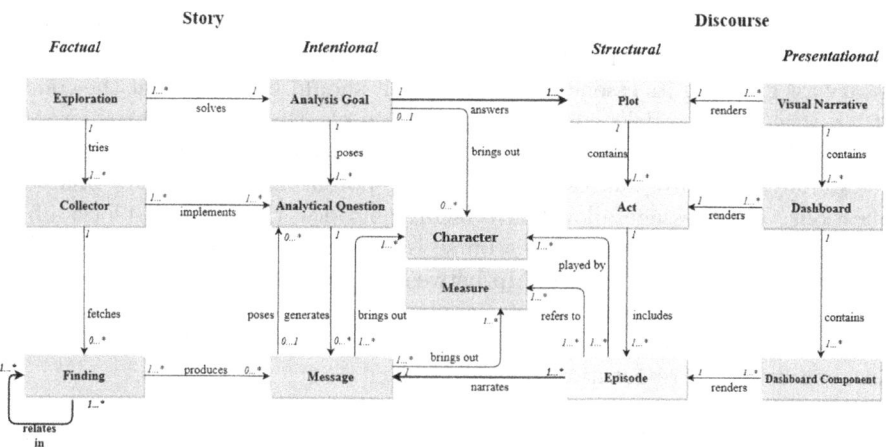

Fig. 1. The conceptual model for data narratives (relations in bold were extended w.r.t. the original version in [12])

their work. This is important for reproducibility, transparency, and linkage, and requires a conceptual model and a process that are both consensual.

The paper is organized as follows: Sect. 2 recalls the key concepts of the conceptual model proposed in [12]. Section 3 reviews the related work for the processes of crafting a data narrative, and Sect. 4 discusses the survey we conducted with data journalists. The proposed process is detailed in Sect. 5. Section 6 concludes and draw research directions.

2 A Conceptual Model for Data Narratives

We recently proposed in [12] a conceptual model of data narratives providing a principled definition of the key concepts of the domain, along with their relationships, and clarifying their role and usage (see Fig. 1).

This model is based on 4 layers following Chatman's organisation [4], who defined narrative as a pair of (a) *story* (content of the narrative), and, (b) *discourse* (expression of it). In our model, the *factual* layer handles the *exploration* of facts (i.e., the underlying data), via a set of *collectors* that allow for manipulating facts with varied tools and fetching *findings*, in an objective way, while the *intentional* layer models the subjective substance of the story, identifying the *messages, characters* and *measures* the narrator intends to communicate, and tracing how they are obtained through *analytical questions*, according to an *analysis goal*. As to the discourse, the *structural* layer models the structure of the data narrative, its *plot* being organized in terms of *acts* and *episodes*, while the *presentational* layer deals with its rendering, that is communicated to the audience through visual artifacts (*dashboards*[1] and *dashboard components*).

---

[1] We use the term dashboard since it is general enough to accommodate various types of visualizations (e.g. a Business Intelligence dashboard, an infographics, a section in a python notebook, a section in a blog or web page).

The interested reader is redirected to [12] for a deeper presentation of the model. Here, we will highlight the main decisions behind the model that are necessary for grasping its essence. Importantly, it should be noted that the concept of *message* is the model's corner stone, which is clearly evidenced by the way we have related message to the other concepts. A specific message is rooted in the facts analyzed, conveying essential findings, potentially raising new analytical questions. The message allows introducing episodes, the building blocks of the discourse. Each episode of the discourse is specifically tied to a message which it aims to convey. The relationship between messages and episodes is the basis for structuring stories that address analysis goals, narrated by structured discourses (with cohesive acts being the backbone of the narrative structure) and dashboards their presentational counterpart.

3   Related Work

In this section, we review the works describing the internals of the data narration process, as well as the tools that automate (part of) the crafting process.

3.1   Global Data Narration Processes

Data narration is a complex process, at the crossroads of several domains: data exploration, data visualization, data management, etc. Despite the many contributions in each of these areas, few works offer comprehensive workflows describing the entire data narration process. The first attempt to model data narration processes come from the visualization community. For example, Kosara and Mackinlay [9] proposed a two-phases process: First, narrators collect information and *explore* their interrelationships, pointing to key facts, and then, they *tie* those facts together into a story. Chen et al. [5] surveyed early proposals and concluded that their crafting processes are composed of two main phases: (a) *visual analytics*, which requires seeing all aspects of complex data, explore their interrelationships, and is supported by multiple coordinated views and sophisticated interaction techniques, and (b) *storytelling*, which is meant to convey only interesting or important information (i.e. findings) extracted through the analysis, presented in a simple and easily understandable way.

To bridge the gap between these two phases, Chen et al. proposed an intermediate one, called *data synthesis* [5]. In this phase, the narrator assembles and organizes the findings to be communicated, to represent explicitly the essential relationships between them, building a compelling narrative. Lee et al. [10] also identified three main phases: *explore data* to retrieve findings, *make a story* to turn findings into a sequence of narrative pieces to build the plot of the narrative, and *tell a story* to materialize the plot in a visual manner. The authors stress the importance for the data narrator to go back and forth between the exploration and the story-making phases. More recently, Duangphummet et al. [6] proposed a protocol consisting of the following phases: *conceptualization* of the data narrative domain, targeted audience and distribution channel, *data preparation* to deliver data that is relevant to the use, *realization* to deliver a storyline with detailed content and an initial form of key visualizations, *visualization design* to redesign the visualizations and create visualization prototypes, and finally, the *visualization development* where technical requirements are defined, and the key visualizations for target devices are developed and deployed.

In addition to [10], many works underline the importance of moving between the data narrative crafting phases. For instance, Wang et al. [19] ran a workshop on data comics, organized by an interdisciplinary team with expertise in data visualization, graphic design, data comics, and illustration. They observed that to create stories, students require to *move back and forth between the story, visualizations, and the data.*

Besides the previously described works proposing global crafting processes, some works describe subprocesses, focusing on the necessary activities to be conducted. Without being exhaustive, we mention here some major contributions. Battle and Heer [1] identified three ways to start a data narrative: having a precise idea in mind, having a vague idea refined during data exploration, or having no idea before exploring the data. Weber et al. also point that the crafting process starts by either an idea, a problem or a question [20]. Notably, many works underline the importance of different story structures and different kinds of interactivity in data narration [13, 20]. In particular, Weber et al. [20] encourage to use non-linear structures and set up interactivity. Many works specifically deal with the phase of structuring the narrative [5, 10, 14, 18].

Finally, very few works highlight the importance of intentional aspects. Thudt et al. [16] stress that subjective perspectives can be introduced at every step of visualization creation: during data collection and processing, visual encoding, and when refining the presentation. In the context of OLAP cube exploration, Vassiliadis et al. [17] propose a set of intentional operators to express high-level analytical intentions and automate their translation to database queries.

3.2 Automated Data Narration

Many recent works addressed the automatic generation of data narratives, providing another source of insights on how this process is perceived.

Wang et al. [18] conducted a qualitative analysis of 245 infographics examples to explore the infographics design space in terms of structures, sheet layouts, fact types, and visualization styles. Based on those, the authors propose a system for supporting a fact sheet generation pipeline consisting of three phases: (i) fact extraction, (ii) fact composition, and (iii) presentation synthesis. Shi et al. [15] proposed Calliope, a system that can automatically generate visual data stories with facts arranged into a logical sequence. It consists of two main modules: (i) the story generation engine, for generating, choosing and organizing the facts that will participate in the narrative, and (ii) the story editor, that visualizes the data story (generated as a series of visualization charts) and allows the users to change it based on their preferences. Shi et al. [14] described the workflow for crafting data videos, consisting of 4 phases: (i) *collecting a series of data facts* around a certain topic, (ii) *constructing a storyline* as an assembly of these data facts into a sequence, (iii) *choosing data visualizations* for the data facts and deciding how to animate them by drawing a storyboard, and finally, (iv) *realizing the storyboard* via a design software in which the narrator edits and combines the animated visualizations until a coherent data video is accomplished.

In CineCubes [7], Gkesoulis et al. detail the process of crafting a data movie in the form of a powerpoint presentation, to answer a specific user's need described by a query. First, an introductory act is built with the initial query, and two subsequent acts are used to put context. These acts contain visualizations highlighting important facts, as well as text and audio describing these facts. A summary act concludes with all the important highlights of the previous acts.

In all these works, the proposed phases are consistent with those described in the previous subsection. Being a mostly automatic generation, the construction is linear in the sense that there is no back and forth movements between phases. In addition, they target a specific domain or data format and organize stories accordingly to pre-established patterns. In particular, we highlight the absence of intentions, that are, at best, modeled via an initial query or a topic.

Lessons Learned. Most of the works describing the data narration process agree on the 3 general phases of exploration (to retrieve findings), structuring (organizing the information gathered into narrative pieces) and presentation (crafting visual artifacts). Automated data narration is still in its infancy, mainly applying rigid patterns and lacking the necessary flexibility of moving between the 3 phases. One of the key findings is that the intentional layer of the model presented in Fig. 1 is largely absent from the works reviewed. This means the substance of the story, i.e., the **composition** of story elements (analytical questions and hypothesis, messages, etc.) as pre-processed by the author's cultural code [4] is ignored. We claim that this absence is regrettable; if data narrations are to be shared, reused, their crafting process documented, then this intentional layer deserves more attention.

4   Data Journalist Practices

A preliminary study, in the form of a survey [3] (in French), investigates the professional practices of data journalists.

The survey consisted of 32 questions[2], answered by 18 data journalists from 14 French regions, who have worked on a big variety of topics, including elections, environment, cinema, terrorism, paradise-papers, real estate. For nearly 50% of them, data narration is at the core of their professional activity, and is occasional or marginal for the others. Concerning training, 56.3% studied social sciences, 18.8% studied sciences and 24.9% graduated from law or journalist schools. One of the journalists works for the International Consortium of Investigative Journalists (ICIJ), 5 of them work for the national press, and the 12

---

[2] https://drive.google.com/drive/folders/1zDzP_ndSlQUJCbtFMVzJDnIbyXK1D2_l?usp=sharing (in French). Note that for some questions more than one answer was possible, and that journalists could leave the questions unanswered.

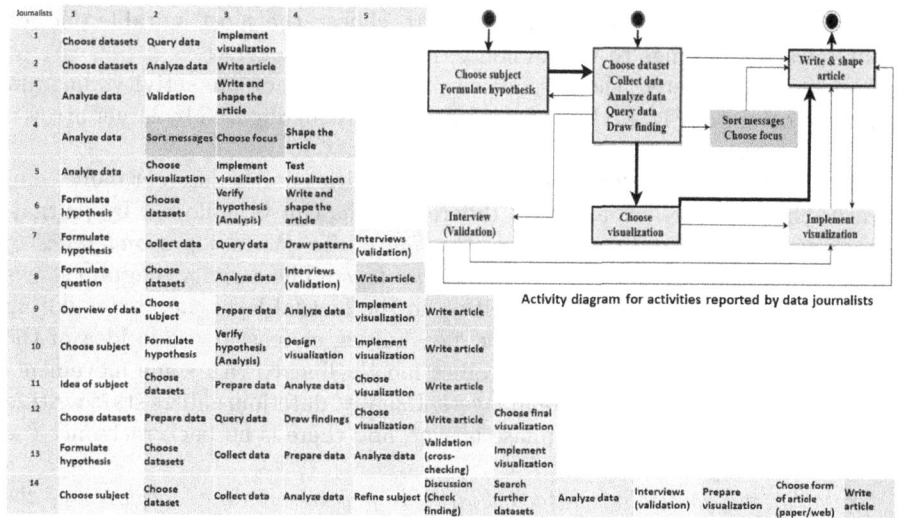

Fig. 2. Sequence of activities reported by journalists (Color figure online)

remaining work for the regional press. Regarding their general working habits, 75% of them work alone. They usually work on open data (72.2%) and more specifically on data from public institutions (44.4%). They consume from minutes to months during the data narration and use different tools during data exploration, such as spreadsheets (93.8%), scripts (50%), notebooks (18.8%), powerBI-like tools (31.3%) and some machine learning tools (28.6%).

Two main questions were asked on their data narration practices. For the first one, "How does a data story's subject emerge?", multiple answers were possible. The answers showed that the goal, or subject, of an article emerges from: an idea to be confirmed by data (68%), a dataset which needs exploration to reveal important facts (68%), a refinement of the subject while exploring the dataset (48%). The second, open question was: "What is the general workflow you apply for data narrative crafting?". Figure 2 sketches the answers provided by 14 of the 18 journalists, where activity names summarize journalists' descriptions of their main activities[3], rows correspond to journalists and column numbers reflect the sequence of activities. We color these activities according to the layers of the conceptual model: factual (pink), intentional (purple), structural (yellow) and presentational (blue). Gray-colored cells indicate that the activity may overlap structural and (more probably) presentational tasks. In addition, activities concerning the checking of findings and the validation of messages (namely interviews, validation or cross-checking), aiming at transforming a factual object into an intentional one, are in between the factual and intentional layer. Similarly, visualizations are used both in the factual layer, to understand data and retrieve

---

[3] Since the question was open, we homogenized the answers and grouped them into few categories.

findings, and in the presentational layer, to choose the most suitable one for communicating findings to the audience in a visual manner.

We have abstracted these sequences in the form of an activity diagram (top-right corner of Fig. 2). Most frequent paths are highlighted by larger arrows.

Lessons Learned. Figure 2 shows that many activities under different names aim towards the same action, and that different paths can be followed by journalists when crafting a data narrative. *The figure also shows a preponderance of activities from the factual and the intentional layer.* The activity diagram shows that journalists enter the workflow either in the factual layer, i.e., by exploring a dataset, or by the intentional layer, i.e., having at least a vague idea of the subject. After this, the workflow becomes mostly linear, with some movements between the factual and intentional layers. Usually, data journalists start writing their articles once the analysis phase is over, and there is no backtrack once the presentational layer is entered.

Notably, the journalists attached little importance to the activities of the structural layer. At the exception of one of them, structuring activities are either hidden in writing activities or even not mentioned explicitly. Precisely, many of them agree that while data exploration usually takes long, visual storytelling can be extremely fast, potentially done on the fly, with some of them actually not even involved in the writing of the article. For those that mention it, the activity "write article" includes several hidden details concerning the organization of messages that should be communicated, the visual presentation and communication of the analysis results.

Overall, we can say that there is a chasm between what practitioners do and what literature suggests – and in fact, there are deficits in both sides. On the one hand, compared to what is reported in the literature, the work of the data journalists is over-emphasizing the intentional part and under-investing on the structural and the presentational part. On the other hand, when it comes to the literature, the presented methodologies overemphasize presentation and (to some extent) structuring, and pay much less attention to the intentional part. A process that gracefully hosts all aspects of narrative construction would facilitate narratives that are more complete and intuitive.

5   A Process for Crafting Data Narratives

From the literature review and the survey with journalists, we synthesize a set of requirements for a comprehensive data narration process and we propose a process that fulfills these requirements.

5.1   Requirements

First of all, we note the absence in the literature of a whole workflow for crafting data narratives, including all the activities identified in Sects. 3 and 4. Figure 3

Analyzing		Structuring	Presenting
Dataset collection	**Goal and question formulation**	**Determine audience**	**Visual narrative setting**
Data preparation [6], preprocessing [5]	Idea precede data search [1,19]	Conceptualize targeted audience [6]	Visualization design [6]
Collect information/data [6,8,13,15]	Question precede data search [19]		Tell story [9]
Collect data Choose datasets	Overview data	**Select messages**	Visualize data story [14]
Search further datasets Prepare data	Refine subject	Choose facts [14]	Presentation synthesis [17]
Collector trial	Choose subject	Sort messages	Realize storyboard [13]
Explore data [4,7,8,9]	Formulate question		Refining presentation [15]
See aspects of complex data [5]	Formulate hypothesis	**Choose narrative structure**	Choose form of article
Analyze data Verify hypothesis		Realization [6]	Shape article
Visualization trial	**Message formulation**	Tie fact into story [8]	
Choose data visualization [13]	Convey interesting information [5]	Make story [9]	**Interactivity choice**
Make data visible [19]	Draw conclusion	Construct storyline [13]	Set up interactivity [19]
Choose visualization		Choose focus	
	Message validation		**Dashboard implementation**
Finding formulation	Cross-checking	**Message mapping**	Visualization development [6]
Fact extraction [17], Choose fact [13]	Validation	Organize information pieces [5]	Convey message through visualization [19], Convey clear message [14]
Explore interrelationship [5,8]	Interviews	Organize facts [14]	Visual encoding [15]
Draw finding Draw pattern		Fact composition [17]	Implement visualization
Findings validation		Organize information pieces [15]	Test visualization
Discussion			Choose final visualization

Fig. 3. The main activities for crafting data narratives identified from the literature and a survey with data journalists (AQ abbreviates Analytical Question) (Color figure online)

depicts the activities as phrased in the literature (in gray boxes) and by journalists (in green boxes). We group those referring to the same task and propose new names (the bold ones in Fig. 3) which are consistent with the conceptual model of Fig. 1.

In more details, most authors [5,6,9,10,14,15,18,19] agree that data narration process includes three main phases: (i) *analyzing*, (ii) *structuring*, and (iii) *presenting*. The survey reveals that the data journalists agreed with the literature, especially on the phases (i) and (iii). In Fig. 3, activities are grouped according to these phases. We remark that activities pertaining to the factual and intentional layers of the conceptual model are mixed in phase (i). In addition, while the literature rarely mentions the activities pertaining to the intentional layer, these activities are often pointed by data journalists. Furthermore, as we explained in [12], the substance of a story, representing the narrator's intention in reporting the story, is a constituent of the data narrative [4]. Conversely, while the journalists did not attach much importance to the activities of the structural phase, this aspect is emphasized in the literature. Finally, as noted in [10,19], the narrator should have the possibility to move freely back and forth between the different phases of data narration. However, this movement should not prevent that different groups of activities could be conducted by different persons with different profiles. These groups of activities, identified by layers in the conceptual model [12], should be as isolated as possible.

To summarize, a comprehensive workflow for crafting data narratives should satisfy the following requirements:

- (R_1) cover the activities and the paths identified by the survey with data journalists, which are depicted in Fig. 2,

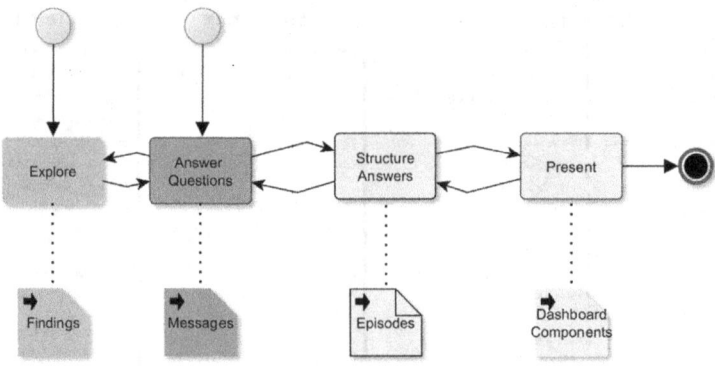

Fig. 4. The process of data narrative crafting (Color figure online)

- (R_2) cover the activities of the three phases identified from the literature,
- (R_3) allow the free back and forth transition between phases,
- (R_4) clearly delineate the different layers of the conceptual model [12] within its activities.

5.2   The Process of Crafting Data Narratives

In this subsection, we propose a comprehensive process for the crafting of data narratives that covers the activities and paths proposed in the literature and reported by journalists (requirements R_1 and R_2), while also being founded upon and coherent with the conceptual model (R_4) and allowing the back and forth movement between its phases (R_3).

The phases of the process are illustrated in Fig. 4. All phases are accompanied by the resulting outcomes, which are exactly the basic constituents of our conceptual model (R_4). We retain the same coloring (pink for factual exploration, purple for intentional question-answering, yellow for the structuring of the answers of the intentional questions into a plot, and blue for presentation). Observe that the factual and intentional layers of the conceptual model are well differentiated here, contrarily to the literature that mix them into one phase.

Consistently with Fig. 2, the process flexibly starts either with the existence of a data set, which is to be explored for findings, or with the emergence of an initiating question, that begs to be answered. This flexibility is important in the sense that prescribing a specific starting point for the collection of findings from the data is not what actually the practitioners do.

The following paragraphs describe the activities pertaining to each phase, including the activities abstracted from the literature and survey results (shown in Fig. 3), and some new activities that intent to cope with missing tasks.

Note that such activities should not be considered as steps to be executed sequentially. Conversely, many activities can be initiated and executed in parallel. Arrows in Fig. 5 indicate a *depends on* relationship. For example, message

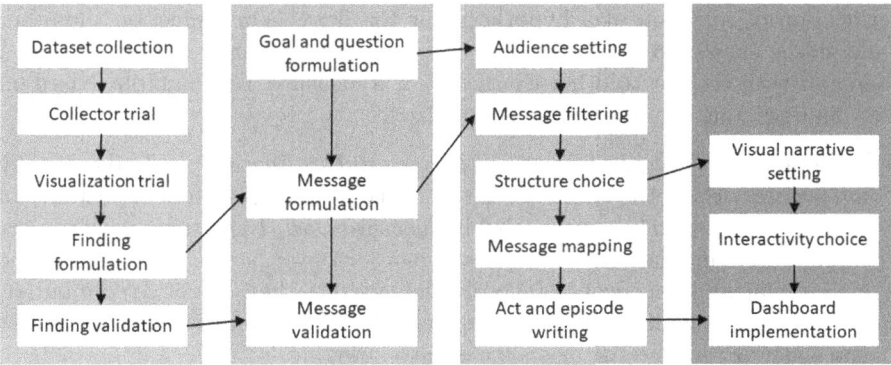

Fig. 5. Activities for data narrative crafting

validation depends on message formulation, as it is necessary to formulate messages before validating them. In addition, at any time, it is possible to come back to previously executed activities (e.g. to rewrite messages or formulate new ones). Backtrack arrows are omitted for clearness.

Exploration. The exploration phase, handling the factual layer, concerns several activities: (i) dataset collection, concerning source selection, data extraction, integration and preprocessing, (ii) trial and reuse of several collectors (i.e. querying, profiling and mining tools) and (iii) trial of diverse visualizations (crosstabs, graphics, clusters, etc.) for collecting findings, then, (iv) finding formulation, concerning the storage of findings and their relationships, and (v) finding validation, which is typically done via statistical tests, but also by discussing and crosschecking with additional data sources (as done by data journalists) and confronting with the state of the art (as done by data scientists [11]). Note that some findings may lead to additional analysis, triggering more collectors and visualisations, or even the collection of more datasets. The exploration phase is time-consuming (data journalists measure it in days or even in months). Then many activities are frequently performed asynchronously.

Question-Answering. This phase, neglected in the literature, handles the intentional layer and concerns activities for (i) formulating goals and questions, (ii) drawing messages from findings, and (iii) validating messages. It supports explicitly the data narrator intention, as its proposed activities help in formulating an analysis goal and a set of analytical questions that reflect their intention.

Furthermore, to cope with literature lacks (evidenced in Fig. 3), we propose a message formulation activity, concerning the derivation of messages from findings, and the identification of characters and measures (the relevant constituents of messages [4]) to be highlighted to the audience.

Remember the distinction of outcomes: A *finding* is a highlight, (or equivalently, a pattern) annotating a dataset, or a subset of it. A finding can be a typical pattern (like e.g., an association rule, or a path of a decision tree) or the verification of rejection of a hypothesis for the data. A *message*, on the other hand, is the answer to the intentional question that exploits a finding to label a character with respect to other characters or a measure. For example, based on data findings, one can answer questions like:

– By comparing *Daily Infections* in *France* to *EU Average*, we find that they are *similar*. Here, the character is the entity *France*, which is an instance of the concept *Country*, and we label its measure *DailyInfections* with respect to another peer character, *EU Average*.
– By correlating the concept *News Authenticity* to the concept *Media outlet*, we find a non-significant correlation, rejecting the hypothesis that the outlet can solely determine the existence of fake news.

The internals of the process, detailed in Fig. 5, allow the flexibility of exploring several paths, that can be chained according to narrator's habits and specificities of the task on hand, alternating data analysis, finding derivation and message writing, but also allowing for the validation of findings and messages, or even the expression of new analytical questions.

In any case, the identification of such messages and their structuring is a task that is practically absent from the related literature, significantly present in the everyday work of practitioners, and structured in our model for the first time.

Structuring. The structuring phase, the missing part in the data journalists processes, handles the structural layer, describing activities for organizing the plot of the narrative in terms of acts and episodes [7] (adopted from the classical structure of plays). Plot setting starts by (i) determining the audience, (ii) eventually selecting a subset of messages for such audience, and (iii) choosing an appropriate narrative structure. Then, (iv) messages are mapped to acts and episodes. In more details, these activities allow the arrangement of the thoughts of the data narrator into different layers: an act which is a major piece of information, and is composed of several episodes that are of lesser importance on their own [12]. Remember that the result of the structuring is an *episode*, which is the annotation of a message (which has a simple structure and labeling) with comments on the context, significance, essence, etc., in other words with the content that makes the message interpretable by human beings.

Also, observe in Fig. 5, the existence of a specific activity to make the actions of writing acts and episodes explicit. Such activities can be performed before or at the same time as choosing visual means.

Presentation. Finally, the presentation phase handles the presentational layer, and includes activities for (i) setting the type of visual narratives, (ii) setting the interactivity mode, and (ii) implementing dashboards for conveying acts and episodes to the audience. Such activities carry on the visualization level and build for each act an associated dashboard and present the narration in a complete visual narrative. Remember that *dashboard components* are representations of episodes in (typically) a visual form of communication, including text, figures, charts, data plots, or any other means to convey the message.

5.3 Discussion

The purpose of this paper is to support data storytelling via a method based on a conceptual model, that fits in all possible domains where storytelling is applicable. One of the main drivers for the method was to bridge the observed gap between literature and practice, with omissions of important parts from both sides. As a result, the method allows the structuring of the overall process in phases and facilitates valid translations between important intermediate results that are necessary to construct a data narrative, in a way that is flexible, realistic and adequately structured.

The intended users of the method are data journalists and data enthusiasts. We ran various preliminary experiments to empirically assess its potential of being adopted by various data workers. We organized a challenge[4] where data enthusiasts (among which journalists, students, social workers) were mixed with data scientists, aiming at producing data narratives using the open data of a French city. Interestingly, all teams started with a vague idea of the topic they wanted to treat, which was refined after many iterations among data collection, data analysis and question formulation. All of them used a unique timeline for structuring their narratives, which were rendered with varied forms. In another experiment, 44 students in BI were asked to craft a data narrative using a dataset they were familiar with, while having no experience in data narratives crafting. Students were observed during crafting, and some of them, especially those less skilled, were asked to indicate the sequence of activities they realized. This helped them to start, particularly having to write down the analytical questions that guided the data analysis, and to write down messages and early think about structuring. Finally, some of the authors of the present paper crafted a data narrative about tuberculosis, targeted for epidemiologists and public health decision-makers in Gabon. The whole crafting process is described in [11]. We highlight the importance of goal setting and message formulation activities, both of them being validated by many experts with different profiles. In particular, in scientific context, messages are not only validated by statistical tests but also confronted to other data sources and similar works of the state of the art and should pass risk assessment tests.

6 Conclusion

In this paper, we proposed a process for crafting data narratives, that covers the whole cycle of data narration, from data exploration to the visual presentation of the narrative. Backed by a literature review and a survey with data journalists, it accommodates a wide range of practices observed on the field, via clearly delineated activities, while being well founded upon a conceptual model of the domain [12]. We believe that these two models, static and dynamic, can serve as a stepping stone for future research in the area of data narration.

Extending the proposal with tool support for guiding the narrator along the process and (semi-)automating some tasks, is a clear path for future work.

---

[4] Sponsored by French CNRS https://www.madics.fr/event/titre1617704707-3351/#madona.

References

1. Battle, L., Heer, J.: Characterizing exploratory visual analysis: a literature review and evaluation of analytic provenance in tableau. Comput. Graph. Forum **38**(3), 145–159 (2019)
2. Carpendale, S., Diakopoulos, N., Riche, N.H., Hurter, C.: Data-driven storytelling (Dagstuhl seminar 16061). Dagstuhl reports (2016)
3. Chagnoux, M.: La datavisualisation, double point d'entrée du datajournalisme dans la PQR (in French). Interfaces numériques **9**(3), 1–7 (2020)
4. Chatman, S.: Story and Discourse: Narrative Structure in Fiction and Film. Cornell University Press (1980)
5. Chen, S., et al.: Supporting story synthesis: bridging the gap between visual analytics and storytelling. In: TVCG (2018)
6. Duangphummet, A., Ruchikachorn, P.: Visual data story protocol: Internal communications from domain expertise to narrative visualization implementation. In: VISIGRAPP (2021)
7. Gkesoulis, D., Vassiliadis, P., Manousis, P.: Cinecubes: aiding data workers gain insights from OLAP queries. Inf. Syst. **53**, 60–86 (2015)
8. Kosara, R.: An argument structure for data stories. In: Kozlíková, B., Schreck, T., Wischgoll, T. (eds.) EuroVis (2017)
9. Kosara, R., Mackinlay, J.D.: Storytelling: the next step for visualization. Computer **46**(5), 44–50 (2013)
10. Lee, B., et al.: More than telling a story: transforming data into visually shared stories. IEEE Comput. Graph. App. **35**(5), 84–90 (2015)
11. Ondzigue Mbenga, R., et. al: A data narrative about tuberculosis pandemic in Gabon. In: DARLI-AP (2022)
12. Outa, F.E., Francia, M., Marcel, P., Peralta, V., Vassiliadis, P.: Towards a conceptual model for data narratives. In: ER, pp. 261–270 (2020)
13. Segel, E., Heer, J.: Narrative visualization: telling stories with data. TVCG **16**(6), 1139–1148 (2010)
14. Shi, D., Sun, F., Xu, X., Lan, X., Gotz, D., Cao, N.: Autoclips: an automatic approach to video generation from data facts. Comput. Graph. Forum **40**(3), 495–505 (2021)
15. Shi, D., et al.: Calliope: automatic visual data story generation from a spreadsheet. TVCG **27**(2), 453–463 (2021)
16. Thudt, A.F., Perin, C., Willett, W., Carpendale, S.: Subjectivity in personal storytelling with visualization. Inf. Des. J. **23**(1), 48–64 (2017)
17. Vassiliadis, P., Marcel, P., Rizzi, S.: Beyond roll-up's and drill-down's: an intentional analytics model to reinvent OLAP. Inf. Syst. **85**, 68–91 (2019)
18. Wang, Y., et al.: Datashot: automatic generation of fact sheets from tabular data. TVCG **26**(1), 895–905 (2020)
19. Wang, Z., Dingwall, H., Bach, B.: Teaching data visualization and storytelling with data comic workshops. In: CHI. ACM (2019)
20. Weber, W., Engebretsen, M., Kennedy, H.: Data stories: rethinking journalistic storytelling in the context of data journalism. Stud. Commun. Sci. **18**, 191–206 (2018)

Forecasting POI Occupation
with Contextual Machine Learning

Alberto Belussi[ID], Andrea Cinelli, Anna Dalla Vecchia, Sara Migliorini$^{(\boxtimes)}$[ID],
Michele Quaresmini, and Elisa Quintarelli[ID]

Department of Computer Science, University of Verona, Verona, Italy
{alberto.belussi,sara.migliorini,elisa.quintarelli}@univr.it,
{andrea.cinelli,anna.dallavecchia,michele.quaresmini}@studenti.univr.it

Abstract. The increasing availability of historical information has emphasized the importance to explore, understand and extract value from it in order to achieve both short-term goals and strategic objectives. Intelligent techniques to handle heterogeneous data, together with user preferences, may be beneficial for end users; among them we can mention recommendation systems, which are able to guide users through huge catalogues of alternative items. This kind of systems represent an invaluable help not only for the users, who can feel disoriented in presence of so many alternatives at their disposal, but also for service providers or sellers, which can benefit from inferred hidden knowledge and guide towards particular items the choices of specific groups of users sharing some common preferences. This influence capability of recommendation systems can be particularly useful in the touristic domain, where the need to control and manage the level of crowding of POIs (Points Of Interest) has become a pressing need in the recent years. In this paper we study the role of contextual information in determining POI occupations and we explore how machine learning and deep learning technologies can help in producing good POI occupation forecasters by enriching historical information with contextual one. Throughout the paper we refer to a real-world application scenario regarding the touristic visits performed in Verona, a municipality in Northern Italy, between 2014 and 2019.

Keywords: Context · POI occupation and recommendation · Crowding forecast · Machine learning · Deep learning

1 Introduction

The collection and analysis of historical data about users' behaviour is the core of any recommendation system which needs to understand the preferences and tastes of users for suggesting them the best next item in a collection. Many literature solutions demonstrate how the use of contextual information can help in defining better and more suitable recommendations, both for single users or group of users [2,3,11]. Recommendation systems are becoming an essential tool for guiding users through huge catalogues of items, since the increasing amount

© Springer Nature Switzerland AG 2022
S. Chiusano et al. (Eds.): ADBIS 2022, LNCS 13389, pp. 361–376, 2022.
https://doi.org/10.1007/978-3-031-15740-0_26

of data is difficult to explore and understand, thus each choice can become a nightmare instead of an opportunity. They could be used also by service providers or suppliers not only for increasing their knowledge about users, but also to guide them towards particular items in their preference list.

Recommendation systems are successfully used in many different application domains, from e-commerce and on-demand TV shows, to the touristic one. As regards to the latter, the ability to guide user choices has become particularly important in the recent years since, due to circumstances such as the COVID-19 pandemic, it has become prominent the need to avoid crowding and restrict the number of people that can access the same POI together. Therefore, a natural extension of currently available recommendation systems is the ability to predict the level of crowding of a given POI and to redirect people to other attractions, which are equally appreciated, but have currently a less amount of visits [8,9].

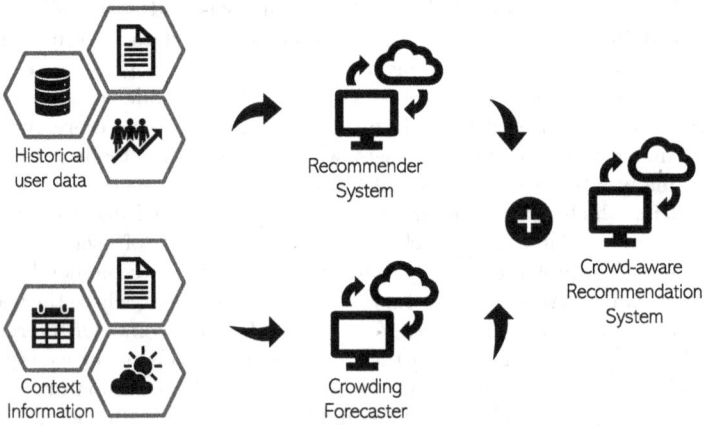

Fig. 1. Architecture of a Crowd-aware Recommendation System (CRS) obtained by combing the results of a classical recommendation system and a crowding forecaster.

In the remainder of this paper, we concentrate on the touristic application domain and we design a solution able to guide users towards the less crowded attractions based on historical and contextual information. More specifically, we will use the term "Recommendation System" for denoting a system that, given the set of available Points of Interest (POIs) $\mathcal{P}$ and the preferences of a user u, is able to determine a subset of $\mathcal{P}$ which contains the most suitable POIs for u. Conversely, we will use the term "Crowding forcaster" for denoting a system that given a POI p and a set of contextual information $C = \{c_1, \ldots, c_n\}$ is able to predict the level of crowding for p in C based on historical contextual information. The combination of a recommendation system and a crowding forecaster can allow to produce better suggestions for users, which essentially reduce waiting time at queue lines, but it can also be an invaluable help for attraction owners who nowadays need to prevent and manage crowding situations, without giving up on pursuing to increase the number of visits in less known or visited POIs. This kind of system obtained by combining a classical recommendation system and a crowding forecaster will be denoted in the following as *Crowd-aware recommendation system* (*CRS*) and its architecture is reported in Fig. 1.

The aim of this paper is twofold: (a) firstly, we will study how contextual information can impact the level of crowd of a given POI. At this regards, we will demonstrate that many different and intertwined factors can have an impact on this aspect. (b) Secondly, we compare the use of machine learning and deep learning techniques for producing the better crowding forecaster system, which is able to take care of all these articulated contextual pieces of information. The aim of this second contribution is to define the better generalized architecture that can be used for POIs with different features, different amounts of available historical information and different availability of contextual information.

The remainder of the paper is organized as follows: in Sect. 2 we discuss some precedent work about contextual recommendations in the tourism domain and crowding forecasting, Sect. 3 introduces the considered application domains, Sect. 4 shows how contextual features influence POIs visits, whereas Sect. 5 describes our proposal to forecast POIs occupation with machine learning and deep learning techniques. Finally, Sect. 6 summarizes the obtained results and presents some future work.

2   Related Work

In the research area of recommender systems there is a strong interest in developing context-aware algorithms for decreasing recommendation errors by enriching available information about items and users with contextual ones [2,3]. In [10,11] the notion of context is used to infer more precise contextual user preferences to be used for suggesting the best sequence of items to group of users by considering the current context they are acting in. In the urban tourist scenario, and in particular in POI recommendations, multiple factors may be considered despite most of the approaches developed in the past mainly exploited three contextual dimensions, namely time, geolocation and social conditions (e.g. POIs visted alone, in groups, with kids) [4,6,14]. In general, additional factors may be included in the recommendation process; indeed, in [7,13] the authors consider weather conditions in POI recommendations. They propose novel recommender systems to suggest next points of interest to visit that match users' preferences (what to visit) and the specific context of the visit (how to visit each POI). However, the contextual information is quite limited and restricted to only a weather feature in [13] and to an hourly weather summary (e.g., cloudy), temperature (e.g., cold) and temporal information such as the time interval related to the visit (e.g., evening) in [7]. Both proposals do not consider other external information like the occupancy rate of each POI, the day of the week, the presence of holidays or other important events in the considered city. Since our main aim is to prevent a high level of crowding in each POI of a certain area and to consequently suggest interesting, but less crowded POIs to tourists, we need to be able to consider all the contextual features that may influence POI visits. In [8,9] the prediction of POI occupancy is obtained only on the base of historical accesses, without enriching data with context, and then it is used to balance travelers without considering their personal preferences. To the best of our knowledge, a proposal that partially tackles a problem similar to ours is [12]; the authors predict parking occupancy in areas where the availability of parking data produced by sensors is limited and needs to be integrated with heterogeneous contextual information, like POIs presence. However, the authors do not include the analysis of user contextual preferences to suggest free parking slots.

3   Touristic Scenario

In this paper we use a real-world dataset containing the visits performed by tourists to POIs located in Verona, a city in the Northern Italy. This dataset contains about 2,1 million records spanning 6 years (i.e., from 2014 to 2019) regarding about 500,000 tourists and 9 different POIs. Each of these records reports an anonymised user identifier, the POI identifier and the timestamp of the visit. The information contained in the dataset has been enriched with some contextual information regarding the weather conditions and some semantic temporal information, like the presence of holidays. In general, such features could be extended in order to include the presence of touristic events in the analyzed area or spatial information (e.g. POIs in the immediate proximity). Table 1 provides an overview of the dataset statistics.

Table 1. Summary of the touristic datasets considered in the paper.

ID	Name	Category	Num. of visits
49	Arena Amphitheatre	Monument	421,490
61	Juliet's House	Museum	375,305
59	Lamberti's Tower	Monument	290,243
71	Castelvecchio Museum	Museum	271,552
54	Church of St. Anastasia	Church	230,352
52	The Cathedral	Church	205,293
42	Archaeological Museum at the Roman Theatre	Museum	145,854
58	Palazzo della Ragione	Monument	111,440
202	Juliet's tomb and Cavalcaselle Museum	Museum	100,701

Definition 1 (Touristic Visit). *Given a set of POIs $\mathcal{P}$ and a set of users $\mathcal{U}$, a visit performed by a user u is represented by a tuple:*

$$v = \langle u, p, t, lat, long \rangle \tag{1}$$

where: $u \in \mathcal{U}$ is a user identifier, $p \in \mathcal{P}$ identifies the POI, t is a timestamp representing the date and time of the visit, and lat and long identify the spatial position (i.e., latitude and longitude) where the POI is located.

The set of all visits performed by users in $\mathcal{U}$ is denoted in the following as $\mathcal{V}$. A tourist visit can be enriched with some contextual information better characterizing the conditions in which the visit has been performed.

Definition 2 (Visit context). *Given a visit $v \in \mathcal{V}$, we say that it is performed in a context C defined as follows:*

$$C = \{ts, doy, dow, hol, pres, wind, rain, temp, hum\} \tag{2}$$

where ts is a predefined timeslot inside the day, doy is the day of the year, dow is the day of the week, hol is a boolean value representing the fact that the visit is performed in a public holiday and/or during a weekend, or not, pres is the atmospheric pressure, wind is the wind speed, rain is the amount of precipitation, temp is the temperature, and hum is the percentage of humidity.

In the following section we study the role of context in influencing the number of tourists visiting a particular POI in a given moment.

4   The Role of Context

Given the dataset presented in the previous section, a preliminary set of analysis has been performed to confirm the role of the context in influencing the level of crowding of the considered POIs. In this section we illustrate some of them.

Dependency on the Day of the Week – In Fig. 2(a) we report the average number of visits for the Juliet's house in February. As you can notice, the number of visits has a dependency with the day of the week: visits concentrate on the weekend and have a peak on Saturday. However, if we enrich the information regarding the temporal contexts, for instance by distinguishing some anniversaries, we are able to better capture the behaviour of tourists and forecast the number of visits. In particular, for the Juliet's house, also known as the house of lovers, the Valentine's day is a very important anniversary. Indeed, Fig. 2(a) reports the fact that even if it happens on a Wednesday, which is usually a very quiet day, the actual number of visits on the Valentine's day triplicates and matches the number of visits of the most busy week days.

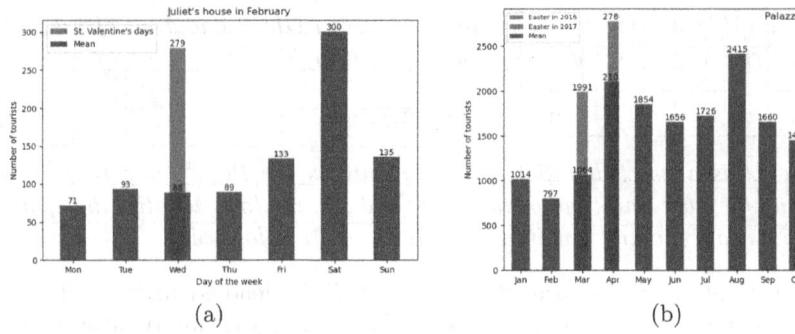

(a)   (b)

Fig. 2. (a) Number of visits to Juliet's house for each day of the week compared to the number of visits at Valentine's day. (b) Average number of visits to "Palazzo della Ragione" subdivided by month.

Dependency on External Events – The analysis of data reveals also that some events can have a great influence on the number of visits: they can change the tourist behaviour on the basis of the period in which they happen. For instance, the Easter holidays are an annual event that happens in a variable moment, mainly in March or April. Figure 2(b) shows in blue the average number of visits performed to the POI denoted as "Palazzo della Ragione". We can notice that March is usually a month with few visits compared to others, but when Easter holidays are in that month, the number of visits drastically increase, whereas when Easter is in April the peak moves to that month.

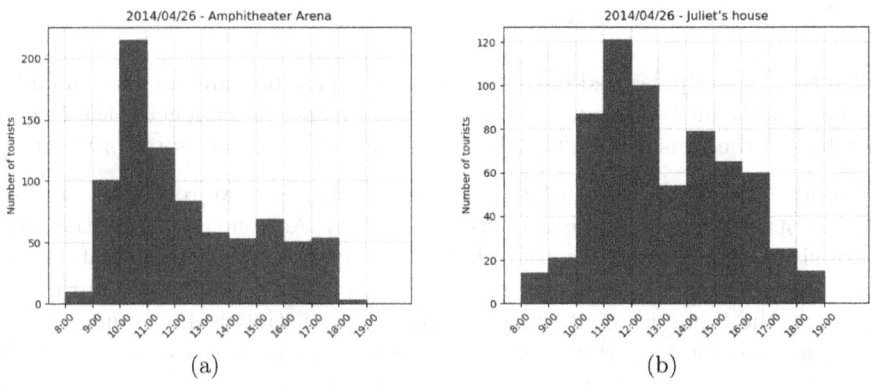

(a)   (b)

Fig. 3. (a) Number of visits to the Amphitheater Arena during April 26th, 2014. (b) Number of visits to Juliet's House during April 26th, 2014.

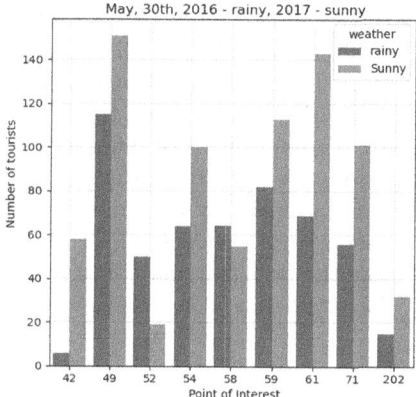

Fig. 4. Number of visits in the same day of year but in presence of different weather conditions: blue line in a rainy day, while yellow line in a sunny day. (Color figure online)

Dependency on the Time Slot – As we have discussed in the introduction, the goal of a crowding forecasting system could also be to drive the choice of tourists, for instance by suggesting a different time-slot for their visit. Indeed, even in the most busy dates, the number of tourists which are present in a given POI is not constant. Figure 3(a) shows the distribution of the number of visits performed by tourists to the Arena and Juliet's House on Saturday, April 26th 2014. This was one of the busiest days, because it is both a weekend day and it is also close to the anniversary of Italy's Liberation, or April 25th. As you can notice, the majority of the visits are concentrate in the morning. In this case a smart recommendation system can suggest tourists to perform the visit between 1 and 2 p.m., anticipating or postponing the launch time, or it can suggest a late afternoon visit before closure.

Dependency on Weather Conditions – Finally, another important contextual feature which can influence the number of visits, independently from the considered day, is represented by the weather conditions. Figure 4 illustrates how in the same day of the year, the number of visits can depend on the fact that it is raining or not. We can notice that some POIs, like 42 and 202, are greatly influenced by this condition, because they are outdoor attractions and in presence of rain there are no visits. Conversely, other POIs, like 49, are influenced but in a less extend, and finally, other indoor attractions (like 52 and 58) could benefit from an adverse weather condition, for instance because they are indoor attractions where tourists can spend some hours in a rainy day.

5   Forecasting POI Occupation with a ML/DL Model

In this section we discuss how the context-aware forecasting occupation problem can be modelled with a machine learning or a deep learning approach. More specifically, we compare the results obtained by considering only the raw historical records about the performed visits (see Definition 1), or their combination, with the ones obtained by including also the contextual information presented in Definition 2. For each POI we try to estimate the level of occupancy in different moments of the day by considering three different time slots (i.e., morning, noon, evening). In both cases (raw and contextual) we compare the accuracy of a machine learning model represented by a random forest, and a deep learning model represented by a deep neural network (DNN). Several configurations are tested in order to determine the best one and study their behaviour in presence of POIs with different characteristics. The implementation of the models has been done in Python by using the Tensorflow [1] and the Keras [5] libraries[1].

Forecasting with Only Raw Data

As regards to the estimation of the number of tourists in each POI with only the raw historical records described in Definition 1, we initially use a Random Forest model and subsequently a DNN model, obtaining the results reported in Table 2 and Table 3 respectively. For the random forest we tried different configurations represented by a different number of trees. As you can notice, the accuracy obtained for the various POIs is quite different and it decreases as the number of available training data decreases: for the default forest with 100 trees, it spans from about 31.4% in the best case to 50.8% in the worst one which is associated with POI 202, namely the one with the smallest amount of historical records. Conversely, the accuracy of the network trained and tested with all the POIs together (i.e., raw ALL) is about 38% and it is not substantially affected by the number of trees in the network.

Table 2. Results obtained by applying a Random Forest model on raw data. The row ALL identifies the network trained with all POIs together. The metric used to evaluate the accurac y is the well-known Mean Absolute Percentage Error (MAPE)

PoI ID	MAPE		
	10 trees	100 trees	1000 trees
49	31.8%	31.4%	31.3%
61	37.9%	38.4%	38.3%
59	33.4%	33.1%	33.0%
71	38.7%	38.6%	38.5%
54	33.3%	33.1%	33.1%
52	33.6%	33.4%	33.5%
42	36.0%	35.7%	35.8%
58	40.5%	40.6%	40.5%
202	51.0%	50.8%	50.9%
ALL	38.0%	37.9%	37.9%

---

[1] The source code and the datasets used in this paper are available at https://github.com/smigliorini/crowd-forecaster.

In case of the DNN model we tried different values for the following hyperparameters that control the architecture or topology of the network: the number of nodes, the number of epochs and the dropout. The last two parameters are used to approximate the best solution without falling into an overfitting. The architecture of the DNN is illustrated in Fig. 5 and it includes: an input layer, a dense layer with n nodes and ReLU activation function, a droupout layer with rate **DP**, another dense layer with n nodes and ReLU activation function, followed by another dropout layer with the same rate **DP**, and finally an output layer with one node and activation function linear. Notice that in Tensorflow, the Dropout Layer randomly sets, at each step during the training time, the input units to zero with a certain frequency **DP**. In this way a certain percentage of randomly chosen units (i.e. neurons) are ignored during the training phase. The percentage of units which are dropped out in the considered DNN model are reported in column **DP** of Table 3. Conversely, the first column represents the number of nodes and column **EP** reports the number of epochs. For each considered PoI, the corresponding cells report the MAPE obtained with the current configuration. In the cells we distinguish the following cases: (1) an overfitting (i.e., when the validation error is significantly greater than the training error) is identified by the presence of an "*" symbol before the percentage, (2) a potential good model is identified by a cell with a gray background, (3) the best found model has a gray background and a bold MAPE value, and finally (4) an unknown fit (i.e., when the validation error is significantly smaller than the training error) is represented by a cell without a background color. If we consider the best DNN models for each POI or for all POIs together in Table 3, they are able to provide

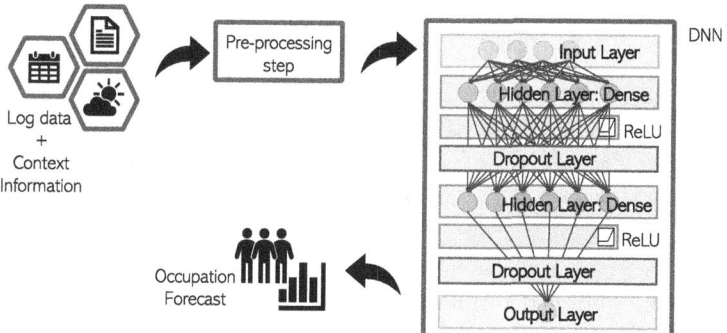

Fig. 5. Architecture of the DNN model. It includes: an input layer, a dense layer with n nodes and ReLU activation function, a droupout layer with rate **DP**, another dense layer with n nodes and ReLU activation function, followed by another dropout layer with the same rate **DP**, and finally an output layer with one node.

a smaller error w.r.t. the corresponding random forest in Table 2. In this case, the errors span from about 24.7% to 49.4%, while the error for the global network decreases to about 36%. As you can notice, there is not a single best configuration for all POIs, but each one could require a different model depending on the number of training records and its behaviour w.r.t. the context dimensions.

Table 3. Results obtained by applying a DNN model on the raw data. The column ALL identifies the network trained with all PoIs together. Each cell reports the obtained MAPE error: an "*" symbol before the value denotes an overfitting model, gray cells identifies good models, white cells represents unknown fit model, and finally gray cells with a bold MAPE value identifies the best configuration.

DNN Par.			MAPE for PoI									
	DP	EP	49	61	59	71	54	52	42	58	202	ALL
32 Nodes	0.0	300	*27.0%	33.1%	*30.4%	*33.7%	28.5%	26.2%	**35.7%**	*34.0%	*48.2%	*36.8%
		500	*27.5%	33.4%	*31.5%	*34.0%	26.7%	25.0%	*33.6%	*34.1%	*50.7%	*37.3%
	0.2	300	25.1%	34.2%	35.8%	39.2%	32.0%	28.4%	39.1%	37.0%	49.6%	41.2%
		500	24.5%	37.6%	34.5%	32.7%	29.9%	29.4%	37.4%	38.1%	52.4%	40.1%
	0.4	300	31.8%	39.9%	37.1%	40.2%	36.0%	33.7%	42.5%	41.4%	48.3%	41.6%
		500	24.5%	39.6%	37.5%	36.4%	35.5%	34.4%	40.3%	41.4%	51.1%	41.8%
64 Nodes	0.0	300	*25.7%	32.7%	*29.4%	33.1*%	26.1%	25.1%	36.7%	*32.3%	*48.7%	*35.3%
		500	*26.0%	**30.3%**	*29.2%	*33.5%	26.6%	**24.8%**	*34.0%	*33.1%	*48.8%	*35.8%
	0.2	300	25.2%	31.4%	32.5%	32.2%	25.4%	26.7%	34.0%	34.4%	**49.4%**	35.8%
		500	**24.7%**	30.8%	**31.0%**	**30.5%**	25.7%	24.2%	34.5%	36.0%	52.4%	36.2%
	0.4	300	24.5%	35.2%	36.0%	30.9%	32.2%	30.4%	35.7%	37.7%	49.0%	38.6%
		500	23.5%	36.9%	37.0%	34.3%	36.5%	30.8%	37.8%	36.8%	47.8%	40.8%
128 Nodes	0.0	300	*25.5%	30.4%	*28.9%	*32.0%	**25.7%**	24.9%	*34.1%	*36.3%	*45.8%	*35.8%
		500	*25.4%	*30.7%	*29.2%	*33.9%	26.0%	25.3%	*32.7%	*33.3%	*48.7%	*35.6%
	0.2	300	25.9%	30.9%	29.2%	30.9%	25.0%	23.7%	31.8%	**32.6%**	44.6%	**35.8%**
		500	25.5%	30.8%	27.9%	*32.4%	25.2%	23.9%	33.7%	33.2%	46.6%	36.0%
	0.4	300	24.6%	30.9%	28.8%	34.0%	25.2%	23.8%	31.6%	34.3%	45.3%	37.5%
		500	24.3%	30.1%	28.2%	29.6%	25.5%	24.7%	32.7%	36.3%	50.1%	35.6%

However, the obtained results and the error improvements provided by the DNN models are not satisfactory. Therefore, we evaluate in the following section the introduction of contextual information together with the raw historical data about visits, in order to improve the forecasting capabilities of both the random forest and the DNN models.

Forecasting with Contextual Information

In this section, we tried to improve the results obtained in the previous one by considering also contextual information during training. We continue to compare both the behaviour of a random forest model with the one of a deep neural network. The obtained results are reported in Table 4 and Table 5, respectively.

Table 4. Results obtained by using a Random Forest on raw and contextual information together. The raw ALL identifies the network trained with all POIs together.

PoI ID	MAPE		
	10 trees	100 trees	1000 trees
49	23.7%	22.7%	22.6%
61	28.4%	28.0%	28.1%
59	26.6%	26.1%	26.0%
71	30.2%	29.1%	29.1%
54	27.7%	26.6%	26.5%
52	28.7%	28.1%	28.0%
42	30.1%	30.2%	30.1%
58	32.7%	31.6%	31.5%
202	41.8%	41.3%	41.2%
ALL	32.8%	32.6%	32.6%

With the addition of contextual information in the training set, the error of the random forest model decreases with respect to the one reported in Table 2. In this case, the MAPE error spans from about 22.7% to 41.3%. In particular, if we consider the global network trained with all the POIs together, we obtain a decrease of the error from about 38% to 32.6%. Moreover, the random forest is able to produce a better prediction also with respect to the best DNN in Table 3 for the ALL case, which has a MAPE of 35.8%. Finally, if we consider how the accuracy changes with the network dimension (i.e., the number of trees), we can notice that also in this case a decrement or an increment of the number of trees w.r.t. the default one (i.e., 100) does not produce any evident effect.

As regards to the DNN model, we vary the network parameters as in the previous case and we report the obtained results in Table 5. In this case, the MAPE error spans from about 20.7% to 37.0%, with the worst case represented by POI 202, which is the one with the lowest number of historical records. In particular, if we consider the global network trained with all the POIs together, we obtain a decrease of the error from about 35.8% to 31.3%.

Table 5. Results obtained by applying a DNN model on raw and contextual information together. The column ALL identifies the network trained with all POIs together. Each cell reports the obtained MAPE error: an "*" symbol before the value denotes an overfitting model, gray cells identifies good models, white cells represents unknown fit model, and finally gray cells with a bold MAPE value identifies the best configuration.

DNN Par.			MAPE for PoI									
	DP	EP	49	61	59	71	54	52	42	58	202	ALL
64 Nodes	0.2	300	*22.2%	*25.2%	25.2%	*27.6%	*28.9%	24.4%	*31.5%	*33.2%	*40.8%	31.8%
		500	*20.9%	*23.7%	27.2%	*27.7%	*28.3%	*23.5%	*30.3%	*30.2%	*39.0%	32.1%
	0.4	300	21.9%	27.5%	28.2%	30.8%	27.0%	27.5%	35.0*%	*32.7%	38.6%	33.9%
		500	25.3%	27.7%	27.2%	26.9%	27.6%	25.1%	*32.8%	*31.6%	38.6%	34.1%
	0.6	300	27.1%	30.0%	30.0%	33.1%	29.3%	27.6%	33.7%	34.2%	41.5%	38.1%
		500	26.7%	30.7%	27.9%	31.9%	29.6%	26.1%	34.8%	34.5%	44.3%	36.4%
128 Nodes	0.2	300	*22.2%	*25.8%	*24.5%	*27.0%	*24.7%	*22.9%	*31.7%	*3.7%	*40.9%	*31.8%
	0.4	300	**20.7%**	25.0%	24.0%	*26.1%	24.2%	27.0%	*29.0%	*32.8%	*39.2%	33.9%
		500	*21.4%	*26.9%	23.3%	*26.4%	*24.0%	23.4%	*28.3%	*31.4%	*39%	31.8%
	0.6	300	21.9%	28.3%	26.6%	27.0%	27.4%	26.4%	**34.2%**	**33.3%**	**37.0%**	35.6%
		500	21.5%	28.7%	28.2%	**25.7%**	26.9%	26.0%	*33.6%	33.9%	*37.0%	35.0%
256 Nodes	0.2	300	*22.6%	*26.0%	*25.7%	*28.9%	26.4*%	*23.6%	*33.1%	*31.6%	*40.3%	*30.9%
		500	*22.3%	*25.4%	*26.1%	*29.0%	25.7*%	*24.3%	*32.0%	*32.0%	*41.6%	*30.4%
	0.4	300	*21.0%	*23.4%	*23.8%	*26.3%	*24.4%	*23.5%	*30.2%	*29.3%	*37.8%	**31.3%**
		500	*21.3%	*23.9%	*24.2%	*26.9%	*24.0%	*22.8%	*38.9%	*29.7%	*37.8%	30.7%
	0.6	300	21.0%	23.6%	23.2%	*26.0%	24.2%	24.6%	*27.7%	*31.4%	*37.0%	32.5%
		500	*20.4%	**23.2%**	23.1%	*25.6%	**23.9%**	**22.2%**	*27.5%	*31.5%	*36.0%	32.0%
512 N.	0.2	300	22.2*%	25.8*%	26.3*%	29.2*%	27.0*%	25.1*%	32.1*%	32.7*%	42.4*%	*30.9%
	0.4	300	*21.2%	*23.7%	*24.2%	*27.6%	*24.9%	*22.7%	*29.8%	*30.2%	*38.4%	*30.7%
	0.6	300	*21.0%	*23.2%	**22.8%**	*25.9%	*23.6%	22.9%	*27.7%	*28.7%	*36.8%	**31.3%**

Table 6 compares the best results obtained with the four models: random forest and DNN trained with only raw data, and random forest and DNN trained with both raw and contextual data. As you can notice, the best accuracy values are achieved with the last model (i.e., DNN trained with raw and contextual data together). The use of a DNN model allows to obtain a initial improvement also on raw data; for instance, if we consider POI 49 (the one with the greatest number of training data points) just the change from a random forest to a DNN model allows to obtain a decrease of the error from 31.4% to 24.7%. However, even without changing the model, but only including contextual information during training, we obtain an important decrement in the error: from 31.4% to 22.7%, confirming the central role of contextual information in crowding forecasting. This behaviour is confirmed also for the networks trained with all POIs together (row **ALL**), were an initial error rate of 37.9%, obtained with a random forest trained only with raw data, decreases to 31.3% when a DNN model trained with both raw and contextual data is used. For these four network we evaluate also the percentage of error in the 95%, 99% and 100% of cases obtaining result in

Table 6. Comparison of best value obtained by the four models Row ALL* reports the MAPE values for the 95/99/100% of test data.

POI	Raw data				Raw + Context			
	RF		DNN		RF		DNN	
	MAPE	Time	MAPE	Time	MAPE	Time	MAPE	Time
49	31.4%	14 s	24.7%	4 min	22.7%	33 s	20.7%	6 min
61	38.4%	14 s	30.3%	4 min	28.0%	35 s	23.2%	7 min
59	33.1%	17 s	31.0%	3 min	26.1%	34 s	22.8%	4 min
71	38.6%	16 s	30.5%	4 min	29.1%	32 s	25.7%	5 min
54	33.1%	16 s	25.7%	2 min	26.6%	30 s	23.9%	6 min
52	33.4%	14 s	24.8%	4 min	28.1%	31 s	22.2%	6 min
42	35.7%	11 s	35.7%	2 min	30.2%	29 s	34.2%	3 min
58	40.6%	11 s	32.6%	2 min	31.6%	26 s	33.3%	3 min
202	50.8%	12 s	49.4%	3 min	41.3%	33 s	37.0%	3 min
ALL	**37.9%**	74 s	**35.8%**	10 min	**32.6%**	331 s	**31.3%**	12 min
ALL*	30.2/35.8/37.9%		27.8/33.1/35.4%		25.1/30.5/32.6%		24.4/29.1/31.4%	

row **ALL***. These values highlight that in the majority of cases the obtained forecast values are even more accurate.

Generalization Capabilities

In this section we evaluate the capability of the networks proposed in the previous section to generalize what they have learned to other different situations. In particular, we consider the four global networks denoted as **ALL** in the previous section and we assume to have a recently added POI for which we have collected a small amount of historical information (i.e., only 6 months in the last year) and we try to estimate the level of crowding in months never seen before. We do that for three different POIs and we report the obtained results in Table 7. If we compare the results in Table 7 with the ones in Table 6, we can observe an increment in the error values. However, in general such increment is acceptable

Table 7. Evaluation of the generalization capabilities for the four global networks trained with all POI data. Each row simulate the presence of a new recently added POI for which we try to estimate the level of crowding for periods not seen before.

POI	Raw data		Raw + Context	
	RF MAPE	DNN MAPE	RF MAPE	DNN MAPE
61	35.4%	33.5%	29.2%	23.6%
59	39.0%	39.2%	34.4%	27.1%
71	40.0%	41.0%	29.6%	26.8%

and the use of contextual information is able to improve the generalization capabilities of the network. For instance, for POI 59, the increment of error is about 8% for the DNN trained with only Raw Data and about 4% when the DNN is trained with both raw and contextual information. This confirms the assumption that the use of a richer contextualized training dataset not only improves the accuracy of the model, but also increases its generalization capability. An interesting case is represented by POI 61, for which the random forest trained with only raw data has an error less than the one reported in Table 6 for the same POI (but trained with only the data points of the specific POI). In this case, the presence of data points related to other POIs decreases the error of the network, confirming the generalization capabilities of the models.

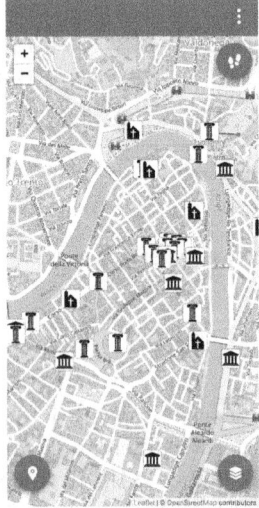 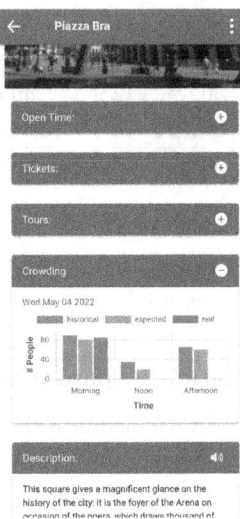

Fig. 6. Mobile phone app presenting some touristic attractions in Verona. On the right the description of a POI with the forecasting of its occupation.

6   Conclusion

In this paper we study the role of the context in forecasting the occupation of different POIs in a given moment through some examples where contextual information greatly influences the amount of visits in a certain POI. Starting from such considerations, we define a machine learning model (i.e., a random forest) and a deep learning model (i.e., a DNN), and we train them firstly with only raw historical data and secondly with both raw and contextual information. We conclude that the contextual models are able to forecast the potential occupation of each POI with a smaller error with respect to the ones trained with only raw historical data, with the DNN outperforming the random forest one. Moreover, the use of contextual information allows to increase the generalization capabilities of the networks, with the DNN providing the best performances. The proposed approach has been included into a mobile phone app, as illustrated in Fig. 6, which provides information about touristic attractions of the Verona city and in particular about their levels of occupation. This app will be available to end users in the next months and will be integrated into a recommendation system able to suggest contextual paths, also considering the distance among POIs.

Acknowledgements. We will thank the touristic office of Verona for providing the datasets of the VeronaCard city pass.

References

1. Abadi, M., et al.: TensorFlow: large-scale machine learning on heterogeneous systems (2015). https://www.tensorflow.org/
2. Adomavicius, G., Tuzhilin, A.: Context-aware recommender systems. In: Recommender Systems Handbook, pp. 191–226 (2015)
3. Chen, G., Chen, L.: Augmenting service recommender systems by incorporating contextual opinions from user reviews. User Model. User-Adap. Inter. **25**(3), 295–329 (2015). https://doi.org/10.1007/s11257-015-9157-3
4. Cheng, C., Yang, H., King, I., Lyu, M.R.: Fused matrix factorization with geographical and social influence in location-based social networks. In: Proceedings of the Twenty-Sixth AAAI Conference on Artificial Intelligence (2012)
5. Gulli, A., Pal, S.: Deep Learning with Keras. Packt Publishing Ltd., Birmingham (2017)
6. Li, X., Cong, G., Li, X., Pham, T.N., Krishnaswamy, S.: Rank-GeoFM: a ranking based geographical factorization method for point of interest recommendation. In: Proceedings of the 38th International ACM SIGIR Conference, pp. 433–442 (2015)
7. Massimo, D., Ricci, F.: Next-POI recommendations matching user's visit behaviour. In: Wörndl, W., Koo, C., Stienmetz, J.L. (eds.) Information and Communication Technologies in Tourism 2021, pp. 45–57. Springer, Cham (2021). https://doi.org/10.1007/978-3-030-65785-7_4
8. Migliorini, S., Carra, D., Belussi, A.: Adaptive trip recommendation system: balancing travelers among POIs with mapreduce. In: 2018 IEEE International Congress on Big Data (BigData Congress), pp. 255–259 (2018)
9. Migliorini, S., Carra, D., Belussi, A.: Distributing tourists among POIs with an adaptive trip recommendation system. IEEE Trans. Emerg. Top. Comput. **9**(4), 1765–1779 (2021)
10. Migliorini, S., Quintarelli, E., Carra, D., Belussi, A.: Sequences of recommendations for dynamic groups: what is the role of context? In: 2019 IEEE International Congress on Big Data (BigDataCongress), pp. 121–128 (2019)
11. Migliorini, S., Quintarelli, E., Gambini, M., Belussi, A., Carra, D.: Sequence recommendations for groups: a dynamic approach to balance preferences. Inf. Syst. **108**, 102023 (2022)
12. Shao, W., et al.: FADACS: a few-shot adversarial domain adaptation architecture for context-aware parking availability sensing. In: 19th IEEE PerCom 2021, pp. 1–10 (2021)

13. Trattner, C., Oberegger, A., Eberhard, L., Parra, D., Marinho, L.B.: Understanding the impact of weather for POI recommendations. In: Proceedings of the Workshop on Recommenders in Tourism. CEUR Workshop Proceedings, vol. 1685, pp. 16–23 (2016)
14. Ye, M., Yin, P., Lee, W., Lee, D.L.: Exploiting geographical influence for collaborative point-of-interest recommendation. In: Proceeding of the 34th International ACM SIGIR Conference, pp. 325–334 (2011)

Smart Contracts for Certified and Sustainable Safety-Critical Continuous Monitoring Applications

Nicola Elia[1]([✉])(ID), Francesco Barchi[1](ID), Emanuele Parisi[1](ID),
Livio Pompianu[2](ID), Salvatore Carta[2](ID), Andrea Bartolini[1](ID),
and Andrea Acquaviva[1](ID)

[1] Università di Bologna, Viale del Risorgimento, 2, 40136 Bologna, Italy
{nicola.elia2,francesco.barchi,emanuele.parisi,a.bartolini,
andrea.acquaviva}@unibo.it
[2] Università di Cagliari, Via Università 40, 09124 Cagliari, Italy
{pompianu.livio,salvatore}@unica.it

Abstract. Monitoring applications are increasingly important to enable predictive maintenance and real-time anomaly detection in industrial and civil safety-critical infrastructures. Typical monitoring pipelines consist of a sensor network that collects and streams IoT data toward a cloud infrastructure that provides storage, visualisation and data analytic capabilities. However, since critical data generated must be often retained for regulatory and tracking purposes, cloud storage requirements become poorly sustainable when dealing with critical infrastructures that have to remain operative for decades while supporting lifelong continuous monitoring. While policies can be applied to remove redundant or outdated information, anti-tamper mechanisms are required to guarantee that data modifications are not driven by malicious intents to alter recorded data. This work presents a blockchain-based framework for continuous monitoring applications enabling certified removal of IoT data in safety-critical databases. The framework allows for the deployment of data-evaluation policies to identify redundant/outdated measurements flowing in the database and, therefore, mark them as eligible for removal. The novelty of our approach stands in the implementation of the data-evaluation policy as a smart contract. Furthermore, the use of a blockchain ensures that critical database operations (like removal) are tamper-proof and compliant with the guideline determined by system stakeholders. We demonstrate the effectiveness of the proposed framework in a real case study using accelerometer data of a bridge monitoring application, and we characterise the overhead of transactions to the blockchain.

Keywords: Database · Smart contract · IoT · Continuous monitoring

N. Elia—Acknowledges support from TIM S.p.A. through the PhD scholarship.

S. Chiusano et al. (Eds.): ADBIS 2022, LNCS 13389, pp. 377–391, 2022.
https://doi.org/10.1007/978-3-031-15740-0_27

1   Introduction

The deployment of sensor networks for safety-critical systems monitoring is becoming widespread in many contexts, from Structural Health Monitoring (SHM) to industrial plants, smart buildings and energy distribution systems. Typical deployments have a layered structure: (i) Sensing IoT nodes performing on-field continuous monitoring, (ii) local network layer orchestrating data collection, and (iii) Cloud components enabling data storage, visualisation and analysis [11].

Some applications, such as SHM, require high sensors sampling rates for their data analysis techniques to be effective, resulting in large data throughput. Monitoring infrastructures that remain in operation for decades pose challenges to data storage capabilities, causing a boundless growth of the database size.

On-the-edge data filtering is not always available, because of embedded systems constraints which limit the leaf acquisition devices capabilities. On-cloud data reducing, when dealing with databases which serve for post-mortem analysis of the monitored safety-critical system, poses the challenge of certifying the integrity of the collected data, to avoid malicious deletions of critical data. In this context, the application of blockchain technology to IoT data is a viable solution. However, the conflicting requirements of data integrity certification and storage space management pose relevant challenges [1,27].

Recently, in the context of SHM applications, early solutions exploiting blockchain technology have been proposed [2,24]. However, *storage space sustainability* and *certification of data-removal policies* issues are still not addressed.

In this work, we propose a novel platform architecture which implements certified data-removal policies to improve the sustainability of cloud databases for monitoring applications with anti-tampering guarantees. The proposed solution exploits blockchain and smart contracts to guarantee that the data-removal policies and algorithms are applied in a certified and tamper-proof manner.

The proposed architecture has been tested in the context of a real case study of an SHM application for rail bridge monitoring, where accelerometer data are collected by an IoT network of sensors and streamed towards a remote database. We performed experiments to demonstrate the feasibility of the proposed approach on the considered case study, showing that we achieve a sustainable SHM database by applying a realistic and certified data-removal policy. Moreover, we perform a characterization of the smart contract performance with respect to deletion policy parameters.

The rest of the paper is organised as follows. In Sect. 2 we discuss the background relevant to our work discussing both structural health monitoring and blockchain technologies. In Sect. 3 we investigate the scientific literature on traditional and blockchain-based monitoring applications. In Sect. 4 we present the design of our system, detailing the architecture components and our smart contracts. In Sect. 5 we apply our framework to the SHM use case and present the experimental results. Finally, in Sect. 6 we conclude our work and present some directions for future works.

2   Background

2.1   Structural Health Monitoring

Ageing infrastructures require frequent checks by companies and governments with a consequent increase in cost. Structural Health Monitoring (SHM) is an innovative field where a continuous assessment of civil structure conditions is performed to determine the required maintenance with a consequent increase in security [12,19,20]. A typical SHM infrastructure includes a set of sensors to be placed on the structure, an in-place computing element to manage the stream of information, and cloud infrastructure for data processing. The core components for such a monitoring system are represented by accelerometers that have to acquire the vibration of the infrastructure to provide data for domain experts' analysis. In the last decades, MEMS (Micro-Electro-Mechanical Systems) capacitive accelerometers have been proposed and experimented within such scenarios. Their extreme low-cost and low-power features allow designing and deploying steady measurement infrastructures for continuous monitoring. These infrastructures can scale up to hundreds of measurement points for a single building [17,23,26]. Types of sensors are not restricted to accelerometers but include inclinometers and crack meters. One of the major open challenges of SHM infrastructures is managing the large amount of data coming from these pervasive sensor deployments, which can hopefully be used for detecting or even predicting anomalies in the structures.

For this reason, sensor networks are often forced to be sparse [19] to reduce the stream and amount of data stored in the cloud. By contrast, sparse monitoring strongly impairs damage detection capability in complex and wide structures such as bridges [21], which demand monitoring a vast space at a fine granularity.

2.2   Blockchain

A blockchain [5] is an append-only data structure handled by a decentralized network of mutually distrusting nodes that rely on a consensus protocol for updating it consistently. Blockchain data is stored within transactions, which usually represent events involving the final users, like, for instance, the transferring of funds between users. Blockchains group transactions in blocks, each containing a list of new transactions, a timestamp and the previous block's hash. While the timestamp certifies the block's production time, the hash ensures the immutability of the published data: any change in a block would also require updating the hash values of all subsequent blocks. Blockchains improve security in decentralized contexts by ensuring entities cannot tamper with or remove data.

The introduction of Bitcoin in 2008 boosted the development of blockchain technologies [6]. Blockchains can serve many use cases, such as documents and processes notarization, supply chain management, healthcare. For instance, shortly after the Bitcoin introduction, several applications began publishing hashes of documents on its blockchain to certify their ownership and timestamp [4]. Later, the development of blockchains that enable users to write software instructions to be executed in a decentralized manner has further simplified the blockchain application to new use cases. Such instructions are referred to as smart contracts [25], and the blockchain that spread their adoption is Ethereum [8].

Both Bitcoin and Ethereum are public permissionless blockchains, that is, systems in which there are no restrictions on the identity of who can join the network. More recently, permissioned blockchains have been proposed, where the identity of participants is well defined. One of the most mature projects in this area is Hyperledger Fabric [3], one of the Hyperledger projects managed by the Linux Foundation. Hyperledger Fabric is a framework for designing and deploying permissioned blockchain: it has a modular consensus protocol and it allows developers to write applications by using standard, general-purpose programming languages.

2.3   Motivations

As explained in previous sections, database size growth is a critical issue in Continuous Monitoring applications. In typical safety-critical applications, stakeholders must rely on a centralized on-cloud data management system for meaningless data removal. This operation, if done incorrectly, can lead to severe consequences that can impact the entire monitoring system and data acquisition campaign. Moreover, data deletion could be prone to be tampered with by malicious users. To the best of our knowledge, we are the first to address this specific problem proposing a system for executing policy-based tamper-proof data deletion based on the blockchain technology.

3   Related Works

This section reviews papers from the literature inherent to real-time IoT continuous monitoring systems, with focus on those based on blockchain.

3.1   Continuous Monitoring

Real-world applications require logging mechanisms for the data storage to be tamper-proof. This is necessary to guarantee that any analysis performed to assess the status of the monitored infrastructure makes use of genuine data. Unfortunately, sensor-to-cloud solutions for monitoring applications such as ExaMon [7] and MODRON [2] do not address this issue. ExaMon [7] features real-time heterogeneous data ingestion from multiple sensing nodes through a MQTT broker but does not have blockchain bindings and cannot provide tamper-proof features. It orchestrates KairosDB and Cassandra databases for its operations. MODRON [2] is a sensor-to-cloud architecture which uses InfluxDB, a database optimized for time series which requires payments to scale on multiple nodes. The latest MODRON version comes with a blockchain plugin to store relevant measurements directly in blockchain [13] but it does not provide the possibility of shared policy instauration to perform critical operations like data removal.

3.2   Blockchain

The idea of applying blockchains to the Internet of Things has been getting considerable attention in recent years, as evidenced by various works in the literature. The majority of the papers either focus on the possible benefits of blockchains for IoT [9,10,22] or propose architectures for specific use cases that show how blockchain technologies can solve some typical problems of the domain addressed, like healthcare [14] and security [15,16,18]. However, many general challenges still need to be solved, such as defining architectures to reconcile the large amount of data generated by IoT systems with blockchain storage and performance limitations. Accordingly, to the best of our knowledge the main difference between the works in the literature and our paper, is that we focus on using smart contracts for recognising obsolete data and removing it.

In [14] the authors propose a smart contracts architecture to improve IoT systems' privacy in the medical field through blockchain and smart contracts. In particular, a private Ethereum-based blockchain is used to improve patients data management and critical events logging. The paper [15] proposes BoSMoS, a blockchain-based system for increasing security in industrial systems by monitoring device software. The system takes snapshots of device software and stores them in a blockchain whose consensus mechanism is configurable. In addition, the authors conduct tests to measure system performance in the event of an intrusion attempt. In [18] the authors extend their previous work [16] and propose an architecture that uses a private blockchain to manage and monitor IoT systems. The authors focus on managing security issues and using the blockchain to keep track of device configuration history.

Overall, with respect to our primary goal, all the papers discussed focus on their specific use case and, even when performing an experimental evaluation, they do not delve into space occupancy issues.

4   System Design

In this section, we illustrate our main design choices. First, in Sect. 4.1 we describe the general requirements of our framework. Then, in Sect. 4.2 we present the cloud layer which involves both database and blockchain. Finally, in Sect. 4.3, we focus on the smart contract. Figure 1 shows framework's high-level architecture.

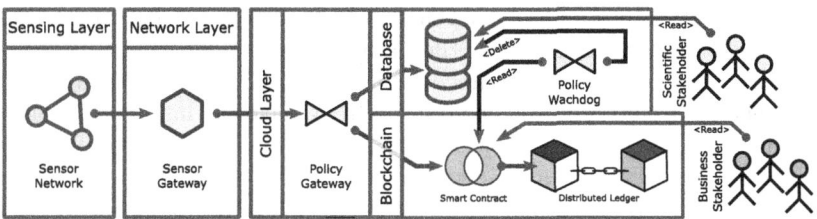

Fig. 1. The proposed blockchain-enabled architecture.

4.1   General Requirements

Different actors may need to interact with Continuous Monitoring systems Cloud Layer for accessing stored data, based on the nature of the monitored infrastructure. For instance, in SHM scenarios, that we will use in Sect. 5 as a use case for validating our approach, it is possible to identify at least four different actors: i) Infrastructure manager, responsible for the building and its maintenance. ii) Company that manages data from measurement systems, thus dealing with data collection and analysis. iii) Company that deals with the installation of measurement systems. iv) Government ministry for transport in the Country.

When disputes occur, often as a result of accidents, all the depicted actors may not trust each other and different actors may need to access stored data, based on the nature of the monitored asset. As a result, the cloud layer is responsible for storing data acquired by the sensors while guaranteeing the following properties: i) Fast access to time-series data is provided for running time-series analysis. ii) Data cannot be altered or tampered with.

A typical cloud layer is built on top of a data ingestion platform, such as ExaMon, which exploits a time-series database to provide fast access to stored time series, thus solving the fast-access needs of the layer itself.

In such a context, non significative data lowers the performance of the whole system, because of two side-effects. The first is that post-processing non significative data leads to calculation outputs which are not meaningful. The latter is that the data storage space spent for storing non significative data is a waste.

Therefore, we propose a system architecture which is able to assign an expiry date to stored data, according to a particular policy. The policy should be able to perform some sort of computations on data coming from sensor network, and apply an algorithm to establish the expiry time-lapse of data itself, i.e. the amount of time after which it may be safely deleted from the data storage to free up data storage space.

4.2   Cloud Layer

The Cloud Layer architecture proposed in this paper will accomplish multiple tasks: i) Ingest and route data from the sensor network; ii) Store data inside a database which provides fast-access for scientific stakeholders; iii) Provide a mechanism to delete non-meaningful data according to a policy shared among business stakeholders, improving the sustainability of the database over time.

With respect to a standard approach, which consists in a database-centric data storage, we propose to assist the database with a blockchain, which will provide the transactions programmed within the Policy Smart Contract. Two custom services have been designed and built to make the interactions between blockchain and database possible. The *Policy Gateway (PGW)* is a service responsible of routing the data coming from the Network Layer to both the database and the blockchain system. It will, indeed, be able to request the execution of transactions to the blockchain, and to write data into the database. The

Policy Watchdog (PWD), on the other hand, is a service which will be able of querying the blockchain and performing delete operations against the database.

To achieve consistency between information stored into the database and blockchain content, the PGW assigns an unique string identifier to each block. It will thus serve as a key for identifying blocks among heterogeneous data storage systems. The PWD will indeed be able to make delete requests targeting the block ids gathered from the blockchain.

4.3 Policy Smart Contract: Requirements and Motivations

A Smart Contract contains a set of suitable methods useful to interact with the distributed ledger. The methods which are intended to be used by the involved actor are exposed as transactions. As per Hyperledger Fabric features, the transactions may be used for updating the World State database, while the history of endorsed transactions is stored within the chain of blocks. In the Hyperledger Fabric framework, the Smart Contract source code is referred to as chaincode.

The World State can be populated with assets, which are sets of key-value pairs. The definition of the asset is made within the chaincode, and only the transactions defined in the chaincode itself are enabled to manage the defined asset. Therefore, each Smart Contract operates on its distributed ledger assets.

A permissioned blockchain can be set up such that the Smart Contracts may be installed on the peers only after the approval of all of them. Therefore, in our context a Smart Contract is suitable for storing and applying the expiry time lapse policy. Since all the transactions that get executed over time are stored in the blockchain distributed ledger, the policy application gets tracked over time in a transparent and immutable ledger. Indeed, considering the architecture represented in Fig. 1, each business stakeholder is required to **endorse** the deployment of the Policy Smart Contract. After its approval, it exposes some transactions which may be triggered both from the PGW, PWD applications and from the business stakeholders themselves.

4.4 Policy Smart Contract: Technical Details

We propose a Policy Smart Contract as an implementation of the Expiry Date Policy within the a Hyperledger Fabric blockchain framework. The Smart Contract has been developed as a Go chaincode, which exploits the Hyperledger Fabric SDK for building a Smart Contract. It defines the asset to be stored by the blockchain, and a set of available transactions.

The **asset** is defined as a data structure which holds a *unique policy string identifier*, which allows to keep track of which policy have been applied to the asset, a *unique asset string identifier*, received along with a data chunk from the PGW, the *md5 hash of the data chunk* time series, computed at data chunk arrival, the asset *expiry time lapse*, expressed as a time interval, computed by the applied policy, and the asset *expiry date*, computed considering the expiry period and the block creation date, which is a property assigned by the blockchain itself.

It is worth noting that sensors data is never stored within the blockchain, avoiding it to grow without tangible benefits. Indeed, it will always be possible to recover the chunk data from the system database using the chunk unique identifier.

Our Smart Contract is intended to provide support for different **policies**, that may run on assets data based on clients requirements. Therefore, a policy is defined as a class which contains a unique string identifier, and a function which receives a chunk of data as input, executes the policy-specific logic and outputs an expiry period. As a consequence, multiple policy instances may be deployed within the Smart Contract, each one implementing its own logic. This gives support to heterogeneous data processing, allowing to apply different policies based on different inputs in a **multi-policy** approach.

A set of atomic **private transactions** are defined to access the World State and the chain blocks. They implement CRUD operations over the World State database, which may be used by the transaction functions to read, create, delete or update the stored assets, and read operations over the chain blocks, which may be used for retrieving data from past transactions. These transactions can be invoked by public transactions, which are on the other hand available to the Smart Contract users for requesting them.

The fundamental **public transactions** are explained in the following lines. **AddChunkWithPolicy** computes a chunk hash, triggers the given policy to establish its expiry time and creates a corresponding asset which is then committed to the World State. **UpdateChunkExpiryDate** updates the given asset with a deterministic expiry date and time, referred to the asset creation time. This approach makes impossible to counterfeit the creation date of the asset, thus making impossible for clients to apply the policy in a fraudulent manner. **GetExpiredChunks** returns all the assets found in World State which are expired to the given expiry date, if the expiry date does not belong to the future.

The full documentation on Policy Smart Contract implementation can be found on the official repository[1].

Given the implemented transactions, our Smart Contract enables different **workflows**, thus enhancing policies flexibility with respect to the system needs. The policy, indeed, may be applied when the asset is created, or assigned in a different moment. Moreover, the policy assigned to an asset may be changed before the asset gets deleted.

The workflow used by the paper authors is represented in Fig. 2. At the end of its **lifecycle**, the asset will no more exist on World State, nor in the database. However, the complete history of transaction will remain stored inside the blockchain blocks. As a consequence, it is always possible to ensure that a missing data chunk has been processed by the Policy Smart Contract.

---

[1] Browse the repository at: gitlab.com/ecs-lab/hyper-watchdog.

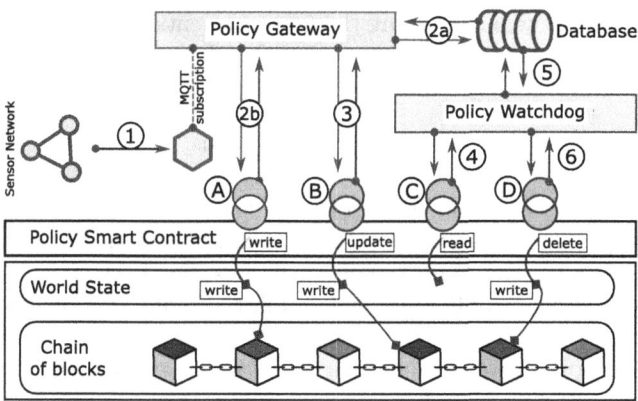

Fig. 2. Proposed **system workflow** and **asset lifecycle**: Policy Gateway receives the data chunk from sensor network (1) through an MQTT broker; it requests the *AddChunkWithPolicy* transaction (A) to store chunk information into the database (2a) and the blockchain (2b). Then, *UpdateChunkExpiryDate* (B) is triggered (3) for making the blockchain compute the chunk expiry date. When the chunk has expired, its unique id will be included in the list returned by *GetExpiredChunks* (C) transaction, therefore the Policy Watchdog will be aware that the block can be deleted (4) and will issue a delete request against the database (5). If data removal is successful, the Policy Watchdog notifies it to the blockchain (6) by triggering the *DeleteChunkIfExpired* (D).

5   Framework Validation

5.1   The Structural Health Monitoring Use-Case

Our use-case is based on a real railway bridge continuous monitoring application. Various elements and services make up the whole architecture, as already shown in Fig. 1. The *Sensing Layer* represents the acquisition system, composed by sensors which produce data at a certain frequency $f_s = 833\,\text{Hz}$. The *Network Layer* takes care of ingesting and pre-processing data: it re-synchronizes the received time series and builds chunks of data characterized by a certain length, N_{chunk}. Therefore, each chunk will contain the time series collected during an amount of time equal to $f_s N_{chunk}$. The chunks are made available to the remaining components of the system by means of an MQTT broker. The *Cloud Layer* takes care of storing data chunks received from the previous layer, making it available to scientific stakeholders for performing data analysis.

5.2   Expiry Time-Lapse Policy

The policy for data deletion should be able to define an algorithm to establish the expiry time-lapse of stored sensor data, i.e. the amount of time after which the data may be safely deleted from the data storage.

Considering our use case, we built an example policy based on mean signal energy of sensors data arrays. Indeed, each 3-axis accelerometer produces three data arrays, each of them containing the measurements along a certain axis. Our policy is able to compute the mean energy of each data array, according to the formula:

$$\overline{E} = \frac{1}{N_{chunk}} \sum_{0}^{N_{chunk}} |x_i|^2 \tag{1}$$

High-energy data is obtained when the bridge gets stimulated by the passage of a train. Our application can discard data coming from sensors when the bridge is not in stress conditions. Thus, the physical meaning of the formula is that low-energy data is most probably non-meaningful, therefore may be deleted sooner.

The resulting values are compared with axis-specific thresholds. Based on the results, the policy gives as output a suitable expiry time-lapse.

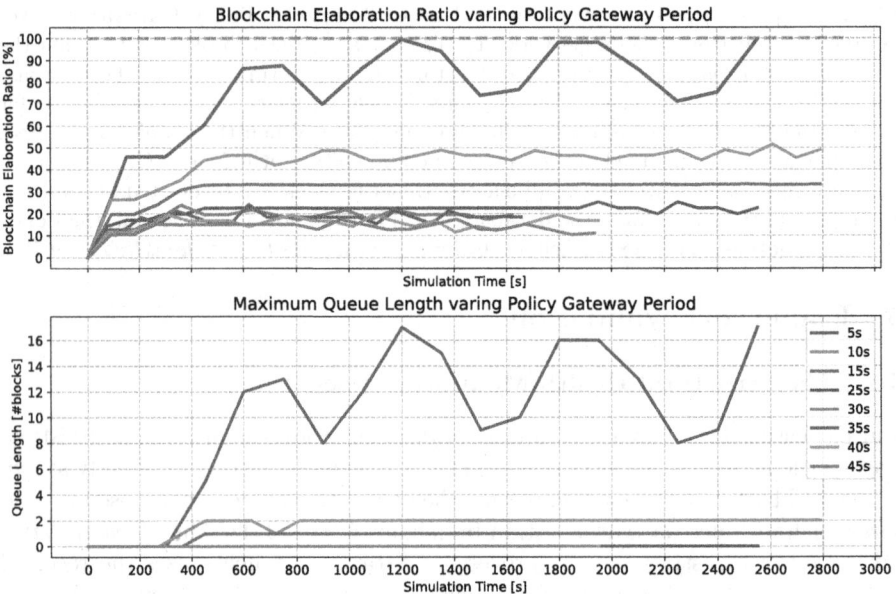

Fig. 3. Blockchain Elaboration Ratio (BER) and queue length versus simulation time (s). Overflow condition BER $>=$ 100% is represented as dashed grey line.

5.3   Test Bench Setup

The system has been tested and validated by deploying the Policy Smart Contract as a Go chaincode running on Hyperledger Fabric Test Network. It includes two peer organizations, each one composed by a peer, and a Raft ordering service consisting in a single node organization. All experiments were conducted using an host machine powered by an Intel^R Core^{TM} i7-10750H CPU @ 2.60 GHz having 64-bit Ubuntu 20.04.4 LTS with 16 GB memory. The Fabric network has been deployed with Docker Compose, with all nodes being containers running under the same Compose network, on the same host machine. As a consequence, network latency was not taken into account.

For testing purposes, the presence of a real database does not affect the performance of the system, because the requests against the database may be made in parallel with the requests made against the blockchain. Therefore, the tests have been conducted with the only presence of the blockchain network and the related services: *Policy Gateway* (PGW) and *Policy Watchdog* (PWD). The latter services have been simulated using a Go application, and the data chunks are injected towards a MQTT broker to the PGW by means of an ad-hoc Python simulator which reads from a database of historical time-series collected within the continuous monitoring application mentioned in the previous section.

All the tests have been conducted implementing the workflow proposed in Sect. 4.3, in a **Blockchain-in-the-loop** fashion. The policies have been tuned such that the expiry times allow the system to actually perform deletions at least a dozen of time within the experiment duration.

5.4   Results

Our use-case dataset contains data coming from a 24h measurement session monitoring. Accordingly, we designed a test to simulate the system in an accelerated way that allows us to explore different parameters in a narrower time range.

More specifically, our goal is to evaluate the system with a series of iterations using different chunk sizes, N. This parameter indeed heavily impacts the timings of transactions as will be shown by the following considerations. Since the sensors acquisition frequency f_s is given, each chunk of data will store measurements for a timespan equal to $t_c = N/f_s$ (s). As a consequence, the frequency at which the PGW is run in real-time context is equal to $f_{\text{PGW}} = 1/t_c$ (Hz).

Provided that in our Blockchain-in-the-loop test environment the chunks are available a-priori, the PGW can be fed-up with a higher frequency than the real-time one, thus we can perform accelerated simulations with a speedup coefficient of $\chi = t_c/t_{\text{PGW}}$. The duration of a transaction, δ_t, will never get affected by the speed of simulation, therefore, between the PGW interventions, the transactions will have less time to be executed than they have in real-time simulations, and they will eventually accumulate. This is why it is necessary to pose a constraint to force the PWD transactions to be run after a suitable amount of PGW transactions, which can be computed as t_{PWD}/t_c, where t_{PWD} is the time that occurs between PWD runs. Thus, the transactions requests will start accumulating if they require more time than t_{PWD} to be executed. Therefore, PWD may not get the actually expired items correctly. This condition may be defined as *overflow condition*, and can be expressed as: $\sum_t^{t+t_{\text{PWD}}} \delta_t > t_{\text{PWD}}$.

Consequently we can define the Blockchain elaboration ratio as:

$$\text{BER} = 100 \frac{\sum_t^{t+t_{\text{PWD}}} \delta_t}{t_{\text{PWD}}} \tag{2}$$

and express the *overflow condition* as: BER $>= 100\%$.

Drawing the evolution of pending transaction requests over time, referred to as *queue length*, we can observe that, as we may expect, lower values of t_c lead to longer queues. As demonstrated by the tests shown in Fig. 3, a lower size of t_c implies an higher amount of transactions, leading to higher BER values.

6   Conclusions

In many different application areas, spanning from structural monitoring to logistics, either private or public actors need to manage data flows with the constraint of being able to store only a portion of them. Leveraging the joint use of two technologies, continuous monitoring cloud infrastructure (e.g. Examon, Modron) and blockchain, we propose a pipeline capable of bounding storage space requirements of a continuous monitoring cloud infrastructure by ensuring the safe removal of portions of data while still safeguarding critical data.

The resulting system, whose pipeline is depicted in Fig. 1, comprises a database that stores all the data coming from the sensor network. The Policy Smart Contract evaluates how long each portion of data should last and memorizes this information in the block-chain, while the PWD filters out data that, according to the policy, do not meet the relevance criteria anymore. Thus, the scientific stakeholders will be provided with instruments to carry out high-performance data analysis, while the business stakeholders will be able to verify data integrity whenever it is necessary, without the need to trust each other. In this way, a tampering detection system is provided.

In Sect. 5, we show the implementation of our policy framework using accelerometric data coming from a sensor network deployed in a railroad bridge SHM use-case, leveraging Hyperledger Fabric [3] as blockchain framework to implement the Policy Smart Contract. According to our validation, the policy is able to detect the stresses on the bridge stimulated by the passage of a train, marking them differently from less-meaningful data. At the same time, the space requirements are kept under control, and are prevented from divergence.

More generally, test results demonstrate that, provided that the policies are well-tuned for the specific use-case, the system is able to free up a significative amount of database space, while improving the overall system security with respect to traditional data storage systems.

Independently from the nature of continuous monitoring scenario, if the policy outputs finite expire time intervals, it's possible to identify an upper threshold of occupied storage space, which will depend by the amount of deleted data with respect to the amount of ingested data. Having an infinite expiry time would lead to permanent storage of some portion of data, thus causing a divergence of occupied space. In the former case, our system allow to bound storage occupation to an upper limit, while in the latter case it helps reducing the speed of data occupation divergence.

In conclusion, our Policy Smart Contract system may be implemented within any continuous monitoring application to improve the sustainability of the database over time without performance loss. Indeed, the blockchain technology exploited at the system core does not affect the data ingestion speed of the database. However, if the number of policies deployed and their parameters result in a high number of transactions, the data deletion system may work slower than expected, leading to inconsistent results.

Therefore, to understand if our system is suitable for a certain monitoring use-case, some simulation tests must be carried out as the one we discussed within this paper: the *BER* value must be evaluated, and the *overflow condition* must never be met. Moreover, simulations must be carried out to predict the storage requirements of the blockchain and the impact of different blockchain deployment configurations, which depend on the underlying blockchain framework chosen by the user.

When using Fabric, considering that the size of a transaction S_{tr} is $\sim 5\,\mathrm{KiB}$, we can predict the blockchain size as $S = S_{tr}t\sum f_i$, where t is the elapsed time and f_i is the frequency of requested transactions of type i. For instance, our use-case requires two transactions for policy evaluation triggered by PGW for chunk creation and expiry date computation, a variable number of transaction, γ, for policy check and data elimination. γ represents the policy behaviour, that in our use-case depends on the PGW and PWD frequencies: $\gamma = \lceil \frac{f_{PGW}}{f_{PWD}} \rceil$. Thus, the blockchain size is equal to $S = S_{tr}t\,(2f_{PGW} + (\gamma + 1)f_{PWD})$.

Considering a scenario in which PGW period is 1 min and PWD period is 3 min, and a worst case in which all the chunks elapse instantaneously, thus $\gamma = 3$, we obtain $S = 0.0446t$ GiB/day. Without data deletion, one year of activity results in 1.7 TiB of database size. With data deletion, implementing the sample policy discussed in previous sections, we would need to store 8 GiB of blockchain transactions with the benefit of securely shrinking the database up to 75%, thus gaining $\sim$163 times the blockchain size.

Further work is required to assess all the trade-offs this infrastructure involves and to explore the impact of more complex policy algorithms, such as discarding too many similar data samples to maintain only subspace prototypes, and reprogrammable policies that can evolve and change in time.

References

1. Adere, E.M.: Blockchain in healthcare and IoT: a systematic literature review. Array **14**, 100139 (2022)
2. Aguzzi, C., et al.: Modron: a scalable and interoperable web of things platform for structural health monitoring. In: 2021 IEEE 18th Annual Consumer Communications Networking Conference (CCNC) (2021)
3. Androulaki, E., et al.: Hyperledger fabric: a distributed operating system for permissioned blockchains. In: Proceedings of the Thirteenth EuroSys Conference (2018)
4. Bartoletti, M., et al.: A journey into bitcoin metadata. J. Grid Comput. **17**(1), 3–22 (2019)

5. Bashir, I.: Mastering Blockchain. Packt Publishing Ltd., Birmingham (2017)

6. Bonneau, J., et al.: SoK: research perspectives and challenges for bitcoin and cryptocurrencies. In: 2015 IEEE Symposium on Security and Privacy. IEEE (2015)

7. Borghesi, A., et al.: ExaMon-X: a predictive maintenance framework for automatic monitoring in industrial IoT systems. IEEE Internet Things J. (2021)

8. Buterin, V., et al.: Ethereum: a next-generation smart contract and decentralized application platform (2014)

9. Christidis, K., et al.: Blockchains and smart contracts for the internet of things. IEEE Access **4**, 2292–2303 (2016)

10. Dai, H., et al.: Blockchain for internet of things: a survey. IEEE Internet Things J. **6**(5), 8076–8094 (2019)

11. Zonzini, F., et al.: Structural health monitoring and prognostic of industrial plants and civil structures: a sensor to cloud architecture. IEEE Instrum. Meas. Mag. **23**(9), 21–27 (2020)

12. Farrar, C., et al.: Structural Health Monitoring: A Machine Learning Perspective. Wiley, Hoboken (2012)

13. Gigli, L., et al.: Blockchain and web of things for structural health monitoring applications: a proof of concept. In: 2022 IEEE 19th Annual Consumer Communications Networking Conference (CCNC) (2022)

14. Griggs, K., et al.: Healthcare blockchain system using smart contracts for secure automated remote patient monitoring. J. Med. Syst. **42**(7), 1–7 (2018)

15. He, S., et al.: BoSMoS: a blockchain-based status monitoring system for defending against unauthorized software updating in industrial internet of things. IEEE Internet Things J. **7**(2), 948–959 (2019)

16. Helebrandt, P., et al.: Blockchain adoption for monitoring and management of enterprise networks. In: 2018 IEEE 9th Annual Information Technology, Electronics and Mobile Communication Conference (IEMCON), pp. 1221–1225. IEEE (2018)

17. Jeong, S., et al.: Sensor data reconstruction and anomaly detection using bidirectional recurrent neural network. In: Sensors and Smart Structures Technologies for Civil, Mechanical, and Aerospace Systems 2019, vol. 10970, pp. 157–167 (2019)

18. Košt'ál, K., et al.: Management and monitoring of IoT devices using blockchain. Sensors **19**(4), 856 (2019)

19. Li, H., et al.: Reviews on innovations and applications in structural health monitoring for infrastructures. Struct. Monit. Maint. **1**(1), 1–45 (2014)

20. Lynch, J., et al.: Structural health monitoring: technological advances to practical implementations. Proc. IEEE **104**(8), 1508–1512 (2016)

21. Reyer, M., et al.: Design of a wireless sensor network for structural health monitoring of bridges. In: 2011 Fifth International Conference on Sensing Technology, pp. 515–520. IEEE (2011)

22. Reyna, A., et al.: On blockchain and its integration with IoT. Challenges and opportunities. Future Gener. Comput. Syst. **88**, 173–190 (2018)

23. Sabato, A., et al.: Wireless mems-based accelerometer sensor boards for structural vibration monitoring: a review. IEEE Sens. J. **17**(2), 226–235 (2016)

24. Sidorov, M., Khor, J.H., Nhut, P.V., Matsumoto, Y., Ohmura, R.: A public blockchain-enabled wireless LoRa sensor node for easy continuous unattended health monitoring of bolted joints: implementation and evaluation. IEEE Sens. J. **20**(21), 13057–13065 (2020)

25. Szabo, N.: Formalizing and securing relationships on public networks. First Monday (1997)

26. Tang, Z., et al.: Convolutional neural network-based data anomaly detection method using multiple information for structural health monitoring. Struct. Control. Health Monit. **26**(1), e2296 (2019)
27. Uddin, M.A., Stranieri, A., Gondal, I., Balasubramanian, V.: A survey on the adoption of blockchain in IoT: challenges and solutions. Blockchain Res. Appl. **2**(2), 100006 (2021)

Outlier Explanation Through Masking Models

Fabrizio Angiulli⬡, Fabio Fassetti⬡, Simona Nisticò$^{(\boxtimes)}$⬡,
and Luigi Palopoli⬡

DIMES Department, University of Calabria, Rende, Italy
{fabrizio.angiulli,fabio.fassetti,simona.nistico,
luigi.palopoli}@dimes.unical.it

Abstract. Given a database and one single anomalous data point, the
Outlying Aspect Mining problem consists in explaining the abnormality
of that data point w.r.t. the data population stored in the input database.
Thus, the problem requires the discovery of the sets of attributes and
associated values that account for the abnormality of a data point within
a given data set. In this setting, the abnormality of the data point at
hand is stated beforehand, e.g., as the result of some outlier detection
techniques (which, for the most part, do not provide information about
why the selected data points are actually anomalous). This paper proposes a solution to the OAM problem exploiting a deep learning architecture. Besides explaining the input data point abnormality by singling
out the smallest set of pairs attribute-value justifying it, our technique
also provides new values for those attributes that would transform the
input outlier into an inlier. Several experiments are also presented that
assess the effectiveness of our approach.

Keywords: Outlier aspect mining · Explainable artificial intelligence ·
Deep learning

1 Introduction

Anomaly detection is an important task in AI and data mining, as witnessed
by the volume of papers appeared on the subject [9,10,24,28]. Accordingly, this
task has many relevant applications in such diverse fields such as finance, cyber-
security, healthcare, fraud detection and others [1,14,16,17,19,23,33].

A relevant problem related to outlier detection is outlying aspect mining
(OAM) (aka, *subspace selection, outlier explanation, object explanation, outlier
interpretation, outlying subspaces detection*): to explain, given a data point known
to be anomalous beforehand, the goal is to find the characteristics that locate it far
from the non anomalous data. The relevance of the task stems from the fact that
most of the anomaly detection techniques do not explain the reasons for a given
object to be recognized as anomalous, notwithstanding that to expose such explanation is often very important in real applications. The task can thus be defined as
that of looking for feature subset(s) (and related values) on which the given data
point is anomalous w.r.t. the population of data objects it belongs to [26].

© Springer Nature Switzerland AG 2022
S. Chiusano et al. (Eds.): ADBIS 2022, LNCS 13389, pp. 392–406, 2022.
https://doi.org/10.1007/978-3-031-15740-0_28

In this paper, we tackle this problem by developing a new approach exploiting adversarial-like neural networks. Our architecture comprises two modules. The former one implements the generative component of the procedure that realizes a generative mechanism different from usual ones in that the goal of the generation is to modify a sample (the given outlier) finding a way to have it become an inlier. The latter module acts as an oracle that predicts if the analyzed point is anomalous or not. The information it provides guides the generative module, as usual in adversarial like approaches, to find the "right way" to modify the sample under analysis in order to "relocate" it close to the points of the reference data set. Summarizing:

- We present a technique, called **MMOAM** , which stands for **Masking Models for Outlying Aspect Mining** to "explain" outliers. To the best of our knowledge, this technique is the first based on deep learning for the considered task.
- We assess the effectiveness of our technique over both synthetic and real data sets.
- We develop a comparison of MMOAM against Subspace Outlier Detection (SOD), an algorithm of reference for the context at hand.

Relevant related literature is briefly recalled next.

Outlier Aspect Mining methods can be grouped into two macro-categories that reflect the type of employed strategy: the former one collects techniques based on **feature selection**, whereas the latter one includes **score-and-search** approaches [30].

The first category includes approaches where feature selection methods, typically used for classification tasks, are applied. [22] solves the OAM on numerical data by converting the problem into a two-class classification and uses the outputs as the starting point to obtain an explanation for sample outliers. Local Outliers with Graph Projection (LOGP), proposed in [12], deals both with outlier detection and outlier explanation by exploiting concepts borrowed from the spectral graph embedding theory. [5] presents a technique that, given a categorical data set and an oulier q, finds the top k attributes associated with the largest outlier score for q. An extension of the approach to numerical data sets is presented in [3].

The latter category collects all the methods which use a measure as a quality metric to choose between sample features. Considering such type of approaches, the paper [34] propose the High-dimensional Outlying Subspace Miner (HOS-Miner) technique. HOS-Miner detects the subspaces in which a given data point turns out to be dissimilar to the other data points. To reach this goal, the Outlying Degree distance function is exploited, which is defined as the sum of the distances between the query data point and its k-nearest neighbours. Density is used as subspace score criterion in [13], where a kernel-density estimation is employed to rank the attribute subspaces and the adoption of such a measure is justified by the fact that the density measure is affected by the growth of the number of dimensions, since points become sparser as the dimensionality increases. The paper [30] presents two dimensional-unbiased measures: the Z-score and the Isolation Path score (iPath). The latter score here is inspired by Isolation Forrest [21], an outlier detection algorithm, based on the idea that anomalies are "few" and isolated from the rest of the data. In addition to these scores, this work also proposes a procedure for searching for attribute subspaces characterizing anomalies. The approach consists in performing a beam search divided into three steps. First of all, to analyze all trivial outlying features, the proposed algorithm inspects all 1-D spaces then, in the second step, it examines all 2-D subspaces through exhaustive research and, finally, in the last stage, it performs the beam search at level l. In [32], starting from [27], a simple grid-based density estimator for outlying aspect mining, called sGrid, is presented. The objective is to make mining algorithms faster, a goal which is actually attained but at the cost of the estimation space unbiasedness. A notable further form of score is the Simple Isolation score Using Nearest Neighbor Ensamble (SiNNE) [25], whose definition is related to an outlier detection algorithm named Isolation using Nearest Neighbor Ensembles(iNNE) [7].

Possible further alternative strategies could be based on **hybrid** approaches [26] where the idea is to combine the strength of both features-selection-based and score-and-search-based approaches. To the best of our knowledge, to date, the only algorithm falling in this category is the one named Outlying Aspect Mining via Feature Ranking (OARank for short) [29]. Here, the hybrid approach is realized by organizing the procedure into two steps. In the first step, the framework uses a feature selection technique and the retrieved features are used as the input for the next step. In second step, the score-and-search algorithm is applied on the features found in the former step to obtain the final output. The goal of the second step is to fine-tune the result of the feature selection phase.

In the rest of the paper and whenever to ambiguities arise we will use the terms "attribute", "feature" and "dimension" basically interchangeably.

The rest of the paper is organized as follows. Section 2 presents the MMOAM technique, Sect. 3 illustrates the experimental results and, finally, Sect. 4 concludes this work.

2   The Proposed Technique

In this section, the MMOAM[1] technique is described. Its foundational idea is to exploit an adversarial approach to explain which characteristics make a certain data point to be located "far" from the given reference data set.

As already mentioned, the architecture we propose realizes a deep learning solution using a generative component coupled with an adversarial component. In particular, the adversarial module tries to predict if the input data point is anomalous or not. The information it outputs is then fed into the generative module that, in its turn, tries to single out the set of changes (specified as a set of attribute-values pairs) needed for the input outlier data point (also referred to as "sample data point" in the following) to be transformed into an inlier w.r.t. the reference data set.

---

[1] https://github.com/simona-nistico/MMOAM.

In deference to the Occam's razor principle, for an outlier explanation to be significant, the transformation that changes the nature of the sample transforming it from anomalous to normal must be *minimal*, which means that the number of features returned as justification of the given data point outlierness must be as small as possible (we notice that this also makes the explanation more easily interpretable by the final user [31]).

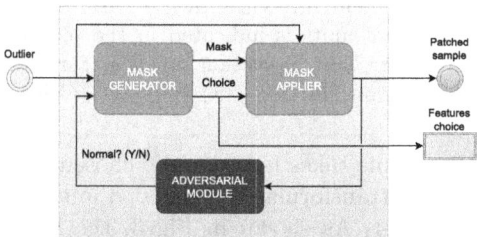

Fig. 1. Proposed architecture. The Mask Generator and the Mask Applier represents the generative module of the MMOAM methodology, the adversarial module acts as an oracle.

The MMOAM architecture is depicted in Fig. 1 and the following sub-sections are devoted to report a more detailed description of the modules included therein.

2.1   The Generative Module

The generative module exploits Masking Models [4] to produce a modified version of the sample data point in order to transform it in such a way that it gets close to the data points belonging to the reference data set. To this aim, MMOAM divides the task into a mask generation step (carried out by the Mask Generator module) and an application step (that is taken care of by the Mask Applier module). The Mask Generator module is a neural network that learns a transformation which tells how to modify the data point to remove the anomaly. The thus computed transformation is encoded into a mask (a sequence of attribute-value pairs, one attribute for each attribute of the original data set). The Mask Applier module then applies the learned transformation by using a suitable combination function (see below).

Figure 2 shows the Mask Generator module architecture: this is a dense neural network with two output branches. The first of these two branches (the one at the bottom in the figure) selects the features to be modified using a dense layer with a sigmoid function as the activation function, so the feature choice is a vector with the same dimension of the sample and values ranging from 0 to 1. The latter one (that located at the top of the figure) outputs the magnitude of the modification to be applied on each involved attribute, this modification are given in output as a real-valued vector with again the same dimension of the input sample. Together, the outputs of these branches represent the transformation that determines the changes to apply by indicating which features need to be modified and how much to modify them.

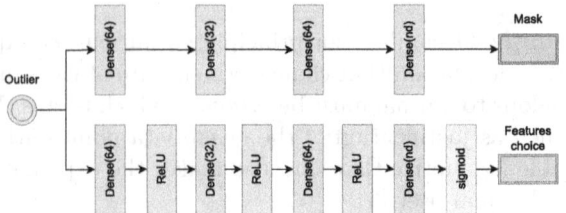

Fig. 2. Mask Generator module architecture. In the green blocks, "Dense" represents a dense layer which number of units is indicated in the round parenthesis. "ReLU" stands for the Rectified Linear Unit activation function $f(x) = \max(0, x)$, "sigmoid" stands for the sigmoid activation function $f(x) = \frac{1}{1+e^{-x}}$. (Color figure online)

The Mask Applier module takes in input the chosen features, the mask and the outlier and applies the transformation (encoded in the mask and the chosen attribute set) to the outlier. As shown in Fig. 3, the information brought by features choice and the mask is combined via an element-wise multiplication. The transformation that results after this operation is summed to the outlier, thus obtaining its "patched" version.

In order to construct an adversarial sample with the desired characteristics, a loss function $l_m()$ is exploited during the training of the Generation Module with the stated objective of changing the input outlier into an inlier using the minimum possible number of transformations. To this end, we adopt a loss built from four different factors, as shown next:

$$l_m(x', y', X_n, m, c) = \alpha_0 fl(0, y') + \alpha_1 \frac{1}{|X_n|} \sum_{s \in X_n}^{n} mse(x', s) + \alpha_2 bce(c, 0) + \alpha_3 ||m||_2^2$$

(1)

where x' is the modified version of x produced by the generation module, y' is the classification obtained from the adversarial module for x', m is the mask produced by the mask generator submodule, c are the chosen features, X_n are the data set normal samples, fl is the focal loss [20], mse is the mean squared error, bce is the binary cross-entropy and α_0, α_1, α_2, α_3 are hyper-parameters.

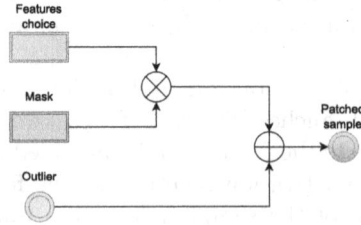

Fig. 3. Mask Applier module structure. In the schema, feature choice and mask are combined through multiplication then the resulting transformation is applied through sum to the outlier to obtain the patched sample.

The first term of the loss function aims at minimizing the anomaly score of the modified sample, the second one serves the purpose of reducing the distance to normal points and, finally, the remaining two are devoted to reducing the number of chosen features and the magnitude of the modification applied to them, respectively. The hyper-parameters α_0, α_1, α_2 and α_3 serve the purpose of controlling the relative weight of each component in the value computed for the loss.

2.2   The Adversarial Module

The Adversarial Module is a neural network trained using a data set consisting in normal and anomalous samples. The task faced by this module is unbalanced classification, as a matter of fact the data set initially have inside it only one anomalous sample that is the sample for which we want an explanation. Then, an iterative strategy where for each iteration the anomalous class is enriched by one new sample, that is the generative model output, is adopted. The goal of this strategy is to improve at every step the explanation, bringing the patched sample as near as possible to the normal data points.

This module outputs a score value (ranging from 0 to 1) obtained from the sigmoid activation function that measures how much a point is anomalous. The loss adopted for training is the focal loss [20], chosen because of the imbalance of the data set. This objective function is a variation of cross-entropy loss, whose peculiar characteristic is to focus its attention on heavily misclassified examples, decreasing in such a way the influence of samples that are easy to classify for the model. Unbalanced classification benefits from this type of strategy. The type of model used in this paper for the Adversarial Module is a Dense Neural Network, about which no further detail is provided here since the core of our proposal is the generation phase, which is its the key-aspect, while the adversarial module is ancillary to it and might be also substituted by any other neural net based anomaly detection module.

The training process employed in the architecture is as follows. Given a data set (X, Y), including a subset of the normal samples, selected using a criterion based on the distance to the outlier, and the anomalous sample under analysis, cumulatively denoted by X, together with their labels, denoted by Y, and an initial copy (X', Y') of this data set, the following steps are performed:

1. The Adversarial Module is trained on (X', Y') for a number e_a of epochs.
2. The Generative Module (thus, both the Mask Generator and the Mask Applier) is trained on (X, Y) for a number e_g of epochs.
3. The patched version $\tilde{o}$ of the outlier o is computed and added to the extended training data set, labelled as anomalous, that is, $X' \leftarrow X' \cup \{\tilde{o}\}, Y' \leftarrow Y' \cup \{1\}$, where 1 is the label assigned to anomalous samples.

This procedure is iterated n_{adv} times in order to obtain the final explanation for the sample outlierness, the explanation given to the user is the result obtained from the last iteration. n_{adv} is an algorithm parameter, as it is reported also in Sect. 3, here we consider it equals to 5 and this brings to good results.

3   Experiments

In this section, the assessment of the proposed methodology is performed considering both a qualitative and a quantitative evaluation. The objectives are to evaluate if MMOAM retrieves the correct features to characterize the outlier and if it is effectively able to mask the sample to make it inlier w.r.t the normal samples.

To this aim, first, we consider a generated data set to assess the robustness of the proposed algorithm to the increase in the number of features and the increase in the number of dimensions characterizing the anomaly. Second, MMOAM is tested on two real data sets to evaluate its behaviour in real contexts.

In both batteries of experiments, also a comparison with the SOD algorithm is performed.

3.1   Synthetic Data

In this section, we consider a situation in which there is a data set with independent attributes and in which there is a sample that is anomalous on one or more of these attributes. To illustrate, data are generated from a normal distribution with mean 3 and standard deviation 0.3; the anomalous point is obtained perturbing one sample by changing one or more of its attributes and having each of these nine standard deviations apart from its original value. Here we test also the scale performances of the method both as the number of data features increases and as the number of perturbed ones increase.

The adopted parameters are as follows:

- the number of considered adversarial explanations is equal to 5;
- the number of epochs carried out at each explanation step is equal to 300;
- for the adversarial module, the number of epochs performed for each step is equal to 150, with a batch size of 16, and
- the number of normal points considered to produce the explanation is equal to 100.

The anomalous sample for the first considered synthetic data sets is obtained by a modification performed on a random dimension. In Fig. 4, the explanations obtained for each test are shown. The figure shows that, in this setting, MMOAM can modify the sample in such a way as to move it within the data distribution also when the number of dimensions increases. Furthermore, the figure content highlights that, in all the tests, the only dimension detected as important is always the correct one: this proves that MMOAM is capable of consistently singling out what makes the given point anomalous.

In the second set of experiments, the number of dimensions modified to obtain an anomalous sample is increased while the total number of dimensions is set to 20. Except for this, the setting is the same as in the previous test. Thus, to create the out-of-distribution point an increasing number of random features are modified. Figure 5 reports the results of this experiment. Again, our approach allows to correctly detect the features that make the sample anomalous, singling out both the modified attributes.

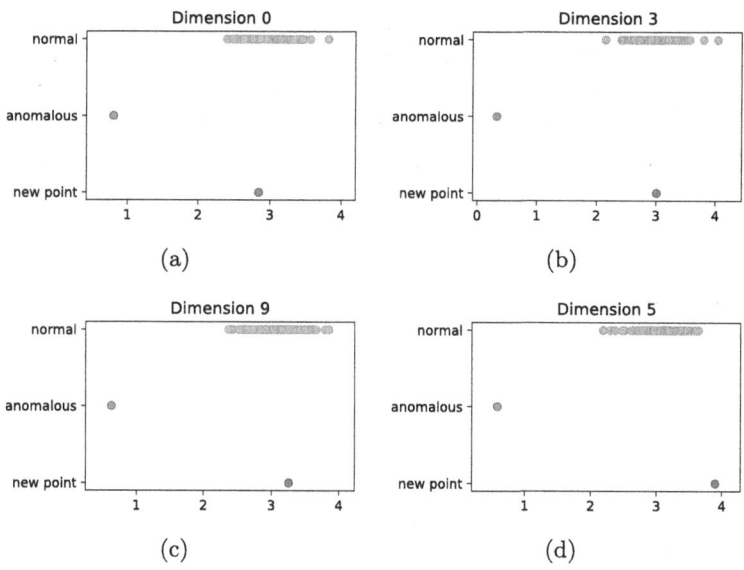

Fig. 4. Explanations produced with increasing number of dimensions. The modified dimension here is 1, the dimension to perturb is chosen randomly. (a) considers the data set with 3 dimensions, (b) considers the data set with 5 dimensions, (c) considers the data set with 10 dimensions and (d) considers the data set with 20 dimensions.

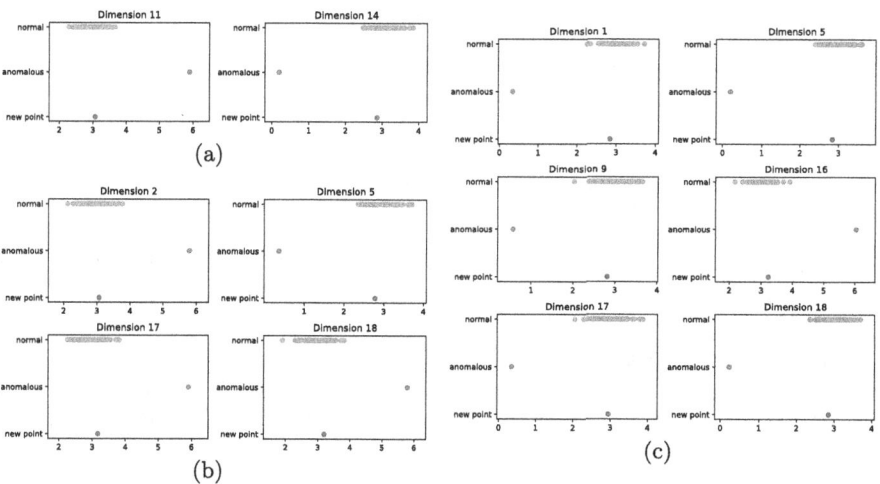

Fig. 5. Explanations produced with increasing number of modified dimensions with a fixed total number of dimensions set to 20. The dimensions modified in the anomalous sample are randomly selected. (a) considers the modification of 2 features, (b) considers the modification of 4 features and (c) considers the modification of 6 features.

Comparison with SOD. In this section, we compare our method with SOD [18], a reference algorithm that carries out subspace outlier detection. We note that, by its nature, SOD allows us to inspect the attributes used for its classi-fication step. For the battery of experiments illustrated next we use the same data setting as that described above for previous experiments. Moreover, we set additional parameters as follows:

- for SOD:
    - the number of neighbours utilized is set to 40;
    - the reference set size is 200;
    - α is set 0.95;
- for MMOAM :
    - the number of utilized points is set to 100;
    - the number of epochs for each discriminator training is set to 150;
    - the discriminator batch size is set to 16;
    - the number of epochs for each generative module training is set 500;
    - the number of adversarial epochs is set to 5;
    - the values of the loss hyperparameters are set to 1.5, 0.6, 0.2 and 0.2, respectively.

Since the ground truth is available, precision and recall measures can be exploited to compare the two algorithms. Thus, let F_r be the set of retrieved features and F_a be the set of features w.r.t. which the sample is anomalous. Then, the precision $P()$ is defined as the portion of retrieved features that are actually anomalous, that is:

$$P(F_r, F_a) = \frac{|F_r \cap F_a|}{|F_r|}$$

while the recall $R()$ is defined as the fraction of anomalous retrieved dimensions, that is:

$$R(F_r, F_a) = \frac{|F_r \cap F_a|}{|F_a|}$$

Precision and recall scores reported in this section are obtained by averaging the value of these scores over 100 runs, each of which uses a different data set generated using the strategy illustrated above.

The situation depicted in the Fig. 6 is associated to tests performed over data sets characterized by a growing number of dimensions and, precisely, with 5, 10 and 20 features. Although both MMOAM and SOD recover all the anomalous characteristics, it is possible to observe that the precision scored by our technique outperforms the one SOD scores. It is moreover worth noting that SOD preci-sion decreases quickly as the number of dimensions increases. On the contrary, notably, this experiment shows that MMOAM reacts better to the increase in the number of features and it maintains high precision and recall scores.

The experiment whose results are summarized in Fig. 7 is one where the number of dimensions is fixed to 20 whereas the number of anomalous features of

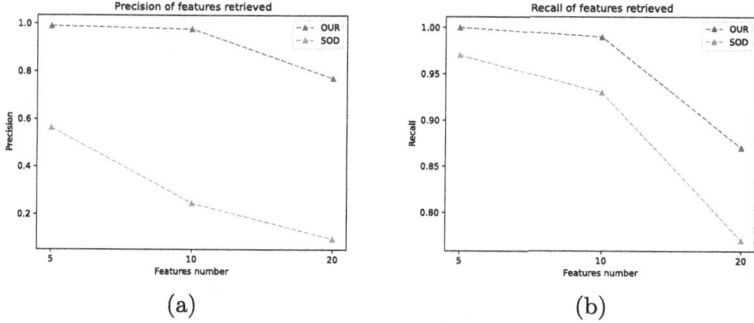

Fig. 6. Mean precision (a) and mean recall (b) reached by SOD and MMOAM on data sets with dimensions 5, 10 and 20. This results are obtained on 100 runs performed on different data sets generated following the same procedure.

the anomalous sample gets larger and larger. Also in this case, the experimental evidence shows that MMOAM markedly outperforms SOD in both in precision and recall. Even if the number of anomalous dimensions is increased, MMOAM remains accurate and, at the same time, retrieves all the correct features, while SOD presents a different trend, in which its precision increases while its recall decreases, so it is no longer able to find all the dimensions to single out.

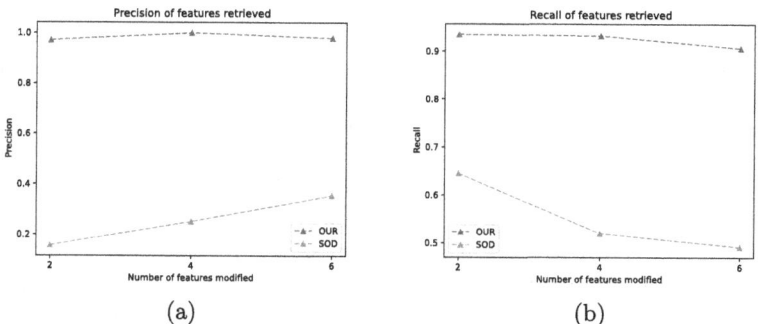

Fig. 7. Mean precision (a) and mean recall (b) reached by SOD and MMOAM on data sets with 20 dimensions and 2, 4 and 6 anomalous features. This results are obtained on 100 runs performed on different data sets generated following the same procedure.

3.2   Real Data Sets

It is crucial to test the proposed technique on real data to understand if it is effective in concrete contexts. The tests of this subsection are devoted to this goal. Here, two different types of evaluations are performed. First of all, we assess the quality of the informative content of the explanation considering a data set whereby in literature is available evidence provided by domain experts. Then, a quantitative evaluation is performed comparing the quality of the subspaces retrieved by our method with the ones detected by SOD. The data sets employed in these experiments are the Iris [15] and the Breast Cancer Wisconsin data sets [8] respectively.

Iris Data Set. The Iris data set [15] contains data regarding three Iris species and, for each species, it stores 50 samples. Each instance is described through 6 attributes: an identifier, the sepal length and width measures, the petal length and width measures and the species to which the flower belongs to, where all the lengths and widths are expressed in centimetres.

For this experiment, the data set has not been considered in its entirety and, actually, only the instances of *Setosa* and *Virginica* Iris flowers are taken into account. To have a setting coherent with the considered task, Iris Setosa samples are chosen to represent the inlier data, while anomalous ones are taken from Iris Setosa samples. The id attribute is dropped since it does not carry any informative content.

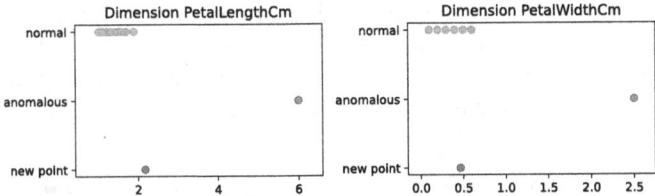

Fig. 8. The explanation for one Iris Setosa sample, the features detected as anomalous are the petal length and width. The value of these features for the patched sample is highlighted by the blue point. (Color figure online)

As shown in Fig. 8, the application of our technique results in singling out petal length and width as the features to act upon (by decreasing their values) in order to transform an Iris Setosa (outlier) sample into an Iris Virginica (normal) one. This finding is confirmed by what is known in botany about these flowers: to distinguish between specimens of these two species only sepal length and sepal width must be analyzed and compared [2].

To conclude with the comparison of our technique with SOD, it is worth noting that, once run on the same data set, SOD fails to deliver same quality results. Indeed SOD singles out almost all data features but leaves out only sepal length, which is, notably, one of the distinctive attributes of the outlier explanation.

Breast Cancer Wisconsin Data Set. Here, the Breast Cancer Wisconsin data set [8] is considered, it is a collection of 30 features computed from digitized images of a fine needle aspirate (FNA) of a breast mass. The samples which form this data collection are annotated as benign or malignant. From this point on, the benign samples are used as normal points while malignant ones are used as anomalous.

Since no ground truth is available for this data set, to compare quantitatively subspaces retrieved by MMOAM with ones retrieved by SOD, the KNN algorithm [11] is used as an evaluator. The experiment carried out is the following: for each sample of the outlier class (so for each one labelled as malignant), MMOAM and SOD algorithms are applied to retrieve a subspace, the results obtained are evaluated training, using only features belonging to these subspaces, a KNN classifier. Then, to rank each point, sum of the distances to its k-nearest-neighbours is used [6], the number of neighbours used is set to 5 since it gives the best results in terms of classification accuracy. The higher is this score w.r.t. the score of the other points, the higher is the sample outlierness. Then, the method efficacy is measured as the fraction of subspaces for which the sample lies in the top-n ranked points. The score obtained ranges from 0 to 1. The more it is near 1, the better is the result.

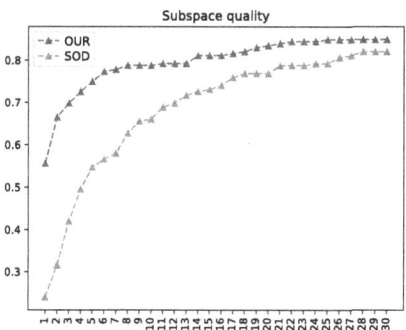

Fig. 9. Comparison of the quality of the subspaces retrieved by MMOAM and SOD, where on the abscissa there is the value of n for the top-n score and on the ordinate the score values.

In Fig. 9, are depicted the results of this evaluation. There, on the abscissa, are reported the values of n used, while on the ordinate, the score values. What is highlighted by the figure is that MMOAM is more accurate to detect subspaces for which anomaly rank is high w.r.t. SOD, which subspaces ranking is lower. It is another proof of the quality of the explanations returned by the proposed methodology.

4   Conclusions and Future Work, Briefly

In this paper, a deep-learning architecture has been illustrated which solves the OAM problem also singling out those attribute values that make the input outlier data point to become an inlier. Several experiments have been presented that demonstrate the general effectiveness of our approach and its capacity to outperform the reference SOD algorithm.

As a future development of this research we envision the possibility to substitute, in our architecture, the classification-based adversarial model with one on the unsupervised kind such as, e.g., an auto-encoder network. Furthermore, we deem it interesting to generalize our approach to make it capable of handling discrete data domains, which will impose a significant remake of parts of our neural architecture (particularly, the generative component of it).

References

1. Abdallah, A., Maarof, M.A., Zainal, A.: Fraud detection system: a survey. J. Netw. Comput. Appl. **68**, 90–113 (2016)
2. Anderson, E.: The species problem in iris. Ann. Mo. Bot. Gard. **23**(3), 457–509 (1936). http://www.jstor.org/stable/2394164
3. Angiulli, F., Fassetti, F., Manco, G., Palopoli, L.: Outlying property detection with numerical attributes. Data Min. Knowl. Disc. **31**(1), 134–163 (2016). https://doi.org/10.1007/s10618-016-0458-x
4. Angiulli, F., Fassetti, F., Nisticò, S.: Finding local explanations through masking models. In: Yin, H., et al. (eds.) IDEAL 2021. LNCS, vol. 13113, pp. 467–475. Springer, Cham (2021). https://doi.org/10.1007/978-3-030-91608-4_46
5. Angiulli, F., Fassetti, F., Palopoli, L.: Detecting outlying properties of exceptional objects. ACM Trans. Database Syst. (TODS) **34**(1), 1–62 (2009)
6. Angiulli, F., Pizzuti, C.: Fast outlier detection in high dimensional spaces. In: Elomaa, T., Mannila, H., Toivonen, H. (eds.) PKDD 2002. LNCS, vol. 2431, pp. 15–27. Springer, Heidelberg (2002). https://doi.org/10.1007/3-540-45681-3_2
7. Bandaragoda, T.R., Ting, K.M., Albrecht, D., Liu, F.T., Zhu, Y., Wells, J.R.: Isolation-based anomaly detection using nearest-neighbor ensembles. Comput. Intell. **34**(4), 968–998 (2018)
8. Bennett, K.P., Mangasarian, O.L.: Robust linear programming discrimination of two linearly inseparable sets. Optim. Methods Softw. **1**(1), 23–34 (1992)
9. Bhuyan, M.H., Bhattacharyya, D.K., Kalita, J.K.: Network anomaly detection: methods, systems and tools. IEEE Commun. Surv. Tutor. **16**(1), 303–336 (2014). https://doi.org/10.1109/SURV.2013.052213.00046
10. Chandola, V., Banerjee, A., Kumar, V.: Anomaly detection for discrete sequences: a survey. IEEE Trans. Knowl. Data Eng. **24**(5), 823–839 (2012). https://doi.org/10.1109/TKDE.2010.235
11. Cunningham, P., Delany, S.J.: K-nearest neighbour classifiers - a tutorial. ACM Comput. Surv. **54**(6), 1–25 (2021). https://doi.org/10.1145/3459665
12. Dang, X.H., Assent, I., Ng, R.T., Zimek, A., Schubert, E.: Discriminative features for identifying and interpreting outliers. In: 2014 IEEE 30th International Conference on Data Engineering, pp. 88–99. IEEE (2014)
13. Duan, L., Tang, G., Pei, J., Bailey, J., Campbell, A., Tang, C.: Mining outlying aspects on numeric data. Data Min. Knowl. Disc. **29**(5), 1116–1151 (2015). https://doi.org/10.1007/s10618-014-0398-2
14. Duraj, A., Chomatek, L.: Supporting breast cancer diagnosis with multi-objective genetic algorithm for outlier detection. In: Kościelny, J.M., Syfert, M., Sztyber, A. (eds.) DPS 2017. AISC, vol. 635, pp. 304–315. Springer, Cham (2018). https://doi.org/10.1007/978-3-319-64474-5_25

15. Fisher, R.A.: The use of multiple measurements in taxonomic problems. Ann. Eugen. **7**(2), 179–188 (1936)
16. Hauskrecht, M., Batal, I., Valko, M., Visweswaran, S., Cooper, G.F., Clermont, G.: Outlier detection for patient monitoring and alerting. J. Biomed. Inform. **46**(1), 47–55 (2013)
17. Hilal, W., Gadsden, S.A., Yawney, J.: A review of anomaly detection techniques and applications in financial fraud. Expert Syst. Appl. 116429 (2021)
18. Kriegel, H.-P., Kröger, P., Schubert, E., Zimek, A.: Outlier detection in axis-parallel subspaces of high dimensional data. In: Theeramunkong, T., Kijsirikul, B., Cercone, N., Ho, T.-B. (eds.) PAKDD 2009. LNCS (LNAI), vol. 5476, pp. 831–838. Springer, Heidelberg (2009). https://doi.org/10.1007/978-3-642-01307-2_86
19. Kruegel, C., Vigna, G.: Anomaly detection of web-based attacks. In: Proceedings of the 10th ACM Conference on Computer and Communications Security, pp. 251–261 (2003)
20. Lin, T.Y., Goyal, P., Girshick, R., He, K., Dollár, P.: Focal loss for dense object detection. In: Proceedings of the IEEE International Conference on Computer Vision, pp. 2980–2988 (2017)
21. Liu, F.T., Ting, K.M., Zhou, Z.H.: Isolation forest. In: 2008 Eighth IEEE International Conference on Data Mining, pp. 413–422. IEEE (2008)
22. Micenková, B., Ng, R.T., Dang, X.H., Assent, I.: Explaining outliers by subspace separability. In: 2013 IEEE 13th International Conference on Data Mining, pp. 518–527. IEEE (2013)
23. Narayanan, V., Bobba, R.B.: Learning based anomaly detection for industrial arm applications. In: Proceedings of the 2018 Workshop on Cyber-Physical Systems Security and PrivaCy, pp. 13–23 (2018)
24. Pang, G., Shen, C., Cao, L., Hengel, A.V.D.: Deep learning for anomaly detection: a review. ACM Comput. Surv. **54**(2), 1–38 (2021). https://doi.org/10.1145/3439950
25. Samariya, D., Aryal, S., Ting, K.M., Ma, J.: A new effective and efficient measure for outlying aspect mining. In: Huang, Z., Beek, W., Wang, H., Zhou, R., Zhang, Y. (eds.) WISE 2020. LNCS, vol. 12343, pp. 463–474. Springer, Cham (2020). https://doi.org/10.1007/978-3-030-62008-0_32
26. Samariya, D., Ma, J., Aryal, S.: A comprehensive survey on outlying aspect mining methods. arXiv preprint arXiv:2005.02637 (2020)
27. Silverman, B.W.: Density Estimation for Statistics and Data Analysis. Routledge, Milton Park (2018)
28. Steinwart, I., Hush, D., Scovel, C.: A classification framework for anomaly detection. J. Mach. Learn. Res. **6**(2) (2005)
29. Vinh, N.X., Chan, J., Bailey, J., Leckie, C., Ramamohanarao, K., Pei, J.: Scalable outlying-inlying aspects discovery via feature ranking. In: Cao, T., Lim, E.-P., Zhou, Z.-H., Ho, T.-B., Cheung, D., Motoda, H. (eds.) PAKDD 2015. LNCS (LNAI), vol. 9078, pp. 422–434. Springer, Cham (2015). https://doi.org/10.1007/978-3-319-18032-8_33
30. Vinh, N.X., Chan, J., Romano, S., Bailey, J., Leckie, C., Ramamohanarao, K., Pei, J.: Discovering outlying aspects in large datasets. Data Min. Knowl. Disc. **30**(6), 1520–1555 (2016). https://doi.org/10.1007/s10618-016-0453-2
31. Wang, X., Yin, M.: Are explanations helpful? A comparative study of the effects of explanations in AI-assisted decision-making. In: 26th International Conference on Intelligent User Interfaces, pp. 318–328 (2021)
32. Wells, J.R., Ting, K.M.: A new simple and efficient density estimator that enables fast systematic search. Pattern Recogn. Lett. **122**, 92–98 (2019)

33. Xu, H., et al.: Unsupervised anomaly detection via variational auto-encoder for seasonal KPIs in web applications. In: Proceedings of the 2018 World Wide Web Conference, pp. 187–196 (2018)
34. Zhang, J., Lou, M., Ling, T.W., Wang, H.: Hos-miner: a system for detecting outlying subspaces of high-dimensional data. In: Proceedings of the 30th International Conference on Very Large Data Bases (VLDB 2004), pp. 1265–1268. Morgan Kaufmann Publishers Inc. (2004)

Author Index

Printed in the United States
by Baker & Taylor Publisher Services